T0396847

BIG IDEAS IN MATHEMATICS

Yearbook 2019

Association of Mathematics Educators

editors

Tin Lam Toh

Joseph B. W. Yeo

Nanyang Technological University, Singapore

Published by

World Scientific Publishing Co. Pte. Ltd.
5 Toh Tuck Link, Singapore 596224
USA office: 27 Warren Street, Suite 401-402, Hackensack, NJ 07601
UK office: 57 Shelton Street, Covent Garden, London WC2H 9HE

British Library Cataloguing-in-Publication Data
A catalogue record for this book is available from the British Library.

BIG IDEAS IN MATHEMATICS
Yearbook 2019, Association of Mathematics Educators

ISBN 978-981-120-536-1

For any available supplementary material, please visit
https://www.worldscientific.com/worldscibooks/10.1142/11415#t=suppl

Printed in Singapore by Mainland Press Pte Ltd.

BIG IDEAS IN MATHEMATICS

Yearbook 2019

Association of Mathematics Educators

Related Titles

Mathematics Instruction: Goals, Tasks and Activities, Yearbook 2018
edited by Pee Choon Toh and Boon Liang Chua
ISBN: 978-981-3271-66-1

Empowering Mathematics Learners, Yearbook 2017
edited by Berinderjeet Kaur and Ngan Hoe Lee
ISBN: 978-981-3224-21-6

Developing 21st Century Competencies in the Mathematics Classroom, Yearbook 2016
edited by Pee Choon Toh and Berinderjeet Kaur
ISBN: 978-981-3143-60-9

Effective Mathematics Lessons through an Eclectic Singapore Approach, Yearbook 2015
by Khoon Yoong Wong
ISBN: 978-981-4696-41-8

Learning Experiences to Promote Mathematics Learning, Yearbook 2014
edited by Pee Choon Toh, Tin Lam Toh and Berinderjeet Kaur
ISBN: 978-981-4612-90-6

Nurturing Reflective Learners in Mathematics, Yearbook 2013
edited by Berinderjeet Kaur
ISBN: 978-981-4447-74-6

Reasoning, Communication and Connections in Mathematics, Yearbook 2012
edited by Berinderjeet Kaur and Tin Lam Toh
ISBN: 978-981-4405-41-6

Assessment in the Mathematics Classroom, Yearbook 2011
edited by Berinderjeet Kaur and Khoon Yoong Wong
ISBN: 978-981-4360-97-5

Mathematical Applications and Modelling, Yearbook 2010
edited by Berinderjeet Kaur and Jaguthsing Dindyal
ISBN: 978-981-4313-33-9

Mathematical Problem Solving, Yearbook 2009
edited by Berinderjeet Kaur, Ban Har Yeap and Manu Kapur
ISBN: 978-981-4277-20-4

Contents

SECTION TWO: CONTENT KNOWLEDGE AND TEACHING IDEAS OF SELECTED BIG IDEAS IN MATHEMATICS

SECTION THREE: PEDAGOGICAL PRACTICES IN TEACHING TOWARDS BIG IDEAS IN MATHEMATICS

Chapter 1

Big Ideas in Mathematics

Joseph B. W. YEO TOH Tin Lam

This chapter provides an overview of all the other 20 chapters presented in this book. The remaining chapters in this book are classified into three main categories. The first section contains five chapters which present to readers general perspectives of Big Ideas in mathematics and mathematics education. The second section consists of eight chapters that discuss the content knowledge and teaching ideas of selected Big Ideas in mathematics. The last section contains seven chapters on general pedagogical practices in teaching towards Big Ideas in mathematics. This chapter concludes with some observations on Big Ideas and implications for teachers.

1 Introduction

Research on the practices of effective teachers has received much attention of mathematics education researchers (e.g. Ma, 1999; Weiss, Heck, & Shimkus, 2004). For example, effective teachers ask effective questions and are able to engage students in co-constructing the mathematics discourse in the classrooms. In terms of instructional materials, these teachers make teaching materials explicit to their students (Leong, Cheng, Toh, Kaur, & Toh, 2018).

Other studies emphasize the importance of teachers having solid content knowledge for teaching. Turner-Bisset (1999) identified content knowledge as one important attribute of expert teachers. There are many other studies which emphasize the importance of content knowledge for teachers (e.g. Chapman, 2005; Kilpatrick, 2001; Schmidt, 2002; Toh, 2017).

While the above are two important aspects of effective or expert teachers, it has been contended that pedagogical practice and mathematics content knowledge of a teacher must be grounded on a set of Big Mathematics Ideas (Charles, 2005).

2 Big Ideas in Mathematics

As early as the beginning of this century, the National Council of Teachers of Mathematics (NCTM) have stressed the importance of Big Ideas of mathematics. According to the *Principles and Standards for School Mathematics* (NCTM, 2000), "[t]eachers need to understand the big ideas of mathematics and be able to represent mathematics as a coherent and connected enterprise (p. 17). Charles (2005) gave a formal definition of a Big idea:

> A Big idea is a statement of an idea that is central to the learning of mathematics, one that links numerous mathematical understandings into a coherent whole (p. 10).

In brief, one can appreciate that Big Ideas are about connections of various ideas across mathematics. Charles (2005) further stressed the importance of Big Ideas as a grounding for mathematics teachers.

In the most recent curriculum revision of the Singapore mathematics curriculum, the Singapore Ministry of Education (MOE, 2018) highlights that one of the three key emphases of this review is to "develop a greater awareness of the nature of mathematics and the big ideas that are central to the discipline" (p. 9) in order to achieve coherence across different topics. In other words, this is similar to Charles' (2005) notion of a Big Idea as one linking various "understandings into a coherent whole".

By using a word count on the 2020 secondary mathematics syllabuses (MOE, 2018), the word "connection" occurs 37 times. It is thus not an exaggeration to state that Big Ideas are about connections. From the perspective of classroom practice, the emphasis of Big Ideas in mathematics can be translated to conscious effort of getting students to view mathematics as a highly connected system of thinking and concepts across various topics, rather than seeing these as isolated unrelated concepts with little or no relation with others.

Readers should note that this emphasis on connection is not entirely new in the Singapore Mathematics curriculum review. In fact, as early as 2006, the syllabus documents (MOE, 2006a; 2006b) have already highlighted that it is crucial that students are able to "[r]ecognise and use connections among mathematical ideas, and between mathematics and other disciplines" (p. 11). Rather, the new emphasis in the latest curriculum revision to emphasize that the recognition and use of connections are based on a sound understanding of the nature (or disciplinarity) of mathematics.

This book, which is a derivative of the lectures and workshops from the annual Mathematics Teacher Conference 2018, builds on the synergy of mathematicians and mathematics educators to provide researchers and mathematics educators a diverse perspective of Big Ideas in mathematics and, more importantly, how these Big Ideas can be translated to effective practices in the mathematics classroom. We believe that the notion of Big Ideas in Mathematics is an opportunity in which mathematicians and mathematics educators are able to offer differing (and often complementary) perspectives as an effort to offer readers a whole full spectrum of understanding of this topic.

This book consists of 21 chapters, of which the first chapter is the introduction of the book while the remaining 20 chapters are classified into three main categories. Section One provides the general perspectives of Big Ideas in mathematics (and statistics) and mathematics education. Consisting of five chapters, this section sets the tone for the next two sections by discussing two broad views of mathematics (and statistics) and the development of Big Ideas in mathematics education from three different perspectives.

Section Two consists of eight chapters that deal directly with the content knowledge and teaching ideas of selected Big Ideas (e.g. equivalence, invariance, functions, models, proportionality) in the 2020 Singapore school mathematics syllabuses, and other Big Ideas such as those that are specific within a branch of mathematics (e.g. calculus).

On the other hand, Section Three, consisting of seven chapters, explores various pedagogical practices that might help teachers teach towards Big Ideas in mathematics. Although some of the authors may illustrate their pedagogy with one or two Big Ideas in mathematics (e.g. measures), the teaching method is generic enough to be applied to other Big Ideas.

3 Section One on General Perspectives of Big Ideas in Mathematics and Mathematics Education

The focus of the five chapters in Section One is on various general viewpoints of Big Ideas in mathematics and mathematics education. The first two chapters present two broad views of mathematics (including statistics) while the next three chapters look at the development of Big Ideas in mathematics education from different perspectives.

This section begins with *Chapter 2*, where Koh and Tay describe three breakthrough ideas in the history of mathematics, framed in the contexts of three "great crises" in mathematics – the discovery of irrational numbers, the foundations of calculus, and the foundations of mathematics. Through the lens of the development of mathematics in history, Koh and Tay hope to enable teachers to appreciate the significance and influence of these breakthrough ideas, so as to facilitate teachers to build connections across these ideas.

On the other hand, Zhu proposes statistical thinking as an alternative Big Idea to understand the world around us: one that is different from the kind of deterministic mathematics described in Chapter 2. In *Chapter 3*, readers will learn to appreciate the differences between statistical and mathematical thinking, some key characteristics of statistical thinking and its importance in guiding our thinking, learning and daily life in this era of big data in the 21st century.

In *Chapter 4*, Wagner looks at the development of Big Ideas in mathematics education from a different perspective: he believes that these ideas develop as responses to human concerns, and so the teaching of mathematics should be positioned as being responsive to community and environmental concerns in order to support responsible citizenship oriented around sustainable development and peace.

Hurst, in *Chapter 5*, hopes to make use of Big Idea thinking to re-conceptualize mathematics education. He highlights the importance of connections within and between Big Ideas as the essence of Big Idea thinking, and the expression of these connections in terms of learning sequences and trajectories. Hurst further classifies Big Ideas into three broad categories of sequential Big Ideas, umbrella Big Ideas and process Big Ideas using an extensive model.

In Singapore, Big Ideas are categorized differently in the 2020 Singapore school mathematics syllabus (MOE, 2018): they are framed around four recurring themes in mathematics (namely, properties and relationships; operations and algorithms; representations and communications; and abstractions and applications). The syllabus document specifies eight clusters of Big Ideas (namely, equivalence, invariance, functions, models, diagrams, notations, measures and proportionality), but it emphasizes that these eight Big Ideas are not meant to be authoritative and comprehensive. In *Chapter 6*, Choy examines the four recurring themes and the eight clusters of Big Ideas in the syllabus document and presents a notion of teaching towards Big Ideas. He also discusses some challenges and suggests what teachers can do to mitigate some of the difficulties surrounding the worthwhile task of teaching towards Big Ideas.

4 Section Two on Content Knowledge and Teaching Ideas of Selected Big Ideas in Mathematics

In Section Two, we zoom in on some Big Ideas in mathematics in the 2020 Singapore school mathematics syllabuses in order to understand more about these Big Ideas and to suggest some teaching ideas. As the eight clusters of Big Ideas specified in the syllabus document are not meant to be exhaustive, this section will also examine some other Big Ideas.

We begin this section with the Big Idea of equivalence in *Chapter 7*. First, Yeo K. K. J. reviews literature on Big Ideas in mathematics, Big

Ideas for teaching and learning, and the Big Idea of equivalence. Then he proposes three activities on equivalence that can be introduced in the primary mathematics classroom.

In *Chapter 8*, Tay also discusses the Big Idea of equivalence but he uses examples from the Singapore pre-university mathematics curriculum. According to Tay, one must have a deep consideration of the nature of mathematics in order to engage in discussion on Big Ideas in mathematics. This chapter also looks at the Big Idea behind solving equations (namely, the technique of 'isolating the unknown') and the Big Ideas of definition and notation.

Closely related to the Big Idea of equivalence is the Big Idea of invariance. In *Chapter 9*, Toh P. C. demonstrates how invariance threads across several topics horizontally across content strands and vertically across levels by giving examples of invariance in geometry at the primary and secondary levels, and number theory at the pre-university level.

The Big Ideas of functions and mathematical modelling in secondary school mathematics are discussed in *Chapter 10*. Yap believes that the introduction of empirical flavour in the teaching of related topics will be able to illustrate to students the essence of these topics. This chapter also provides two possible lesson plans incorporating the ideas that he has proposed.

In *Chapter 11*, Seshaiyer and Suh also advocate the use of mathematical modelling as a Big Idea in teaching mathematics to solve real world tasks that involve not only mathematics but other disciplines as well. They give an example of a modelling task given to a class in an elementary school during a lesson study and discusses the solutions from some students. They believe that this kind of tasks can provide students with the opportunity to engage in the four pillars of 21st century skills: communication, collaboration, critical thinking and creative problem solving.

Next, Yeo J. B. W. unpacks the Big Idea of proportionality in *Chapter 12* by examining how the concepts of proportion, ratio, rate and variation are connected to one another, so that primary and secondary school teachers are better equipped not only to explain to their students the similarities and differences among these concepts but also to connect these ideas into a coherent whole and to appreciate how proportionality is used in real life.

Chapter 13 also deals with Big Idea of proportionality but from another perspective. In this chapter, Chua suggests presenting secondary school students with different real-world situations in order for them to identify those that depict proportionality. He then discusses the underlying idea behind the concept of proportionality and its connections to other topics such as gradient, trigonometry and statistics.

All the above chapters in Section 2 deal with Big Ideas across topics. But there are also Big Ideas within a specific branch of mathematics, e.g. calculus. In *Chapter 14*, Toh T. L. paints a portrait of how pre-university calculus lessons would appear for a teacher who has a deep understanding of "Big Idea of calculus education". The examples on building up students' repertoire of concept images are drawn from the current pre-university mathematics curriculum on calculus.

5 Section Three on Pedagogical Practices in Teaching towards Big Ideas in Mathematics

In Section Three, we look at some pedagogical practices that might help teachers teach towards Big Ideas in mathematics. This section begins with the use of an approach called teaching through problem solving in *Chapter 15*. Although Lim illustrates this pedagogy using the concept of area in the primary school classroom, and in the process brings in the Big Ideas of measure and invariance, the reader can potentially apply the teaching method to another topic to draw out other Big Ideas. In this teaching approach, the teacher acts as a facilitator during students' presentation of their solution to the given problem and subsequent class discussion.

But mathematical vocabulary plays an important part in connecting mathematical ideas during teacher instruction or student discussion. So, in *Chapter 16*, Kaur, Wong, Toh and Tong analyzed data from a large-scale research in Singapore to study how widespread the emphasis among secondary school teachers on developing mathematical vocabulary in their students was. They also use Marzano's six steps for effective vocabulary instruction to illustrate, through examples drawn from the research study, how teachers may facilitate the development of mathematical vocabulary during their instruction.

Having the correct mathematical vocabulary is the first step to connecting and communicating mathematical ideas. But in *Chapter 17*, Cheng and Koh have found that primary school students still have difficulty in expressing themselves. The focus of this chapter is on the design and use of journal writing prompts to support students in developing mathematical concepts, reasoning and justifying, and in communicating their ideas. In this chapter, Cheng and Koh provide a model as a guideline on how to design such journal prompts, guided by Big Ideas and big process ideas. They also discuss three kinds of journal prompts (content, process and affective prompts), with examples from their work with teachers in a primary school.

Another way to develop students' reasoning and justifying skills is to use typical textbook problems to generate new problems. In *Chapter 18*, Dindyal explains how to use standard secondary school textbook problems to form conjectures and to generate counterexamples. He also illustrates how typical problems can be modified to elicit inductive reasoning (so as to generalise) or deductive reasoning (e.g. justifying and proving).

In contrast, Leong, in *Chapter 19*, uses typical problems from one page of a secondary school textbook to sift out a 'small idea', then extends it into a Big Idea across neighbouring topics in the same chapter. Continuing with this method of inquiry, Leong extends the Big Idea beyond the chapter to a 'bigger idea' among other topics in the textbook. Finally, he extends the 'bigger idea' to an overarching mathematical idea of representation in mathematics. Leong concludes his chapter by encouraging teachers to use this method of inquiry for other 'small ideas' to sift out new Big Ideas.

On the other hand, in *Chapter 20*, Seto, Choon and Pang focus on the organizing of the teaching of a topic around Big Ideas by engaging teachers to construct a web of interconnected ideas (WICI) for the topic and to draw up the Content Representation in Mathematics (M-CoRe) in relation to the WICI. M-Core is helpful to elicit Big Ideas in the content and Big Ideas in teaching to develop students' understanding of that content. Seto et al. also provide illustrations of their WICI and M-CoRe tools through the teaching of mixed numbers at Primary 4 and simple probability at Secondary 2.

Last but not least, Low and Wong advocate a whole school approach to connect mathematical topics and Big Ideas, not just horizontally within a level, but vertically across levels. In *Chapter 21*, Low and Wong give an example of finding the height of the school flag pole, which can be implemented across three levels (namely, Secondary 1 to 3). In each level, students will use a different method based on the topics that they have learnt for that level, but the unifying underlying principle behind all three methods is actually the Big Idea of proportionality. Low and Wong conclude the chapter by discussing some issues and challenges that teachers may face when teaching for vertical connection.

6 Concluding Remarks

As we read all the chapters in this book, we notice that the authors have diverse interpretation of Big Ideas in mathematics. Despite the differences in opinions among the authors, we observe two common trends that run across the chapters in this book: (1) Big Ideas are about connections across the various mathematical ideas; and (2) with an emphasis on Big Ideas in the mathematics curriculum across all levels of school mathematics, it necessitates teachers to re-consider their understanding of mathematics (school mathematics in particular) and the nature of mathematics.

References

Chapman, O. (2005). *Constructing pedagogical knowledge of problem solving: preservice mathematics teachers*. In H. L. Chick, & J. L. Vincent (Eds.), *Proceedings of the 29th Conference of the International Group for the Psychology of Mathematics Education* (Vol. 2, pp. 225 – 232), Melbourne: PME.

Charles, R. I. (2005). Big ideas and understandings as the foundation for elementary and middle school mathematics. *Journal of Mathematics Education Leadership*, *7*(3), 9-24.

Kilpatrick, J. (2001). *Adding it up: helping children learn mathematics.* Washington, DC: National Academies Press.

Leong, Y. H., Cheng, L. P., Toh, W. Y., Kaur, B., & Toh, T. L. (2019). Making things explicit using instructional materials: A case study of a Singapore teacher's practice. *Mathematics Education Research Journal, 31*(1), 47-66.

Ma, L. (1999). *Knowing and teaching elementary mathematics: Teachers' understanding of fundamental mathematics in China and the United States.* Mahwah, NJ: Erlbaum.

Ministry of Education. (2006a). *Mathematics syllabuses: Primary.* Singapore: Author.

Ministry of Education. (2006b). *Mathematics syllabuses: Lower secondary.* Singapore: Author.

Ministry of Education. (2018). *2020 secondary mathematics syllabuses (draft).* Singapore: Author.

National Council of Teachers of Mathematics (NCTM). (2000). *Curriculum and Evaluation Standards for School Mathematics.* Reston, VA: NCTM.

Schmidt, W. (2002). The benefit to subject-matter knowledge. *American Educator, 26*(2), 18 – 22.

Toh, T. L. (2017). On Singapore prospective secondary school teachers' mathematical content Knowledge. *International journal of Mathematics Teaching and Learning, 18*(1), 25-40.

Turner-Bisset, R. (1999). The knowledge bases of the expert teachers. *British Educational Research Journal, 25*(1), 39 – 55.

Weiss, I. R., Heck, D. J., & Shimkus, E. S. (2004). Mathematics teaching in the United States. *Journal of Mathematics Education Leadership, 7*, 23-32.

Chapter 2

Some Great Breakthrough Ideas in Mathematics

KOH Khee Meng TAY Eng Guan

There are numerous breakthrough ideas in the rich history of mathematics. In this chapter, we shall present some of them pertaining to the Singapore mathematics curriculum content. These include number systems, trigonometric functions, Cartesian coordinates, limit, and the Fundamental Theorem of Calculus. These are framed in the contexts of three "great crises" in mathematics – the discovery of irrational numbers, the foundations of calculus, and the foundations of mathematics. By putting school mathematics content within the lens of the development of mathematics in history, it is hoped that teachers can appreciate how significant and influential these ideas were, and in turn be able to use the connections to motivate students learning mathematics.

1 History as a Big Idea

Charles (2005) defines a Big Idea in mathematics as "a statement of an idea that is central to the learning of mathematics, one that links numerous mathematical understandings into a coherent whole." (p. 10) Although the historical development of a mathematical concept is itself not "a statement of an idea", it does fulfill the role of "linking numerous mathematical understandings into a coherent whole". To this end, this chapter employs a perspective via a history of mathematics to help readers think about mathematics pedagogy in the Singapore mathematics classroom.

Kleiner (2010), a mathematician, explains that although he admired the elegance of the formal structure of mathematics, he was dissatisfied with the exclusive focus on the formal definition-theorem-proof mode of instruction. He found that the history of mathematics boosted his "enthusiasm for teaching by providing perspective, insight and motivation ... for example, when [he] taught calculus, [he] was able to understand where the derivative came from, and how it evolved into the form we see in today's textbooks" (p. viii).

There are numerous breakthrough ideas in the rich history of mathematics. In this chapter, we shall present three breakthrough ideas that are pertinent to the Singapore school mathematics content. These are framed in the contexts of three "great crises" in mathematics – the discovery of irrational numbers, the foundations of calculus, and the foundations of set theory. Our exposition takes its starting point from the "first crisis" described in Eves (1983a, p. 43) and follows through by condensing the narratives in the two books of Eves (1983a, 1983b).

2 The First Crisis – Discovery of Irrational Numbers

The Theorem attributed to Pythagoras is one of the very first great theorems in mathematics. The result is great because it is the first result linking numbers, algebra and geometry.

The followers of Pythagoras formed a school, both literally and philosophically. One of their main philosophical doctrines is encapsulated in the phrase "Everything is a number". To them, all things in the world could be expressed through natural numbers and their ratios. In particular, to the Pythagoreans, all the points on the number line would be rational, i.e. each could be expressed in the form a/b, where a is an integer and b is a non-zero integer. One can imagine their distress when it was discovered (according to folklore, by one Hippasus of Metapontum) that $\sqrt{2}$ could not be expressed in the required form a/b, and thus there exists a number that is not rational.

Figure 1 shows how $\sqrt{2}$ can be constructed and placed as a valid point on the number line. The well-known proof that $\sqrt{2}$ is not rational is one by contradiction. Suppose $\sqrt{2}$ is rational, i.e., it can be expressed as a/b, where a and b are non-zero integers and a and b have no common prime factors. Squaring both sides, we have $2b^2 = a^2$. Since 2 divides the Left Hand Side of this equation, 2 must divide the Right Hand Side, a^2. Since 2 is prime, 2 must divide a, thus $a = 2k$ for some integer k. This results in $b^2 = 2k^2$, and following the same argument as before, we have that 2 must divide b. But a and b were chosen such that they have no common prime factors. This is a contradiction that implies the complement of the assumption, i.e., $\sqrt{2}$ is not rational.

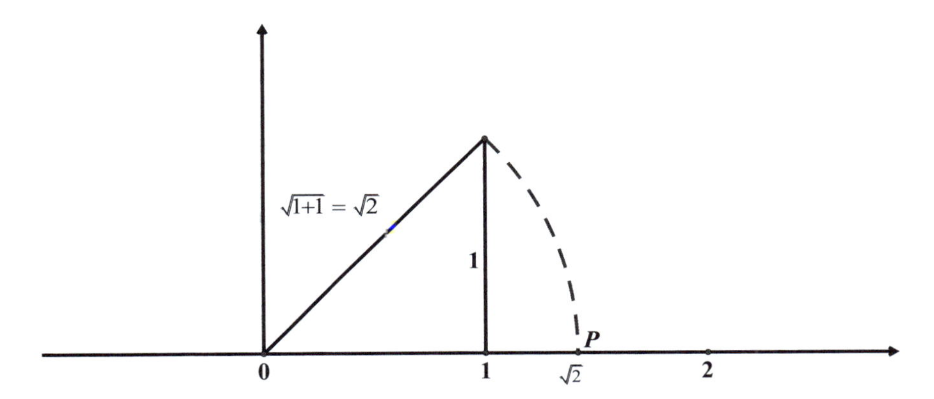

Figure 1. The square root of 2 is not rational

A popular myth has it that the Pythagoreans literally killed the messenger, sending poor Hippasus to the bottom of the sea because of his discovery. The discovery of the irrational numbers shocked the Pythagoreans and caused the first crisis in the foundations of mathematics because the latter refuted the main philosophical doctrine of the Pythagoreans.

As often happens in mathematics, when one irrational number was discovered, an infinite number of irrationals followed. For example, if n is not a k-th power of a natural number, where k is an integer not less than 2, then the k-th root of n is irrational. Following from this, the set of numbers was expanded beyond the rationals to include the irrationals.

The rationals together with the irrationals form the set of real numbers whose representation on a number line is a complete continuum without any further gaps.

Much later, the complex numbers were conceived by Gerolamo Cardano around 1545 in response to the problem of square roots of negative numbers appearing as solutions of cubic equations. Work on the problem of general polynomials ultimately led to the Fundamental Theorem of Algebra, a great result that owes everything to the existence of complex numbers. This is because with complex numbers, exactly n solutions (some repeated) exist for every polynomial equation of degree n.

With regard to polynomials (with complex coefficients), we are now equipped with the complex number system which is sufficient to capture all possible roots. Indeed, the first crisis in mathematics allowed the breakthrough from the constraints of rational numbers to the discovery of the 'complete' complex number system. Figure 2 shows the relationships between the numbers in a Venn Diagram.

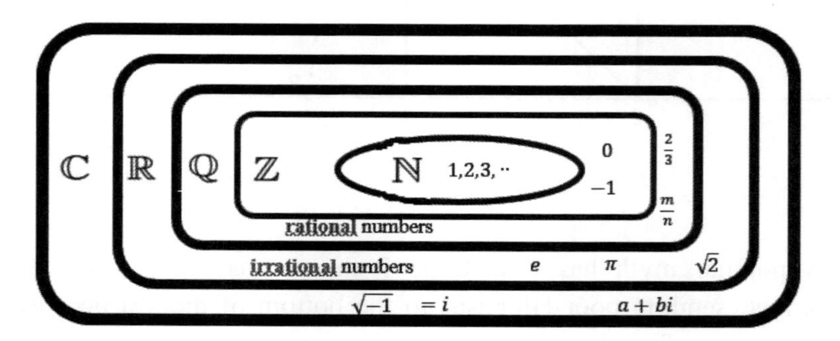

Figure 2. The 'complete' number system

3 The Second Crisis – the Foundations of Calculus

The 1-dimensional real line (see Figure 3) provides a 1-1 correspondence between the real numbers and points on the line.

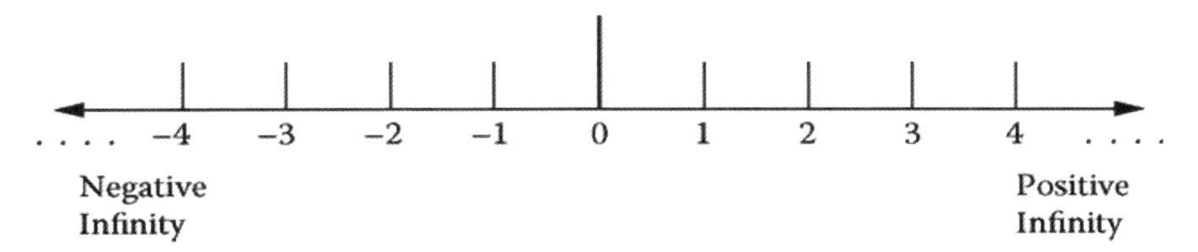

Figure 3. The 1-dimensional real line

René Descartes (1596-1650) was a French writer, mathematician, physicist and philosopher (he is often cited for his argument that his existence was real – "I think, therefore I am"). Legend has it that while lying in bed one day, he spotted a fly on the ceiling and thought about giving a precise location to it. Figure 4 shows a 2-coordinate position notation that would have achieved his purpose.

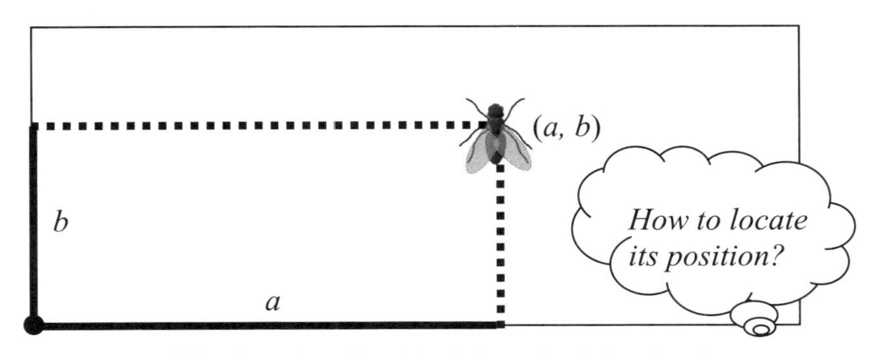

Figure 4. The fly on the wall and the 2-dimensional Cartesian plane

Descartes invented the coordinate plane, and it is now called the Cartesian plane in his honour. The Cartesian plane provides a 1-1 correspondence between the set of ordered pairs of real numbers and the set of points on the plane.

This was a great breakthrough because the Cartesian plane became a tool to merge algebra and geometry. The field of Analytic Geometry was born to these two illustrious parents. Graphs of polynomials gave an extra visual perspective to the study of roots. The natural logarithm and exponential functions could be further studied and applied to problems in archaeology, finance, and natural growth and decay. The study of ordinary differential equations and partial differential equations via calculus naturally followed later. Construction of mathematical models using graphs was made possible.

As an example of a great achievement in analytic geometry, Joseph Fourier (1768–1830) showed that any periodic function could be decomposed into the sum of a set of sine and/or cosine functions. The initial purpose was to solve the heat equation in a metal plate. Prior to Fourier's work, no general solution to the heat equation was known but particular solutions could be obtained if the heat source behaved in a simple way, in particular, if the heat source was a sine or cosine function. With the Fourier series method, given a more complicated function, one needs only to decompose that function into the sum of sine and/or cosine functions, solve the simple functions term by term, and then superimpose them back again to form a solution for the original function. The technique used to solve the heat equation could now be applied to a wide range of problems, especially those involving linear differential equations with constant coefficients, for which solutions are available for 'simple' sinusoidal equations. To this end, the Fourier series has many applications in electrical engineering, acoustics, optics, signal processing, image processing, quantum mechanics, econometrics, etc. and so is a compulsory topic in many undergraduate engineering, science and economics programmes.

A more modern application of analytic geometry is in the Global Positioning System (GPS) that all of us use today. The following simplified explanation of the GPS is based on the article by Kalman (2002).

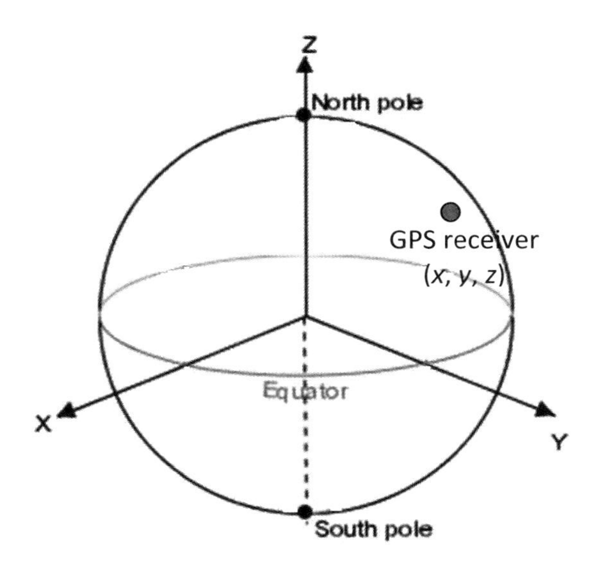

Figure 5. A GPS receiver on the surface of the earth

Figure 5 shows a GPS receiver at your position (x, y, z) on the surface of the earth. The aim of the subsequent calculation is to obtain the values of x, y and z so as to locate your position expressed as (x, y, z) on the 3-dimensional Cartesian space.

Figure 6 shows four satellites, S_i, positioned at (x_i, y_i, z_i) respectively, for $i = 1, 2, 3, 4$. These satellites transmit signals continuously.

At time t_0, you receive transmissions from the four satellites containing their positions (x_i, y_i, z_i) and the times of transmission t_i, for $i = 1, 2, 3, 4$. Taking into consideration the time difference and the speed of light, it is possible to work out the distance d_i of Satellite i, for $i = 1, 2, 3, 4$ from your position.

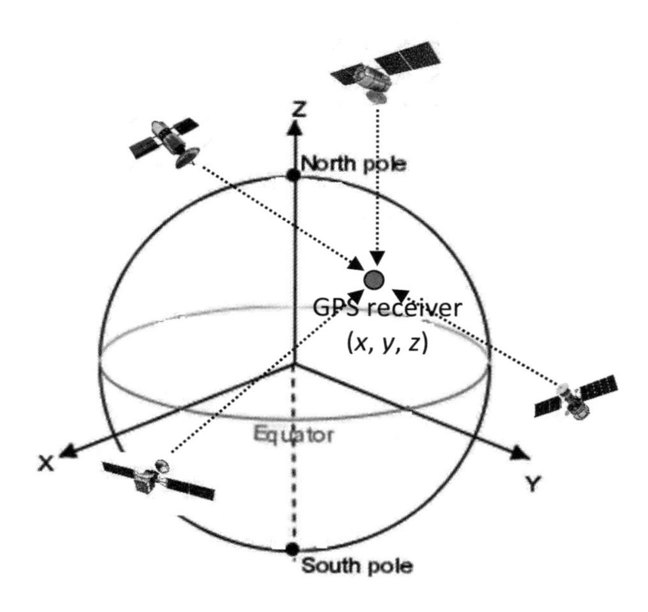

Figure 6. Four satellites

The distance between two points (a, b, c) and (p, q, r) is given by $\sqrt{(a-p)^2+(b-q)^2+(c-r)^2}$. With this formula, we are able to obtain a system of 4 equations for the distance of each of the 4 satellites to your position as below:

$$(x-x_1)^2+(y-y_1)^2+(z-z_1)^2=d_1^2 \qquad (1)$$

$$(x-x_2)^2+(y-y_2)^2+(z-z_2)^2=d_2^2 \qquad (2)$$

$$(x-x_3)^2+(y-y_3)^2+(z-z_3)^2=d_3^2 \qquad (3)$$

$$(x-x_4)^2+(y-y_4)^2+(z-z_4)^2=d_4^2 \qquad (4)$$

Next, we operate on the equations as follows: (1) – (2), (1) – (3), (1) – (4). The aim of taking the differences of equation pairs is to obtain a system of 3 linear equations in 3 unknowns x, y, and z. This is easily solved to give your position (x, y, z).

In the 17th and 18th centuries, calculus was developed on the back of analytic geometry. Differential calculus is a branch of mathematics concerned with the determination, properties, and application of derivatives and differentials. Integral calculus is a branch of mathematics concerned with the determination, properties, and application of integrals. Figures 7 and 8 show their respective foci of concern.

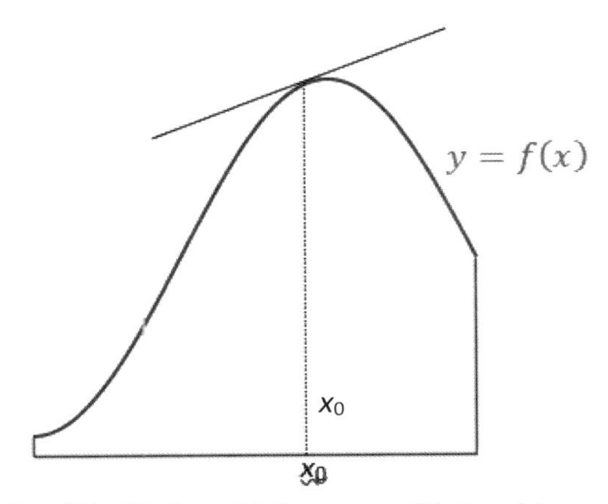

Figure 7. The concerns of differential calculus

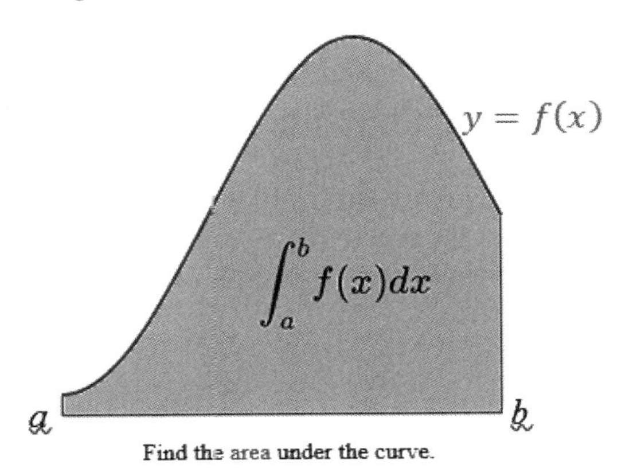

Figure 8. The concerns of integral calculus

The two giants of mathematics who are acknowledged as the pioneers of calculus were the Englishman Isaac Newton (1643-1727) and the German Gottfried Wilhelm Leibniz (1646-1716). Newton began his work with derivatives and Leibniz with integrals but astonishingly, both of them arrived independently at the same conclusion regarding the relationship between derivatives and integrals, i.e., that essentially each is the inverse of the other. This great result in mathematics is called the Fundamental Theorem of Calculus.

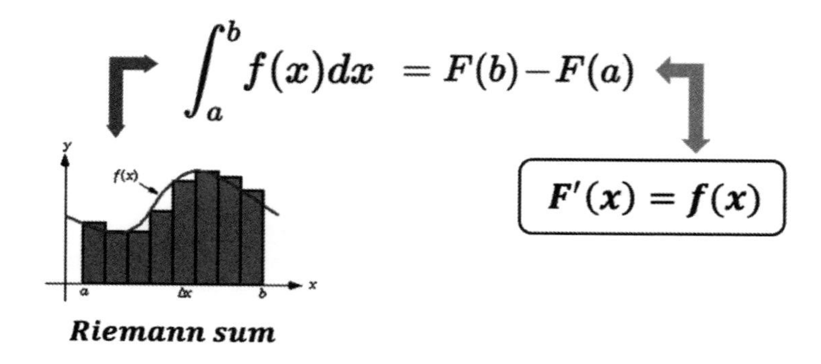

Figure 9. The Fundamental Theorem of Calculus

Students in high school seldom get the chance to grasp the enormity of this theorem and thus often work on derivatives and integrals without having a sense of why one is the inverse of the other. For this purpose, we would like to insert here a short proof of the Fundamental Theorem of Calculus that should be accessible to such students so that they can appreciate Calculus better as an interplay between derivatives and integrals.

Let f be a differentiable function of x. We define the real-valued function F on the real domain such that $F(a)$ is the area bounded by the curve $y = f(x)$, the x-axis and the line $x = a$.

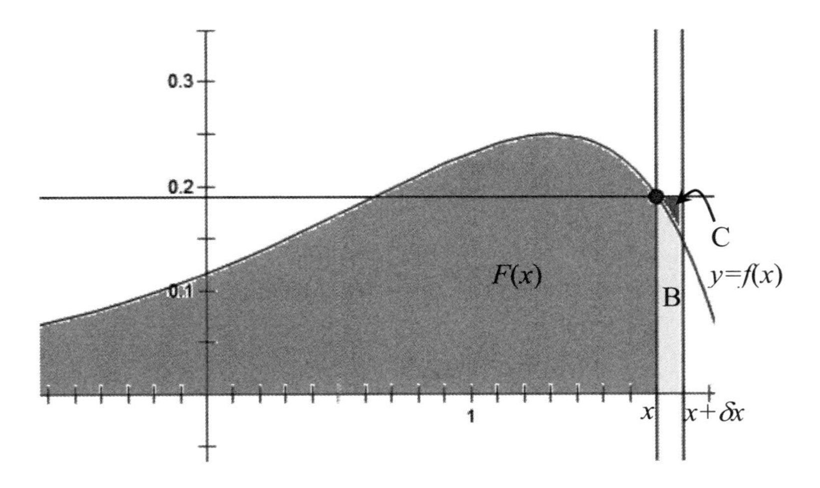

Figure 10. Proof of the Fundamental Theorem of Calculus

From Figure 10, we can see that for a small change δx of x, the increase in F is the area B. From the definition of the derivative, we have

$$\begin{aligned} F'(x) &= \lim_{\delta x \to 0} \frac{F(x+\delta x) - F(x)}{\delta x} \\ &= \lim_{\delta x \to 0} \frac{\mathrm{B}}{\delta x} \\ &= \lim_{\delta x \to 0} \frac{\delta x \cdot f(x) - \mathrm{C}}{\delta x} \\ &= \lim_{\delta x \to 0} \left(f(x) - \frac{\mathrm{C}}{\delta x} \right) \\ &= f(x) \qquad \text{since C is of the order } (\delta x)^2. \end{aligned}$$

Thus, f is the derivative of F. Since the area under the curve from $x = a$ to $x = b$ is denoted by $\int_a^b f(x)dx$ which by definition is $F(b) - F(a)$, we have shown that F is the anti-derivative of f.

Calculus found significant applications in engineering and science. However, basic concepts, such as limits, continuity and series, were initially not rigorously defined. For instance, the Riemann integral depends on the sum of an infinite number of rectangles but what is the meaning of the following sum of an infinite series?

$$a_1 + a_2 + a_3 + \ldots + a_n + \ldots$$

As an example, we consider

$$1 - 1 + 1 - 1 + 1 - 1 + \ldots$$

Which of the following would then be true?

i. $(1-1)+(1-1)+(1-1)+\ldots = 0+0+0+\ldots = 0.$

ii. $1+(-1+1)+(-1+1)+(-1+1)+\ldots = 1+0+0+\ldots = 1.$

iii. Or do we take the average of (i) and (ii) and obtain $\frac{1}{2}$?

This kind of conundrum prompted George Berkeley, an Irish philosopher, to critique the foundations of calculus in his publication *The Analyst* (Berkeley 1734). There, he acknowledged that the results of calculus were true but that calculus could not claim to be more logically rigorous than religion. He questioned whether mathematicians "submit to Authority, take things upon Trust" (Berkeley 1734 p.93) just as religious people were then criticized to be doing.

The reasonable criticism of Berkeley on the breakthrough work of Newton and Leibniz was indeed a crisis regarding the validity of mathematical truths. Fortunately, subsequent work by Cauchy, Bolzano and Weierstrass resulted in a theory of limits that successfully restored the prestige of mathematical validity. For example, the limit of an infinite sequence is defined as follows: A sequence of real numbers

$\{a_n\}_{n=1}^{\infty}$ converges to a limit L if for every $\varepsilon > 0$, there exists an N such that $|a_n - L| < \varepsilon$ for all $n > N$.

Let us thus return to our conundrum: 1 – 1 + 1 – 1 + 1 – 1 + … = ?

By defining the limit of a series to be the limit of the sequence of its partial sums, we have the following sequence of partial sums:

1, (1-1), (1-1+1), (1-1+1-1), …, which is 1, 0, 1, 0, ….

From the N-δ definition of the limit of a sequence, we can see that the sequence of partial sums above does not converge. Hence, none of the proffered options (i)-(iii) are true.

The teacher who is clear about the development of rigour regarding calculus will first be able to understand the difficulty students have with ideas of convergence, limits and infinitesimals. Perhaps, by relating to the students how great mathematicians themselves struggled with these concepts, students will gain empathy and hopefully, take enough time and persistence to work on understanding fully these difficult notions. Secondly, that the derivative is the inverse of an integral as shown in the Fundamental Theorem of Calculus is a very important exemplar towards seeing the Big Idea of Relationships in mathematics.

4 The Third Crisis – the Foundations of Mathematics

About 300 B.C., the Greek mathematician Euclid produced his monumental work *Elements*. This was the first great application of the axiomatic method in mathematics. Just as chemical elements are the basic units on which all chemical objects are built, the *Elements* posited a number of axioms (or postulates) as fundamental truths of Geometry that need not be proved, and on which many theorems could be logically deduced.

In the concept of Greek axiomatics, an axiom expresses a property of some basic objects (such as points and lines in Geometry) and this expression is accepted as self-evident, somewhat because it has a strong correlation to similar items in the real world. Thus, Greek axiomatics is sometimes called *material axiomatics*. This concept of the self-evidence of axioms was shaken in the nineteenth century by the 'discovery' of a non-Euclidean geometry (where parallel lines need not meet only at infinity) and of a noncommutative algebra.

Mathematicians thus had a more careful look at the axiomatic method, and wanted to make it more rigorous. David Hilbert and Bertrand Russell were among those who made great contributions to the evolution of the so-called *formal axiomatics* which contrasted with *material axiomatics* in that an axiom in *formal axiomatics* is now a basic assumption about some undefined primitive terms. From this perspective, Geometry is a purely abstract study that does not take its cues from physical reality or even imagery.

In constructing mathematics using formal axiomatics, one needs to list a collection of symbols for the primitive terms and then list another collection of statements about these primitive terms as axioms. However, there is more than arbitrarily forming these two collections of primitive terms and axioms. The two collections should form a system with the following desired properties:

- **Consistency** Contradictory statements must not be implied by the collection of axioms. (If otherwise, some 'axioms' should be discarded.)
- **Independence** An axiom in the collection should not be logical consequences of some other axioms in the collection. (If otherwise, the 'axiom' is superfluous and can be discarded.)
- **Completeness** Adding another axiom to the collection of axioms while preserving the consistency and independence of the new collection will necessitate adding new primitive terms to the original collection of primitive terms. (If otherwise, adding new 'axioms' may help build a more encompassing branch of pure mathematics using the same number of primitive terms.)

In pursuit of a set of axioms and inference rules in symbolic logic from which *all* mathematical truths could be proved, Alfred North Whitehead and Bertrand Russell wrote the three volume *Principia Mathematica* in 1910, 1912 and 1913. It was heralded as a masterpiece of mathematics and philosophy. Although the logicist approach was important philosophically, mathematicians in general did not feel that it appropriately 'defined' mathematics. For example, it took 362 pages of the *Principia Mathematica* to reach the logical conclusion that 1 + 1 = 2.

Nonetheless, Hilbert continued pushing his grand design for viewing mathematics as a study of axiomatic systems. Although formalists have been accused of trying to reduce mathematics to symbol manipulation, in truth, Hilbert's aim was to ensure that the foundations of mathematics was sound, which to him meant that axiomatic systems should all be consistent and complete.

We now reach the third crisis in the history of mathematics. Kurt Godel proved in 1931 that Hilbert's grand design was an impossibility. Godel's First Theorem is stated as follows:

> For any consistent formal system *F* which contains the natural number system, there are undecidable propositions in *F*; that is, there are propositions *S* in *F* such that neither *S* nor not-*S* is provable in *F*.

The theorem implies that any collection of axioms for the natural number system must be incomplete to remain consistent. Any consistent collection of axioms will have statements *S* about the natural numbers for which we *cannot* prove if it is true or not. Perhaps the Goldbach conjecture that "every even number greater than 2 can be expressed as a sum of two primes" is one of them – after all, it has been worked on since 1742!

Godel's first theorem showed that Hilbert's attempt to find complete axiomatic systems was an impossible endeavour. Thus, the *Principia Mathematica* was shown to be a vain attempt to establish a set of axioms and inference rules in symbolic logic from which *all* mathematical truths could be proved. Could Hilbert at least then hope to prove the consistency of mathematics through proving that relevant axiomatic systems are consistent? Godel's Second Theorem destroyed that vision as well:

> For any consistent formal system *F* which contains the natural number system, the consistency of *F* cannot be proved in *F*.

Godel's two theorems shattered the attempt of mathematicians to establish mathematics as a branch of human knowledge that is impeccable and beyond reproach with regard to its validity. Since the axiomatic method cannot guarantee the consistency of important fields of mathematics nor can it determine the truth or falsity of all propositions in these fields, mathematics would respond graciously to this crisis by climbing down from its perch of infallibility. De Sua (1956) links back to the religious aspect of the second crisis with the following comment: "Suppose we loosely define a *religion* as any discipline whose foundations rest on an element of faith, irrespective of any element of reason which may be present. Quantum mechanics for example would be a religion under this definition. But mathematics would hold the unique position of being the only branch of theology possessing a rigorous demonstration of the fact that it should be so classified." (p.305)

The teacher who is clear about the development of rigour regarding the foundations of mathematics will first be able to understand that mathematics is the field of human knowledge and study which possesses the 'highest' level of verifiability and yet is not 'beyond doubt'. Perhaps, by relating to the students how important proof is to mathematics, students will buy into the disciplinarity of mathematics and hopefully, ask questions of themselves and the teacher towards probing as far as possible the veracity of a mathematical proposition. Students will learn also not to simply accept statements and solutions from textbooks as if they were an unimpeachable authority.

5 Conclusion

In this chapter, we briefly present the history of mathematics by highlighting three crises that profoundly affected and changed the course of mathematics. These three crises are the discovery of irrational numbers, the controversy regarding the foundations of calculus, and the controversy regarding the foundations of mathematics. The first resulted in an expansion of the number system leading to an openness regarding the boundaries of mathematics. The second resulted in new areas of mathematics that strengthened the rigour of mathematical theory and assertions. The third crisis brought mathematics back down to earth to realise its own limitations.

Knowing the history of mathematics can be an invaluable tool for teachers in motivating students to learn the subject. In the context of Big Ideas, the history of mathematics helps to link together seemingly disparate topics, as for example, the Pythagoras theorem and the irrational numbers through the discovery of $\sqrt{2}$ as a number from the right-angle triangle with shorter sides of length 1 and the proof of its irrationality by contradiction.

Finally, in the telling of these three crises, one can see the greatness of our discipline through its pursuit of truth and proof, and its humility through the same approach which ultimately limited its (still substantial) reach.

References

Berkeley, G. (1734). *The analyst: A discourse addressed to an infidel mathematician.* London: Wikisource.

Charles, R. I. (2005). Big ideas and understandings as the foundation for elementary and middle school mathematics. *Journal of Mathematics Education Leadership*, 7(3), 9-24.

De Sua, F. (1956). Consistency and completeness - a resume. *American Mathematical Monthly, 63,* 295-305.

Eves, H. (1983a). *Great moments in mathematics (Before 1650).* USA: Mathematical Association of America.

Eves, H. (1983b). *Great moments in mathematics (After 1650).* USA: Mathematical Association of America.

Kalman, D. (2002). An underdetermined linear system for GPS. *The College Mathematics Journal, 33,* 384-390.

Kleiner, I. (2010). *Excursions in the history of mathematics*. NY: Springer.

Chapter 3

On Characteristics of Statistical Thinking

ZHU Ying

Statistical thinking is crucial for understanding the world around us by providing a different way of thinking from deterministic mathematics. This chapter aims to elaborate key characteristics of statistical thinking and underlying core issues. Having a good understanding of the key characteristics of statistical thinking can help us think and evaluate results critically, make sound decisions and guide our learning and daily life.

1 Introduction

Statistical thinking, as a big idea in mathematics, is crucial for understanding the world around us. Emerging from the rejection of deterministic thinking of mathematical scientists, statistical thinking grew among economists, kinetic gas theorists, and the biometricians from 1820-1900. It was stated that "[s]tatistical thinking concerns the relation of quantitative data to a real-world problem, often in the presence of *variability and uncertainty*. It attempts to make precise what the data has to say about the problem of interest." (Mallows, 1998)

In the 1990s there were strong beliefs that the development of statistical thinking would be the next step in the evolution of the statistical discipline and that people's conception of statistical thinking would alter their understanding of reality (Provost & Norman, 1990; Snee, 1999). It was stated by Provost and Norman (1990) that "the 21st century will place even greater demands on society for statistical thinking throughout industry, government, and education." In response to a strong call from some prominent statisticians to develop students' statistical thinking (Moore, 1990), the American Statistical Association set the agenda for the future of statistics education. An emphasis on statistical thinking, using more data and concepts, fostering active learning were included.

Over the past decades, the desire to imbue students with statistical thinking has led to a statistics education reform with a paradigm shift in statistics instruction from statistical techniques, formulas, and procedures to developing statistical thinking, reasoning and conceptual understanding. In the 2018 Singapore H1 Mathematics Syllabus, the importance of statistical thinking has been emphasized: "A basic understanding of mathematics and statistics, and the ability to think mathematically and statistically are essential for an educated and informed citizenry and in other fields of study. For example, social scientists use mathematics to analyze data, support decision making, model behaviour, and study social phenomena." (Ministry of Education, 2018).

This chapter discusses the key characteristics of statistical thinking and underlying core issues.

2 Statistical Thinking and Determinism

Determinism had brought great success to mathematicians and scientists. In particular, the marvelous achievement of Newtonian mechanics convinced many people that there existed law and order in this universe that could be discovered and well represented by mathematical formulas (Kline, 1971). Such deterministic view once dominated our thoughts and guided our actions for long time.

Unfortunately, determinism may only apply to certain areas in the world. For most of the phenomena in the world, particularly related to human society, determinism seems powerless.

With amazing success of mathematical statistics in many areas, such as actuarial science, molecular motion, economic operation, population problem, distribution of intelligence and measurement error, statistical methods and the theory of probability have produced many thoroughly reliable laws that are simply unattainable by deterministic mathematical methods (Ben-Zvi & Garfield, 2004). People who are formerly convinced of determinism began to consider nondeterminism.

More significantly, people recognized the existence of an underlying irregularity and disorder even in phenomena formerly considered lawful. For instance, after many years of supreme effort astronomer Johannes Kepler (1571-1630) found that the path of Mars was an ellipse path (Kline, 1971). However, today we know from theory and more accurate observations that the true path of Mars deviates from an ellipse due to all sorts of perturbations arising from the gravitational attraction of the other planets. In fact, Kepler's laws happened to be the descriptions of the average behavior of the planets (Porter, 1986). Strictly speaking, they do not hold today. Moreover, the fate of Kepler's laws has been the fate of all scientific laws. They hold for a time and then some refinements are shown to be necessary with the general increase in scientific knowledge and enhancement of human cognitive abilities (Kline, 1971).

People thus formed a statistical way of perceiving the world. The mathematical laws of science amounted to no more than summaries of the average effect of irregular and disorderly occurrences (Kline, 1971; Porter, 1986). This is just like how kinetic theory of molecules states that temperature is simply the mass effect of irregular motions of billions of molecules. Since temperature is essentially a statistical quantity as an average measurement, temperature of a single molecule is somewhat meaningless. This attitude toward nature world and its law is known as the statistical way of perceiving the world, and it has made significant achievement in the reform of human thought in 20th century. Over the past two decades in statistics literature, statisticians have been vigorously promoting the idea that the development of a statistical way of thinking must be central in the education process and the variation is a fundamental element in statistical thinking (Pfannkuch, 1997; Ben-Zvi & Garfield, 2004). Though the variation element has not been addressed by educationists until recently, it is a powerful underlying conception that allows us to relate observed behavior to the abstract ideas of pattern, exceptions, and randomness.

3 Characteristics of Statistical Thinking

3.1 *Conduct research through method of mass observation*

The method of mass observation is a method peculiar to statistics. The interpretation of the "mass observation" may be two folded: (a) to make observations of general population under study that was composed of numerous individuals, and the "mass" here refers to large sample size; (b) to perform the same experiment a large number of times, and the "mass" here refers to large number of repetitions.

Mass observation method is based on the law of large numbers (LLN) whose essence is that in a mass of individual phenomena the general *regularity* is revealed more accurately than that is embraced by an individual observation. Therefore, in the basis of a statistical research, a mass observation of the facts always lies.

There is close relationship between mass observation method and LLN. We know in statistics the "population" is assumed to be a homogeneous class in terms of the quality. However, the quantity of the essence present in each and every individual of population is assumed to vary from one to the other. To use those variable individual values to find quantitative statistics that is representative of general characteristic of population (for example, mean, variance, etc.), sufficiently large number of individual observations are required as a sample. Only through such kind of mass observation, individual contingent idiosyncrasies may cancel one another so as to enable the inevitable overall regularity to be revealed. For instance, English-language letter frequencies vary quite a bit across different authors and different subjects. However, accurate average letter frequencies can be obtained by observing and analyzing a large amount of representative English text from a variety of sources, which is simply the effect of LLN. Figure 1 shows relative frequencies of English letter that are based on a sample of 40,000 words from *Cryptological Mathematics* by Lewand (2000). Letter E occurs most often with a frequency of 12.7%, about over 172 times more common than Z (0.074%) used in English text.

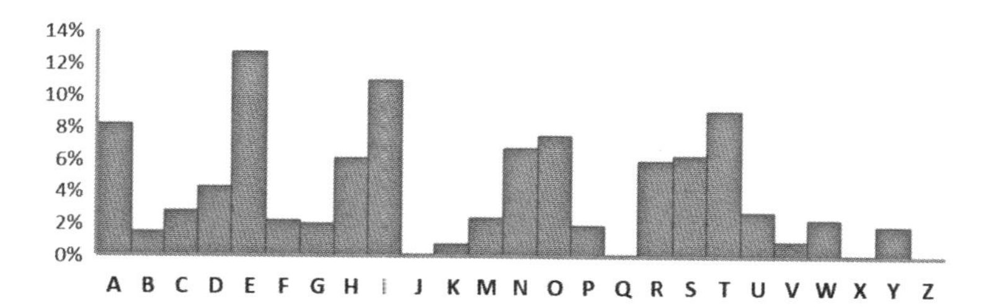

Figure 1. Relative frequencies of letters in the English language

Belgium mathematician and statistician Quetelet first introduced the theory of probabilities to statistical science. He had an excellent discussion on relationship between numerous individuals and population.

> "A person examining too nearly a small portion of a very large circle would see in this detached portion merely a certain quantity of physical points, grouped in a more or less irregular manner, and so, indeed, as to seem as if they had been arranged by chance. But, placing himself at a greater distance, the eye embraces of necessity a greater number of points, and already a degree of regularity is observable over a certain extent of the segment of the circle; and by removing still farther from the object, the observer loses sight of the individual points, no longer observes any accidental or odd arrangements amongst them, but discovers at once the law presiding over their general arrangements, and the precise nature of the circle so traced." (Quetelet, 1842)

Quetelet thought that the relationship between individuals and population is just like the relationship between points and circle. Statisticians consider population features such as mean and variance as indicators of stable properties of variable individuals that become evident only in the aggregate mass. Such stability can be regarded as the certainty in situations involving uncertainty (Ben-Zvi & Garfield, 2004).

3.2 *Care about both general characteristics and individual variability*

Through mass observation people noticed that individual variations are effaced and that the general characteristics of a population tend to be sharply defined. Various laws and theories of nature and human society amounted to no more than summaries of the average effects of irregular and disorderly occurrences, which are simply impermanent descriptions.

In a period of peace and prosperity, a nation as an entity usually displays a stable state with regular behaviour pattern. For instance, people from all walks of life live and work in peace and contentment; Men and women get married and raise families; People obey laws and engage in regular social activities; Elections are held on schedule and the winners take office; Principle of equitable and mutually beneficial relations are upheld between nations. If we did not know any more about the nation than just these facts, we may attempt to assert that the behaviour of nations and even life itself was determined by some super-being and constrained to follow this invariable pattern.

An advocate for the statistical view, however, would urge us to look closer at the behaviour of individuals themselves. What do we discover? Some people do not go to work, and they idle away their time and even steal; Quite a few people do not marry, or marry but have no children; At election time only a fraction of people vote, while of the rest, some do not care and some play false; Some people disregard the law and engage in anti-national activities; In international communication, some countries pursue hegemony, but some forfeit their sovereignty. Does the group behaviour merely describe its general mass effects? The statistical way of thinking reminds us to pay attention to all sorts of destabilizing factors causing irregularities and disorder that are often hidden behind the stable group behaviour, and be aware of possible social instability and revolution as a potential result (Kline, 1971).

The statistical thinking not only cares about general characteristics of population and its overall effect, but also recognizes the variability and even haphazardness of individual actions. It specifically allows for the possibility of alteration in the general characteristics and mass effect. One must consider both types of variability in order to understand scientific and social phenomena. *Variability* is fundamental in developing statistical thinking (Cobb, 1992; Moore, 1998). An important reason for focusing students' attention on the nature and sources of variability is not to encourage the practice in terms of procedures but instead to encourage their statistical thinking in order to understand the reality behind the data.

3.3 *Allow for errors*

In deterministic mathematics, the results turn out to be either correct or incorrect, either this or that. However, things are entirely different in statistics. Statistical hypothesis testing shows typical statistical way of thinking for decision making, and thus plays a fundamental role in statistics literature, particularly in statistical inference. The following example shows its unique way of thinking.

In traditional Chinese culture, full moon is appreciated by people in praise of a good sign and good fortune, even in some form of moon worship. However, in some other countries' culture, people believe that the full moon has a reputation for trouble and claim the moon as a sinister force that controls human behavior. Medical science researchers examined the existence of "lunar effect" influence on higher suicide occurrence and higher incidence of psychiatric disorder (Blackman & Catalina, 1973).

The number of psychiatric patients admitted to the Emergency Service of North Richmond mental health centre, Virginia was obtained before, during and after the full moon from August 1971 to July 1972, as shown in the table below.

Table 1

Number of patients served in psychiatric emergency room of the North Richmond mental health centre from August 1971 to July 1972 in three different moon phases

Year/Month	Before full moon*	During full moon	After full moon**
1971 August	6.4	5.0	5.8
1971 September	7.1	13.0	9.2
1971 October	6.5	14.0	7.9
1971 November	8.6	12.0	7.7
1971 December	8.1	6.0	11.0
1972 January	10.4	9.0	12.9
1972 February	11.5	13.0	13.5
1972 March	13.8	16.0	13.1
1972 April	15.4	25.0	15.8
1972 May	15.7	13.0	13.3
1972 June	11.7	14.0	12.8
1972 July	15.8	20.0	14.5
Average	10.92	13.33	11.46

Source: Blackman and Catalina (1973)
* Data refer to mean number of patients seen in the 10-day period before the full moon.
**Data refer to mean number of patients seen in the 10-day period after the full moon.

The data in Table 1 showed that more patients were seen during the full moon phase at the Emergency Service of North Richmond mental health centre than in other moon phases. Can we use one random sample to tell the existence of "lunar effect"? In other words, if the null hypothesis is that there is no "lunar effect", could we conduct a hypothesis test and decide if we should reject the null hypothesis or not on the basis of sample data? As one sample is just like one example, we cannot use a single example to prove a universal statement. However, a single counterexample is sufficient to disprove a universal statement (or reject a conjecture). A decision making in hypothesis testing centres on the null hypothesis. Using the court trial analogy, as the trial begins a defendant is assumed to be innocent. All evidence presented in a trial is to contradict the innocent hypothesis and hence obtain a conviction. If there is enough evidence against innocence, the court will declare the defendant guilty. Otherwise, the court will find the defendant not guilty. But this does not mean that the defendant is innocent. In a similar way, the null hypothesis in hypothesis testing is assumed to be correct, and a researcher conducts a study showing evidence to make a decision between "reject null hypothesis" and "fail to reject it". It is worth mentioning that "failure to reject null hypothesis" does not imply that null hypothesis is true, but means that there is insufficient evidence to reject the null hypothesis and we thus retain the null hypothesis. We may reject the null hypothesis when we have enough evidence to show that sample data are not consistent with the null hypothesis. In practice, people sometimes use expression "accept null hypothesis" to represent "fail to reject null hypothesis". However, we should note that "fail to reject" is not synonymous with "accept it".

A hypothesis test was conducted to test the existence of the "lunar effect". Our null hypothesis (H_0) is that there is no "lunar effect" and the alternative hypothesis (H_1) is that there exists "lunar effect". A paired sample *t*-test was performed on the differences between the number of patients seen on the day of full moon and the mean number seen in the 10-day period before the full moon, and it gave a *t*-test score of 2.13 (df = 11, *p*-value = 0.0284). As the *p*-value is less than 0.05, could the "lunar effect" we see in the sample just be an accident due to chance (that is, H_0 is true and a relatively rare event has occurred), or is it good evidence that there exists "lunar effect" in the population (that is, H_1 is true)? Since a rare event is considered to be very unlikely to occur, if it does happen, it indicates that there may exist certain unknown non-accidental factor influencing the outcome. So most likely we would not agree with the notion that the rare event has occurred by chance. It thus would make more sense to reject H_0 and in favor of H_1 that there exists "lunar effect", and it would be meaningful to make further investigation on the "lunar effect".

A similar test was conducted on differences between the number of patients on the day of full moon and the mean number seen in the 10-day period after the full moon. It yielded a *t*-test score of 1.55 (df = 11, *p*-value = 0.075). With the *p*-value greater than 0.05, we do not reject H_0

and conclude that there is no significant difference in the number of patients during the full moon day and the period afterwards. This finding may be due to a residual full-moon effect that was gradually dissipating.

Hypothesis testing allows us to use sample data to make inferences regarding population characteristics. Since in practice we could never examine the entire population and decision makings are based on sample data, there is always a possibility of making errors. In reality, a statistical hypothesis (the null hypothesis, H_0) may or may not be true, and we could make a decision of rejecting or failing to reject H_0 based on sample information. In statistical hypothesis testing, there are four possible outcomes as given in Table 2. Two situations lead to correct decisions that true H_0 is accepted and false H_0 is rejected, whereas the other two are called two types of errors in hypothesis testing that lead to incorrect decisions. A type I error occurs if one rejects H_0 when it is true. Conversely, a type II error occurs if one fails to reject H_0 when it is false. It is important to understand that a statistically significant result may happen and yet the null hypothesis is true (Type I error) because unlikely results are not impossible (Batanero, 2000).

Table 2
Possible outcomes of hypothesis testing

	H_0 True	H_0 False
Fail to reject H_0	Correct Decision	Type II error
Reject H_0	Type I error	Correct Decision

When we make a statistical decision, it is inevitable to make some level of error because fundamental *uncertainty* lies in a statistical inference procedure. Thus in statistical thinking, errors are allowed. It is impossible to make a decision without taking risks in making errors. However, we could control the risks of making wrong decisions and keep both Type I and Type II errors at acceptably low levels.

3.4 *Not the best, but for the better*

When we use statistical methods to solve problem, we could always give a better solution, but can never achieve the best solution. Table 3 shows life expectancies for men in 53 European countries in 2011.

We could use statistics method to find that the mean life expectancy is 74 years for men in European countries in 2011. Apparently, this is not an accurate value of the life expectancy, but is a point estimate of the mean life expectancy of European men. Some people will live longer than 74 years, some may live not as long as 65 years, but most people can expect to live for about 74 years. Furthermore, we could find the lower quartile and upper quartile to give a range between 69 and 79 that represents middle 50% of the data for estimating the life expectancy of European men. The estimate using interquartile range is more informative than the point estimate, but may not work so well because some people live longer than 79 while some live no longer than 69. If the life expectancy of European men follows a normal distribution, we could use confidence interval to estimate that at the 99% confidence level, the mean life expectancy of European men is between 72 and 76. In statistical interpretation, we say that in the long run, 99% of our sample confidence intervals will contain the true mean life expectancy of European men.

An interval estimate is often considered as a better estimate than a point estimate since it allows more information about a population characteristic. For a fixed population and a fixed confidence level, as we increase the sample size, the width of the confidence interval becomes narrower showing lower variability. Thus our uncertainty decreases and we would have greater precision. This makes sense because larger samples are closer to the true population. Given the same sample size, the width of the confidence interval increases as the confidence level increases. Thus a 99% confidence interval is wider than a 95% one. A higher confidence level means more accurate as it is more likely to contain the true value.

In this way, we can always choose a better estimate to improve the precision and accuracy of the parameter estimate, but we will never achieve the exactly correct value of the parameter of our interest. The famous quote by Box (1979) that "all models are wrong, but some are useful" expresses the quintessential statistical way of thinking. Though we could never find the best solution to a statistical problem, we could always look for a better one among various different solutions. How to decide which one is better may rely on many factors, such as experience, understanding and preference of the decision-maker. Statistical thinking could always guide us toward a better and useful solution.

Table 3

Life Expectancies for Men in European Countries in 2011

73	79	67	78	69	66	78	74
71	74	79	75	77	71	78	78
68	78	78	71	81	79	80	80
62	65	69	68	79	79	79	73
79	79	72	77	67	70	63	82
72	72	77	79	80	80	67	73
73	60	65	79	66			

Source: "WHO life expectancy" 2013

3.5 *Focus on correlation*

In our daily lives, we think and see things so often in causal terms that we may believe causality can easily be shown. When we see two events happen one after the other, our minds tempts to assume that the latter event is a consequence of the former one. The following true story written in Agnès Poirier's book (Poirier, 2018) shows us such an example.

In 1943, a renowned French philosopher Jean-Paul Sartre (1905-1980) wrote his first major work in philosophy, *Being and Nothingness.* His publisher, Galimard, knew that its commercial prospects were slim and thus produced a first printing of 1000 copies. To their amazement, the 700-page philosophical tome flew off the shelves, selling unexpectedly well. Both publisher and Sartre thought it must be because Sartre's thoughts on freedom and responsibility resonated with Parisians that the book was especially popular. To make his work more impactful,

Sartre made several amendments and supplements overnight, and submitted it to publisher for additional printing. Strangely, the newly printed books were returned and the bookshops requested for the exact same version as the first printing. How could that be? Then the truth dawned. It was during World War II, the price for copper kitchen weights rose significantly due to a serious shortage of raw metals. Kitchen balance and weights were essential to housewives' daily routine. It was found accidentally that the first printed *Being and Nothingness* weighed exactly one kilogram. People then rushed to buy the book to use it as a perfect and cheap substitute for standard copper weight, which had been sold on the black market or melted down to make ammunition. It was simply ironic that the reason for the popularity of *Being and Nothingness* was its weight, not its content.

In the real world, causality can rarely if ever be proven, only shown with a high degree of probability. Sticking too much to causality is perhaps one of the most stubborn fallacies of the human mind. In many circumstances, if we give up the persistence in seeking causality and turn to analyze correlations among different things, we may obtain unexpected results. The following story (Mayer-Schönberger & Cukier, 2013) that happened in America gives an example of how thinking in terms of correlation leads to success.

One day, an angry man stormed into a Target store (the American discount retailer store) in Minnesota to see a manager. "My 17-year-old daughter got this in the mail!" he shouted. "She's still in high school, and you're sending her coupons for baby diapers and cribs? Are you trying to encourage her to get pregnant?" When the manager called the man a few days later to apologize, however, the voice on the other end of the line was conciliatory. "I had a talk with my daughter," he said. "It turns out there's been some activities in my house I haven't been completely aware of. She's due in August. I owe you an apology."

In fact, Target has relied on predictions based on big-data correlations for years. Basically, its method is to analyze vast amounts of customers' data and let the correlations lead their work.

Target's analytics team reviewed the shopping histories of women customers who signed up for its baby gift-registry. They noticed that these women bought lots of various unscented lotion at around the third month of pregnancy, and that a few weeks later they tended to purchase supplements like magnesium, calcium, and zinc. The team ultimately uncovered around two dozen related products that enabled the company to calculate a "pregnancy prediction" score for every customer, and created a shopping model for pregnant customers based on their purchasing patterns. The correlation-based purchasing model enables the retailer to screen out those customers who have high "pregnancy prediction" scores and even enables the retailer to estimate the due data within a narrow range, so as to send relevant coupons for each stage of the pregnancy. Based on correlation analysis, this highly targeted promotion improved sales performance significantly.

The results obtained from statistical analysis may only explain the correlation connection between two things, but cannot establish causal relationship. This is one of the important characteristics of statistical thinking. With big data sweeping through the entire world, the importance of correlation has reached an unprecedented level. Viktor Mayer-Schönberger, Professor of University of Oxford and author of *Big Data*, told us that growing respect for "correlation" perhaps is the most fundamental change brought about by big data. It transforms how we understand and explore the world from a way of thinking in terms of causality to a way of thinking in terms of correlations. In the entire human history to date, people around the world have been conditioned to look for causes, searching for "why". However, the continuing quest for causality may lead us down to the wrong paths. So we suggest that, in a big-data world, in many situations, knowing what not why is good enough. For instance, machines generally do not break down without giving any form of warming. Armed with sensor data, correlational analysis can help to identify subtle changes and specific patterns during machine vibrations that typically crop up before something breaks. Spotting the abnormality in an earlier stage enables us to fix the problem before the breakdown actually occurs. This is called "predictive analysis" or "predictive maintenance". The cost of collecting and analyzing the data that indicates when we should take early action is much lower than the cost of an outage. The predictive analysis may not explain the cause of a problem, but knowing what is a problem is good enough for us to fix the problem.

Although correlation does not imply causation, in many situations knowing correlation alone could help to do many things. Moreover, correlations are not only valuable in their own right, but also point the way to causal investigations. By telling us which two things are potentially associated, correlations allow us to investigate further whether a causal relationship is present.

4 Conclusion and Discussion

We live in a world of uncertainty. We often have incomplete information when we need to make a decision. Statistical thinking provides a different way of understanding the world from deterministic mathematics and has its unique characteristics. In a world where every moment is determined by a complex set of uncertain factors, when a large number of observations are gathered and studied, overall regularity can be revealed. Given omnipresent nature of variability, statistical thinking not only cares about general characteristics of population, but also recognizes the individual variability in order to understand scientific and social phenomena. Statistical inference, as a typical statistical way of thinking, allows for errors when we make a decision. Controlling the risking of making errors could help us make a sound decision on whether a pattern is meaningful or just a coincidence. It is important to understand the two types of errors and do not interpret statistical test results in a deterministic way. With a statistical way of thinking, we always look for a better solution in the context of a problem, though it may not be the best one. In the era of big data, it is important to realize that analyzing correlations alone among different things rather than seeking causality could lead to great success.

In this big data era, big questions may come in different disciplines. Hence, new ability to analyze the data is required. Successful application of statistical science lies on embedding statistical thinking in the process. The challenge thus comes from both the specific areas of application and developing methods and concepts that will be broadly applicable. However, the fundamental tenets of good statistical thinking have not changed. At the heart of these statistical thinking are core issues about *uncertainty* and *variability* that have both a permanent value and a continuing challenge that is conceptual, mathematical and computational for the 21st century (Cox & Efron, 2017).

In recent years, much attention has been paid to simulating interest in statistical thinking. Many literature are often aimed at empirical prediction from large noisy data sets rather than probing the underlying interpretation of the data or understanding the nature of the data measurement process (Ben-Zvi & Garfield, 2004). Therefore, more emphasis should be placed on raising students' awareness about the characteristics of statistics thinking, developing teaching strategies to enhance students' statistical thinking in a variety of contexts and situations. Having a good understanding of those key characteristics of statistical thinking and the underlying core issues can help us understand the world around us and guide our thinking, learning and daily life in the 21st century.

References

Batanero, C. (2000). Controversies around the role of statistical tests in experimental research. *Mathematical Thinking and Learning, 2*(1-2), 75-97.

Ben-Zvi, D., & Garfield, J. (Eds.) (2004). *The challenge of developing statistical literacy, reasoning, and thinking.* Dordrecht, The Netherlands: Kluwer Academic Publishers.

Blackman, S., & Catalina, D. (1973). The moon and the emergency room. *Perceptual and Motor Skills, 37*, 624–626.

Box, G. E. P. (1979). Robustness in the strategy of scientific model building. In R. L. Launer & G. N. Wilkinson (Eds.), *Robustness in Statistics: Proceedings of a workshop* (pp.201-236). New York, NY: Academic Press.

Cobb, G. W. (1992). Report of the joint ASA/MAA committee on undergraduate statistics. In *American Statistical Association 1992 Proceedings of the Section on Statistical Education* (pp. 281-283). Alexandria, VA: American Statistical Association.

Cox, D. R., & Efron, B. (2017). Statistical thinking for 21st century scientists. *Science Advances, 3*, e1700768.

Kline, M. (1971). *Mathematics in Western Culture*. New York, NY: Oxford University Press.

Lewand, R. (2000). *Cryptological mathematics.* Washington, DC: The Mathematical Association of America.

Mallows, C. (1998). 1997 Fisher Memorial Lecture: The Zeroth Problem. *American Statistician, 52*(1), 1-9.

Mayer-Schönberger, V., & Cukier, K. (2013). *Big data: A revolution that will transform how we live, work, and think*. Boston, MA: Houghton Mifflin Harcourt.

Ministry of Education (2018). *2018 pre-university H1 mathematics syllabuses*. Singapore.

Moore, D. S. (1990). Uncertainty. In L. A. Steen (Ed.), *On the Shoulders of Giants* (pp. 95-173). National Academy Press.

Moore, D. S. (1998). Statistics among the liberal arts. *J. Am. Stat. Assoc., 93*, 1253–1259.

Pfannkuch, M. (1997). Statistical thinking: One statistician's perspective. In F. Biddulph & K. Carr (Eds.), *Proceedings of the 20th Annual Conference of the Mathematics Education Research Group of Australasia: People in mathematics education* (pp. 406-413). Rotorua, New Zealand: MERGA.

Poirier, A. (2018). *Left bank: Art, passion, and the rebirth of Paris, 1940-50*. London: Bloomsbury Publishing Plc.

Porter, T. M. (1986). *The rise of statistical thinking, 1820–1900.* Princeton, NJ: Princeton University Press.

Provost, L., & Norman, C. (1990). Variation through the ages. *Quality Progress, 23*(12), 39-44.

Quetelet, M. A. (1842). *A treatise on man and the development of his faculties*. Edinburgh: W. & R. Chambers.

Snee, R. (1999). Discussion, development and use of statistical thinking: A new era. *International Statistical Review, 67*(3), 255-258.

WHO life expectancy. (2013). Retrieved from http://www.who.int/gho/mortality_burden _disease/life_tables/situation_trends/en/index.html

Chapter 4

Situated Mathematics: Positioning Mathematics Ideas as Human Ideas

David WAGNER

Lists of 'big ideas' in mathematics education tend to highlight important mathematics concepts. This chapter places these concepts inside a bigger mathematical idea – the idea that mathematical acts and processes arise within contexts. Any mathematical idea develops as a response to a human concern or set of concerns. This view of mathematics underpins the teaching of mathematics as responsive to community and environmental concerns. This chapter considers how mathematics education literature that lists big ideas positions people. The chapter then identifies an alternative positioning in which mathematics can be responsive to community concerns, and thus supporting responsible citizenship oriented around sustainable development and peace. The chapter describes how some educators have worked at and might work at designing mathematics teaching to be responsive to community and environmental concerns. The chapter also identifies challenges of doing this.

1 Introduction

I begin with a thought experiment that can help us think about what is most important in mathematics.

> Hahn and Tai were flipping a coin. Hahn bet $100 on heads and Tai bet $100 on tails. They agreed that the first to get five would win the

$200. Hahn and Tai's game was interrupted and could not be continued with Hahn ahead—four heads to one tails. They are frustrated. They ask you how much of the $200 each person should get. And of course they want you to explain your reasoning.

Before continuing reading, I encourage readers to think about the answer that you would give to this situation. Some possible answers include the following. Perhaps you will recognize your answer in them:

- Hahn gets $200 because he was ahead when the game was almost done.
- They both get nothing, and the $200 goes to charity because it was a failed game.
- They each get their $100 back because the game was incomplete.
- Hahn gets $187.50 and Tai gets $12.50 because the probability of Tai winning is 1/16 assuming a fair coin. 1/16 of $200 is $12.50. (For Tai to win, the next 4 flips would have to be tails and probability of that is $\frac{1}{2} \times \frac{1}{2} \times \frac{1}{2} \times \frac{1}{2} = 1/16$.)
- Hahn gets $199.68 and Tai gets $0.32 because the coin that they were using shows heads 4/5 of the time and thus the probability of Tai winning is 1/625. That probability multiplied by $200 is $0.32. (For Tai to win, the next 4 flips would have to be tails and based on the history of this coin the probability of that is $1/5 \times 1/5 \times 1/5 \times 1/5 = 1/625$.)

The problem describes a situation that seems to invite mathematics. It seems to be a mathematics word problem. However, one could answer without mathematics, as in the first three answers given above. If one answers with mathematics, what is the rationale for using mathematics as opposed to one of the first three responses? It is impossible to use mathematics to justify the use of mathematics. We know that a mathematical answer would likely be favored because it appears objective and people usually want objectivity to govern difficult decisions. However, the apparent objectivity is a smoke screen because one needs to use a non-mathematical reason to choose a mathematical answer, and if one chooses a mathematical answer it is still necessary to use non-mathematical reasoning to identify which mathematics to use.

I doubt that anyone would choose the fifth answer above but I included it here to illustrate the problem of choosing within mathematics. The fifth answer would be insulting to Tai and it goes against the common idea that coin flips are fair. However, the choice among possible mathematical approaches is not so clear in other situations in which people use mathematics. Consider, for example, the choices of how to measure pollution in a climate treaty. Or consider the choices of how to model climate change.

Yet in school mathematics, the focus is usually on *how* to do mathematics. It is not on *when* to do mathematics. Lists of big ideas that govern or guide curriculum reflect and promote this focus of attention on procedure and processes.

2 Big Ideas in Mathematics

There are many lists of 'big ideas' in mathematics education, some discussed in scholarship but more often in professional literature, sometimes in official curriculum documents and sometimes by educators aiming to guide mathematics teachers. When we identify the big ideas of our discipline, we point at aspects which we consider to be especially important to the discipline, perhaps *the* most important ideas, and at the same time we are trying to promote a conception of mathematics that we think needs more attention.

In this chapter, I consider how the conceptualization of big ideas—our choices about what to include in the list, and our choices about how to write about them—position people, particularly learners of mathematics. Lists of big ideas tend to be organized by concepts or processes. This chapter places these ideas inside a bigger mathematical idea, which is the idea that mathematical acts and processes arise within contexts. This idea is juxtaposed against its apparent complement, which is the idea that mathematics is independent of human activity.

The chapter considers how literature that identifies big ideas may promote one of these two ways of thinking about mathematics. Does the literature position mathematics as human activity or independent of human activity? After reviewing the way the literature positions big ideas in mathematics, the chapter takes a next step to consider the implications for teaching mathematics oriented to sustainable development and peace.

With mathematics positioned as a human activity, it can be a tool for humans to be responsive to community concerns, and thus supporting responsible citizenship oriented around sustainable development and peace. I describe how educators can work at designing mathematics teaching to be responsive to community and environmental concerns.

3 Positioning Mathematics

In order to understand how mathematics is positioned by articulations of the big ideas in mathematics, it is necessary to be clear about what is meant by 'positioning.' Van Langenhove and Harré (1999) described positioning as the ways in which people use action and speech to arrange social structures. In other words, when we speak or write, our words put others into social positions. In this chapter, I will be commenting on the way an articulation of big ideas can positions students and mathematics. What does the word choice say about the role of people (students), the role of mathematics, and the relationship between them?

Beth Herbel-Eisenmann and I noted that many references to positioning in mathematics education identify how people are positioned in relation to mathematics (Wagner and Herbel-Eisenmann, 2009). We showed that such analysis does not fit the theory very well because the theory, as developed in the book edited by Harré and van Langenhove (1999), focuses on interactions among people while ignoring things transcendent to the interaction (things outside the interaction, things like mathematics). However, Herbel-Eisenmann and I explained how it is relevant to consider how mathematics is positioned. We showed that mathematics, though a transcendent force, manifests itself in local interactions, mediated through the people and artefacts involved in the interaction. In this way, it is appropriate to consider the way mathematics is positioned by articulations of big ideas in mathematics.

3.1 *Big ideas and how they position mathematics*

It is quite popular to identify big ideas in mathematics, especially in curriculum documents. However, not everyone identifies the same big ideas. Further, the way these ideas are articulated is significant. In this section, I describe how big ideas position students and mathematics in relation to each other.

Big ideas lists often comprise a listing of strands or major topics. Most traditionally, when big ideas are part of curriculum documents, these strands are things like *number*, *measurement*, *patterns*, *data*, etc. When the big ideas are presented in this simple way, there is no mention of people, and most certainly not of people making choices. Thus mathematics is positioned as independent of people. I will refer to this positioning as *mathematics alone.*

While a root list of big ideas may ignore the role of people, the way a strand (a big idea) is elaborated may recognize the role of people. For example, the province of British Columbia, in Canada has traditional strands as big ideas, but elaborates them in ways that identify human action. Under the heading of *number*, which is a big idea, the curriculum says, "Number represents and describes quantity. Quantities can be decomposed into smaller parts." This statement uses a passive voice but identifies action. I will refer to this positioning as *mathematics acts alone* because there is no reference to people in the description but there is reference to action.

In contrast, under the big idea of shape and space, the British Columbia curriculum says, "We can describe, measure, and compare spatial relationship." Here there is action, and the people doing the action are identified—'we'. I will refer to this positioning as *mathematics available for action*. This category name puts mathematics first because it is the organizing idea, and the name recognizes that people are identified as doing the mathematics.

Whatever the level of recognition of human activity in a list of strands as big ideas, mathematics is positioned as the focus of attention. Mathematics procedures and concepts are the big ideas. Thus people (students) are positioned as people who need to learn this mathematics. When writing this paragraph, I noticed how this idea that students are supposed to learn mathematics seems natural and appropriate, but I am going to problematize this idea below by opening up the question of what it means to learn mathematics.

A different kind of positioning appears in Small's series of books on big ideas in mathematics (e.g., Small, 2010). These books were written to support teaching for understanding. Here are two examples of Small's big ideas, each of which applies to many mathematical concepts, which she called strands. One of the big ideas is about comparison and the other about representation:

- Comparing mathematical objects/relationships helps us see that there are classes of objects that behave in similar ways.
- Mathematical objects/relationships can often be represented in multiple ways; each of those ways might make something about those objects more obvious.

In these presentations of big ideas, the focus is on action because the labels of the big ideas comprise actions: the big ideas are comparison, representation, etc. However, these big ideas are presented in the passive voice, and thus mask the humans who would be doing these actions presumably. I say this approach *masks* humans, as opposed to saying that it *hides* or *ignores* humans, because the presence of humans as active subjects is implied but not explicitly identified. In the first one there is a reference to people, but it is important to notice that it seems as if mathematics acts on people instead of people acting with mathematics (acting for themselves or acting on or for others) — "mathematics [...] helps us." I will refer to this kind of positioning as *mathematics acts on us*. In the second one on representation there is also reference to people, but in this case the people take action using mathematical acts. I will refer to this kind of positioning as *mathematics available for action*, as with the example from British Columbia curriculum. In both of these kinds of positioning, *mathematics acts on us* and *mathematics available for action*, the mathematics takes primacy because the big idea is centered on an action that is identified as characteristically mathematical.

Some published sets of big ideas have mixed lists, including in the same list ideas that are actions as well as ideas that are things. For example, Charles (2005) identified 21 big ideas for younger grades. He used labels such as *Numbers*, *Equivalence*, and *Comparison*. Technically, these are all nouns (things), but two of these are nominalisations—nouns made out of something else. *Equivalence* is the noun version of *equal*, which is an adjective (a description). *Comparison* is the noun version of *compare*, which is a verb (an action). Morgan (1998) and others have identified the prevalence of such nominalisation in mathematics and remarked that such nominalisations mask humans as actors. For example, if we use the verb *compare* we need a subject of the sentence, someone doing the comparison, but the word *comparison* can sit in a sentence without a person acting—for example, "A comparison of the two triangles reveals …"

Charles (2005) explained that a big idea cannot be a single word. Those are just convenient labels: "For ease of discussion each Big Idea [...] is given a word or phrase before the statement of the Big Idea (e.g., Equivalence). It is important to remember that this word or phrase is a name for the Big Idea; it is not the idea itself" (p. 10). The idea has to be a sentence. For Charles, "A Big Idea is a statement of an idea that is central to the learning of mathematics, one that links numerous mathematical understandings into a coherent whole" (p. 10). Sentences are more likely to reveal mathematical action as compared to a single word name of an idea. This distinction is important for the way mathematics and students are positioned. However, despite the intentions of the educators who make a list of big ideas, my experience working with teachers tells me that they and others will refer to them by their single-word labels.

To demonstrate how the full sentence rendering of a big idea shows action, we can consider the big idea that Charles labeled as *Proportionality*. That title alone does not suggest action or people, so it has a positioning of *mathematics alone*. However, the sentence describing this big idea shows action: "If two quantities vary proportionally, that relationship can be represented as a linear function" (p. 18). With this sentence it is an example of positioning in which *mathematics acts alone*.

3.2 *Positioning is a choice*

Curriculum writers have choices about what to say is a big idea, and how to write about these ideas. Table 1 presents an overview of the kinds of positioning identified above, identifying the role of people, as absent, active or passive. The last line in the table refers to a kind of positioning that will be introduced later in this chapter—*we do mathematics*—but it is included in the table now to complete the set of positionings.

Table 1

Positioning structures – relationships of mathematics and people

Positioning	Focus	Action	People
mathematics alone	Mathematics	no	absent
mathematics acts alone	Mathematics	yes	absent
mathematics acts on us	Mathematics	yes	passive
mathematics available for action	Mathematics	yes	active
we do mathematics	people doing mathematics	yes	active

This section uses the description of these positioning to revisit some of the big ideas given as examples in the section above. This is to illustrate how a big idea changes in the way it is presented. These distinctions are important for curriculum writers who have initiating influence on the way a set of big ideas is presented. It is also important for teachers and teacher leaders. They may not have a choice about which big ideas are in force because they have to follow curriculum, but they do have a choice about how to write and speak about the big ideas. In this way, teachers and teacher leaders may have the biggest impact on the roll out of big ideas.

Table 2 shows how each of the big ideas given in the above section could be written with different positionings. For each example, one of the positionings represents the quotation, repeated from above, and I wrote the other three to illustrate how the big idea is different with a different positioning in mind. It is important to note that the positioning that is represented for teachers and for students will significantly impact the way they perceive the mathematics they do.

The examples in the table include a big idea that is descriptive (with a focus on properties of objects/numbers/operations), a big idea that focuses on objects (with a focus on shape and space), and a big idea that is a process (with a focus on comparison). It is clear that some of the statements of the big ideas are more awkward than others. This is because process-based big ideas are easier to associate with action than are concept-based big ideas. Conversely, concept-based big ideas are harder to associate with action. For example, I could not conceive of a way to write about comparison in the *mathematics alone* positioning. This is because comparison is an action, and the positioning is defined by the absence of action. Thus the table says N.A. in that place.

The table only works with the first four positionings because the fifth one—*we do mathematics*—will be explained later in this chapter and because that positioning is a significant departure from the other four.

The exercise of trying to write an idea using different positionings is instructive. I recommend this exercise for curriculum writers in particular, but also for mathematics teachers. For an example of how the exercise gave me insights, I draw attention to the move from the less active to more active positionings in any of the examples. Even the way these sentences refer to mathematics needs to be different. To illustrate, for the properties example, I found it difficult to use the word *properties* in the *mathematics available for action* positioning. Properties are static and thus difficult to associate with action. Trying to write that big idea with action forced me to think about the social function of mathematical properties. How do mathematical properties serve our human purposes?

Table 2

Examples of wording big ideas for different positioning

Focus	Positioning	Big Idea
Properties	mathematics alone	Mathematical objects are defined by unique properties.
	mathematics acts alone	"For a given set of numbers there are relationships that are always true, and these are the rules that govern arithmetic and algebra." (Charles)
	mathematics acts on us	Mathematical rules guide the way we negotiate number and prediction.
	mathematics available for action	We can use agreed upon number relationships to communicate effectively and dependably.
Shape and Space	mathematics alone	Geometric objects have identifiable characteristics.
	mathematics acts alone	Geometric objects are defined by the rules of geometry.
	mathematics acts on us	Geometry allows us to design and organize our spatial environments.
	mathematics available for action	"We can describe, measure, and compare spatial relationship." (British Columbia curriculum)
Comparison	mathematics alone	NA
	mathematics acts alone	Mathematics is the relationship among numbers (or shapes), and the relationships among these relationships.
	mathematics acts on us	"Comparing mathematical objects/relationships helps us see that there are classes of objects that behave in similar ways." (Small)
	mathematics available for action	We can use mathematical knowledge to communicate comparisons among objects or numbers.

3.3 *Situating mathematics*

The four positioning structures explored above reflect the way big ideas are written in typical mathematics education contexts. As Charles (2005) noted, for convenience it is usual to use a word or phrase to name the big idea while the idea itself is usually elaborated with a complete sentence. I support his advice because it promotes a view of mathematics as action. In all the examples I found of big ideas in mathematics or in mathematics education, the names (or labels) referred to mathematical objects, characteristics or processes. This positions mathematics as most prominent. The names I chose for the positioning structures begin with mathematics to reflect this focus on the mathematical objects, characteristics and processes.

However, it is important to look at what is missing in each kind of positioning. Illustrations can help us see what is missing. I have drawn diagrams to illustrate the positioning in each of the classifications from Table 1 above. They are presented in order from the most static to the most active, but it is instructive to see them all in relation to each other. This comparison among the diagrams helps show what is missing.

When the big idea does not describe action or people it appears that mathematics exists on its own, not interacting in any way. This positioning, which I have named *mathematics alone* is illustrated in Figure 1.

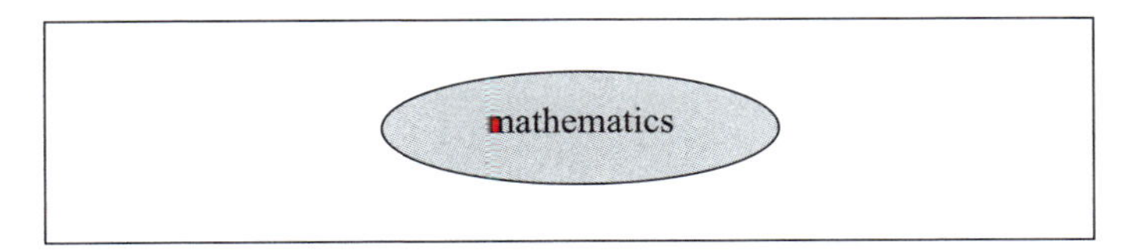

Figure 1. Illustrating positioning – mathematics alone

When the big idea describes action but not people, who is doing the action and to whom? Or what is doing the action to what? It appears from the big ideas in this category that mathematics acts on itself. There is mathematics, the mathematics does things (which would be mathematical action) and this develops further mathematics. This positioning, which I have named *mathematics acts alone* is illustrated in Figure 2.

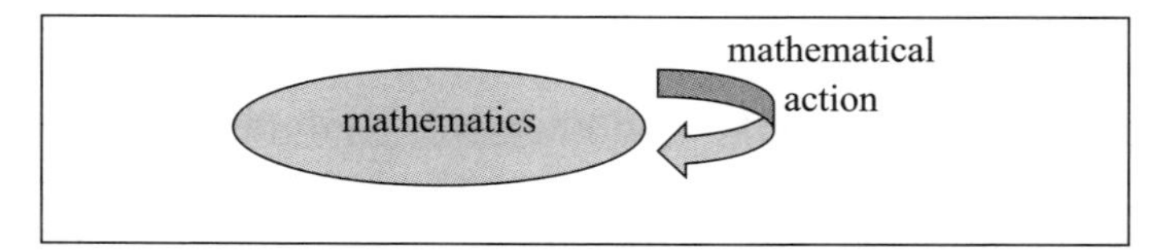

Figure 2. Illustrating positioning – mathematics acts alone

When the big idea describes action and puts people in the passive role, there appears to be mathematical action impacting or guiding people. The strange thing is that it is unclear who is initiating the action. Perhaps it is mathematics, similar to Figure 2. Nevertheless, this positioning, which I have named *mathematics acts on us* is illustrated in Figure 3. I was not sure whether to include another bubble above the arrow to show mathematics doing mathematical things to people. I chose not to include that bubble because the elaborations of big ideas that fit this category typically describe mathematical action and people but do not describe anyone in the position of actor. This trick is done using the passive voice when writing the big ideas.

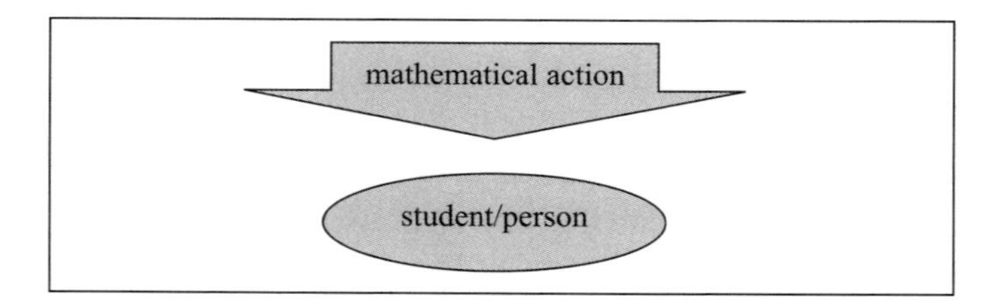

Figure 3. Illustrating positioning – mathematics acts on us

When the big idea describes action and puts people in the active role, the illustration seems to make more sense (to me). This positioning, which I have named *mathematics available for action* is illustrated in Figure 4. The right hand part of the illustration is included in response to the new positioning that I am promoting in this chapter. That new positioning comes out of an intentional concern for making mathematics education a part of our common human agenda to live at peace with each other and with our cohabiters in our environment. I asked myself how these positionings relate to community concerns (which include concerns about the environments in which communities are situated).

The question was this: How do these big ideas position students in relation to their environments? However, the big ideas that are characterized within the first three categories of positioning in Table 1 do not describe students as actors. Their actions in relation to community concerns cannot make sense, even in some possible future, because there are no actions that they take. Mathematics seems to do the acting. In contrast, for this fourth positioning structure, the student is positioned as an actor. However, following the big idea descriptions in this category, there is no mention of community concerns. Thus it seems to be implied that student action in their learning environments may be positioned as equipping students for possible future actions in relation to their communities (again, including the communities' environments).

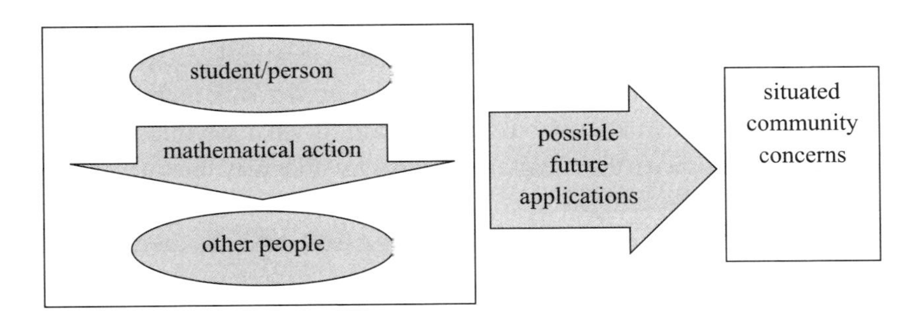

Figure 4. Illustrating positioning – mathematics available for action

Finally, I present Figure 5, which introduces a fifth form of positioning, which describes the kind of mathematics that I promote in the next section of this chapter. I have named this positioning *we do mathematics* (also appearing in Table 1) because it is focused on action, and it foregrounds the role of the person doing mathematics. In this structure, the learning environment includes consideration of community concerns so students are already addressing important concerns. It is not left as a possibility for the future. Of course, engaging with community concerns while learning mathematics also prepares students for future action in this vein.

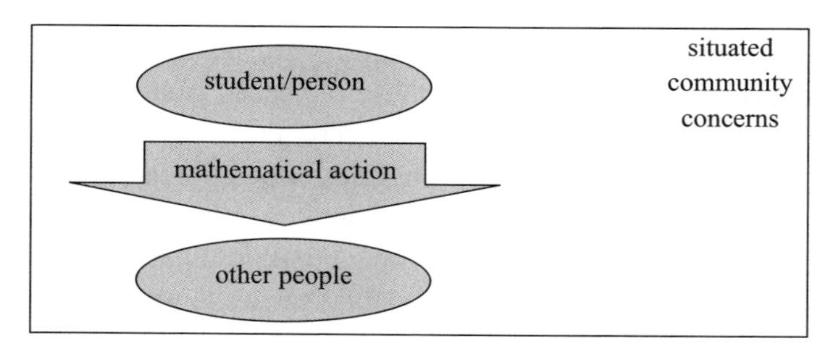

Figure 5. Illustrating positioning – we do mathematics

4 Mathematics for Sustainable Development and Peace

The section above analyzes big ideas in mathematics education. Lists of big ideas reflect the kind of foci and the kind of teaching that have dominated mathematics learning for some time. This way of teaching and learning positions mathematics in certain ways and, reciprocally, it positions students in certain ways. In other words, our imagination of what mathematics can be and our imagination of what we might do with or in relation with mathematics is guided by the way mathematics is presented in our schooling.

Meanwhile it is becoming increasingly clear that we, the people of the world, need to work actively at finding ways to live in more peaceful and more environmentally sustainable ways. I present here a vision for teaching mathematics for peace and sustainability. This vision fits the kind of positioning described above as *we do mathematics*. I say it is a bigger idea than any of the ideas in the lists of big ideas I have found. The bigger idea is that mathematics can and should be developed and used to support sustainable and peaceful living—concisely stated thus:

Bigger idea: We do mathematics to address needs in our community.

It may be appropriate still to use a list of big ideas, which I suggest to be written to highlight activity as described in the above section. All these big ideas, however, should be placed under the bigger idea that mathematics serves the needs of people in particular contexts. Mathematics is positioned as situated in contexts.

This vision requires a clear understanding of what it means to live peacefully and with sustainability. I will use the Sustainable Development Goals (SDGs) negotiated and articulated by the United Nations (UN) in 2015 as part of its 2030 Agenda for Sustainable Development (United Nations, 2015). This set of goals is well-articulated and a warranted base for education because it was constructed in negotiation among the diverse peoples of the world. These goals focus attention on common needs across nations, societies, and cultures for practices that ensure viable environmental and social structures with consideration of people's needs for economic viability. I emphasize that sustainable development as described in the SDGs includes the goals of peace and good citizenship. This goes beyond some conceptions of sustainability, which focus on the environment. The UN resolution motivates attention to these concerns but should do so in ways that reflect these concerns.

It is hard to imagine someone being against the SDGs. Who wants war? Who wants to pollute drinking water? Who wants to give some cultures more prominence than others? Who wants the concerns of men and boys to eclipse the concerns of women and girls? A look at the world around us may lead us to think that there are people who want war, who want to pollute drinking water, etc. However, I believe (or at least I would like to believe) that it is not that people want war or polluted water or these other terrible things but rather that they weigh one goal against others and decide, for instance, "I will pollute this water because it brings me some wealth" (or perhaps "it brings jobs to my community").

4.1 *Competencies for responsible mathematics*

Teaching with the bigger idea in mind, that we do mathematics to address needs in our community, one may wonder which takes precedence—the mathematics or the sustainable and peaceful life. One way to answer this question is to ask which is more important to us. Do we want our children to have strong mathematical knowledge and skill? Do we want our children to contribute to a peaceful and sustainable society? (It is worth thinking about what we want both for our students and for the children in our families.) For me the second wish is stronger, the wish for my children and my students to contribute to peace and sustainability. However, these goals are not mutually exclusive. The two goals support each other. Mathematics develops through a focus on community needs for sustainability and peace—mathematics itself develops this way and so does a learner's mathematical skill and understanding. And community needs can be supported with mathematics.

A useful way to think about this is with mathematical competencies. Consider again the thought experiment with which this chapter opens—the interrupted coin-flip game. There are mathematical skills that one can use to answer the problem but it is also necessary to justify the use of mathematics as the appropriate way to answer the problem.

It is quite common to differentiate between skill and understanding in mathematics. Skemp (1978) famously differentiated between instrumental and relational understanding. I extend this distinction to include other important kinds of knowing. All of these competencies are important under the rubric of the bigger idea—that people do mathematics to address community needs—and its corollary that mathematics teaching should engage students in situated mathematics:

- *knowing how*: Students need to have skill in performing procedures. They need to know how to do important mathematics. I think this form of knowing is over-emphasized in school mathematics but it is still important. Skemp (1978) referred to it as 'instrumental understanding.' In the coin toss example, this could be the skill of multiplying fractions to calculate the theoretical probability, and the skill of applied ratios to connect that probability to the monetary award.
- *knowing why*: Students need to know why mathematical procedures work, and also why mathematical properties are

true/dependable. Skemp (1978) described this as 'relational understanding.' In the coin toss example, it is necessary for the student to understand probability in order to know to multiply the fractions to calculate the theoretical probability, and to understand ratios in order to know how to apply the probability to the monetary award.

- *knowing which*: Students need to know which mathematics to use in which situations. It is important to know which approaches and which procedures. This requires good knowledge of a spectrum of mathematics, and thus relates to what Mason & Spence (1999) referred to as 'knowing-to-act'. It means knowing to use particular knowledge or skills in a given situation. In the coin toss example, this means knowing that the problem could be interpreted as a probability situation.
- *knowing when*: I think of this as critical understanding. Students need to know when it is appropriate to use mathematics and how reliant a decision should be on the mathematics. Such reflection is important for mundane situations such as deciding on a gift purchase for a loved one (e.g., Wagner, 2011) and for significant social issues such as deciding on immigration policy. In the coin toss example, this competence governs the decision to use mathematics or not, and enables justification for one's choice in this regard.
- *knowing to*: Students need to have confidence to use mathematics to address personal and community needs when it is appropriate to do so. This can be referred to as mathematical agency. Mason & Spence (1999) described the difference between passive knowing and active knowing in their description of 'knowing-to-act.' They described classroom contexts, in which students need an active knowing. I extend this to point out the necessity of applying such active knowing to situations outside school, situations where others are not considering mathematical approaches. There is a developing rich tradition of critical mathematics education that gives examples of how this kind of critical agency can be developed in school (e.g., Gutstein & Peterson, 2005: UNESCO-MGIEP, 2017). This competence

may not apply to the coin toss example. However, If we imagine ourselves happening upon an argument between Hahn and Tai, we can imagine ourselves choosing to intervene using mathematics to help calm the dispute in a way that they may accept as fair. That would be an example of this competence at work.

- *knowing about*: I think of this as understanding the discipline. It is meta-understanding. Good mathematics students see how mathematical knowledge and action impacts one's perspective, and how pervasive mathematical practices (in school and in the press) influence social norms. In the coin toss example, this would mean understanding the seductiveness of a mathematical answer and weighing the appropriateness of suggesting mathematics as a solution when we know that mathematics may be favoured due to social norms.

4.2 *Teaching mathematics for responsible citizenship*

For teachers who want their students to develop holistic mathematical competencies (the full list of ways of knowing, given in the above section), let us consider how to do this. The most important step for teachers who want their students to develop mathematics for responsible citizenship, is to engage themselves with their community and environment and to do so with attention to mathematics and potential mathematics. I encourage mathematics educators to 'walk mathematically' and notice what I call "the roots and fruits of mathematics."

The *roots* refer to nature and the mathematics it inspires. Pay attention to natural phenomenon and ask how these inspire our mathematical obsessions. For example, a tree often inspires us to think about right angles. No tree forms an actual, perfect right angle with the land around it, but the idea of perpendicularity governs the growth as the tree reaches for the sun and gravity flattens the soil. Another example, is the human hand, which motivates us to base counting systems on fives and tens. The key question for the roots is this: How do natural phenomena inspire our mathematical obsessions?

The *fruits* refer to human constructions and how mathematics influences the way we design things. This can refer to physical constructs such as a bridge or a street plan, and it can refer to conceptual constructs such as the organization and administration of a hospital and the medical procedures done therein. For example, with a street plan, we may notice a rectangular organization of streets. We would then ask why streets are organized like this? Perhaps a hexagonal organization, like a beehive, would be more efficient. The key question about fruits is this: How do our mathematical obsessions shape the things we construct in the world?

Walking mathematically in this way will enable a teacher to choose contexts in which to situate mathematics learning. Ideally, the teacher should focus on community concerns, especially ones that relate to sustainable development tensions. For example, teachers in my workshop at the Mathematics Teachers Conference 2018, *Mathematics Task Design for Responsible Citizenship*, noticed that Singapore's leaders have be interested in the distribution of ethnic groups. The leaders wished to promote multiculturalism and avert ghettoization. Decisions the city made impacted peaceful living and this had significant impact on environmental harmony.

The next step is to identify the mathematics that people do in the context and the mathematics that might be done to address the concerns in the context. With the example of civic demographics, statistical work is necessary for understanding the current distribution, and modelling work can be used to consider new policy. We know that such statistics and modeling was already at play in Singapore, but we did not yet have time to investigate how it was done. To develop classroom mathematics tasks, mathematics could be used to understand the work already being done by Singapore's leaders and mathematics could be used to consider alternatives.

The next step is to ask which mathematics fits the capability of one's students. In the workshop, we did not take this step in relation to civic organization because we did not have the resources to complete the previous step. Nevertheless, participants said that statistical methods and proportionality could be used to understand the current demographic distribution better. For example, students could gather their own statistics in parts of the city and compare those to published statistics on distribution. It was also suggested that some modeling could be done to consider how policy changes may impact redistribution.

My observation as a teacher and a researcher is that teachers and people in general too often underestimate the capabilities of others. People especially underestimate the capabilities of someone younger. When students engage mathematically with contexts that are meaningful to them, already they will have better understanding than they do with mathematics that is done outside of meaningful contexts. Furthermore, students will be more motivated to explore the context and the mathematics when mathematics is introduced in meaningful contexts.

Finally, the teacher can develop the context into some classroom-ready prompts for student work. For this step it is necessary to emphasize that students should be asked to do mathematics that people actually do in the context or that people might do in the context. When I have worked with teachers and textbook writers who want to develop mathematics for sustainability, I find that it is very easy to fall back to familiar mathematics questions, which involve calculations that have nothing to do with the concerns in the situation.

The five steps described above, given as a list, are summarized here for convenience. It would be helpful to meet regularly with other teachers who have also chosen to follow the advice for teaching responsible mathematics. The last four steps are especially amenable to group work, but even walking mathematically is a rich activity to do with colleagues as you will be inspired to notice new things by the observations of your peers.

- *Walk mathematically*: Pay attention to the mathematics in your community and environment.
- *Choose a context* that involves local concerns that relate to tensions involving sustainability (environmental sustainability, or sustainability of peaceful coexistence).
- *Identify the mathematics* that people do in that context and that they might do.
- *Decide which mathematics* is within the capability of your students (do not forget that they may impress you).

- *Develop questions/tasks* for classroom action, in which the students do mathematics that people really do in such contexts.

4.3 *An example of mathematics teaching for responsible citizenship*

I close with an example of a teacher *walking mathematically*. Everywhere she goes, she looks at her city through mathematical eyes and eyes that care for the community. One day, she walks through a market and takes pleasure in the wide variety of food available, and the many people involved in bringing the food here. She thinks of the organization and the physical work involved and remembers reading something about some food production taking a heavier environmental toll than other foods. She has now *identified a context* for engaging her students in mathematics for responsible citizenship.

She goes home and does some investigation on the internet, finding some articles on the environmental costs of foods. However, these articles are from Europe, the UK and the USA. These articles focus on foods less common in Singapore. It would be good to know about local foods that she and her students eat regularly. She reasons that even the same foods would have different environmental impacts in different locations due to the costs of transportation and production differences in different climates.

She *considers the mathematics* that must be done to bring food to her people. There would be mathematics done by farmers and fishers, but she does not know much about that activity. She makes a mental note to continue researching this, but she wants to engage her students in the food issues already. There is also mathematics done by the vendors she meets at the market and their suppliers. The cost of the foods would depend on the distance the food is transported and on any refrigeration needs along the process.

She *decides to focus* on the cost of transportation, specifically the distances the food travels to come to her market. She will talk with her colleagues about the mathematics involved in food production and its transportation but she already plans to gather some local information. She decides to ask her students to keep track of the food their families eat for one week. She wonders if she should tell them how to gather the data or if she should let them figure it out. Why should she do all the mathematics? Let them do it. They will learn more mathematics this way. The next day she tells her students of her plan to use their local food practices to gather data that they will be able to use to understand their community's needs better and to learn more mathematics. She gives the students time to organize themselves and make decisions about how to keep track so that everyone collects comparable data. This requires some intervention on her part.

The teacher notices that this activity will engage her students in the full range of mathematical competencies. When gathering data, her students will have to make decisions about which data is most important and how to record it. Thus, they will be using algorithms (showing that they *know how* to do mathematics). In their decision-making they will have to *know which* algorithms to use, which requires them to *know why* these algorithms work (knowing how they work). Once they use their data to understand their community's food needs better, they will have to apply this knowledge to the situation and balance it against other concerns, including the importance of people's choices and pleasures regarding food. People do not choose foods on the basis of efficiency alone. These questions will have them thinking about *when* to use their data to make arguments. Finally, they will become more aware of the impact of using mathematics to make their arguments; they will *know about* their mathematics better. She wants them to know that they are expected to use their mathematics to understand their world better and to make it a better place. They will *know to* use their mathematics.

5 Conclusion

A resource that I would recommend for the development of classroom tasks is a guidebook for textbook authors who want to make peace and sustainability the center of their texts (UNESCO-MGIEP, 2017). It is published by the Mahatma Gandhi Institute of Education for Peace and Sustainable Development (MGIEP), which is an arm of UNESCO (United Nations Educational, Scientific and Cultural Organization). I wrote the mathematics chapter in that book along with Masami Isoda, Parvin Sinclair, and Antonius Warmeling. The guidelines for textbook authors in the mathematics chapter can be applied easily to developing classroom tasks. The chapter also points to other helpful resources.

The fast-changing world around us comprises significant change in social structures and in the environment. There is no time to waste with mathematics teaching that positions students as passive. I encourage a focus on big mathematical ideas that foreground human action, and to see these under the rubric of a bigger idea—we do mathematics to address needs in our community. I hope that the advice I give here is helpful for teachers who wish to take seriously the necessity of teaching mathematics for responsible citizenship.

Acknowledgement

I am thankful for the meaningful collaborations I have had with other mathematics educators interested in developing mathematics resources oriented around sustainable development and peace. I am particularly thankful for international collaborations in this vein. In particular, I acknowledge the leadership of Shankar Musafir and Yoko Mochizuki of the Mahatma Gandhi Institute of Education for Peace and Sustainable Development (an arm of UNESCO). I am honoured to be called upon by them to engage with people around the globe in this important work.

References

Charles, R. I. (2005). Big Ideas and understandings as the foundation for elementary and middle school mathematics. *Journal of Mathematics Education Leadership, 7*(3), 9-24.

Gutstein, E., & Peterson, B. (2005). *Rethinking mathematics: Teaching social justice by the numbers*. Milwaukee, WI: Rethinking Schools, Ltd.

Harré, R., & van Langenhove, L. (1999). *Positioning theory: Moral contexts of intentional action*. Blackwell, Oxford.

http://unesdoc.unesco.org/images/0015/001524/152453eo.pdf

http://www.un.org/ga/search/view_doc.asp?symbol=A/RES/70/1&Lang=E

Mason, J., & Spence, M. (1999). Beyond mere knowledge of mathematics: The importance of knowing-to-act in the moment. *Educational Studies in Mathematics*, *38*, 135-161.

Morgan, C. (1998). *Writing mathematically: The discourse of investigation*. London: Falmer.

Skemp, R. (1976). Relational understanding and instrumental understanding. *Mathematics Teaching, 77*, 20-26.

Small, M. (2010). *Big ideas from Dr. Small, Grade 9-12.* Toronto: Nelson Education.

UNESCO-MGIEP. (2017). *Textbooks for sustainable development: A guide to embedding.* New Delhi: United Nations Educational, Scientific and Cultural Organization Mahatma Gandhi Institute of Education for Peace and Sustainable Development.

United Nations. (2015). *Resolution adopted by the General Assembly on 25 September 2015.* New York: United Nations.

Van Langenhove, L., & Harré, R. (1999). Introducing positioning theory. In R. Harré & L. van Lagenhove (Eds.), *Positioning theory: Moral contexts of intentional action* (pp. 14-31). Blackwell: Oxford.

Wagner D., & Herbel-Eisenmann, B. (2009). Re-mythologizing mathematics through attention to classroom positioning. *Educational Studies in Mathematics, 72*(1), 1-15.

Wagner, D. (2011). Warm bodies using cold mathematics. *Antistasis, 1*(2), 7-9.

Wagner, D. (2017). Reflections on research positioning: Where the math is and where the people are. In H. Straehler-Pohl, N. Bohlmann & A. Pais (Eds.), *The disorder of mathematics education: Challenging the sociopolitical dimensions of research* (pp. 291-306). New York: Springer.

Chapter 5

Big ideas of Primary Mathematics: It's all about Connections!

Chris HURST

Big idea thinking provides an opportunity to re-conceptualize how we view and teach primary mathematics. The real value of big ideas lies in interpreting the mathematics within them. Big ideas are those that connect mathematical understandings into a coherent whole, and are central to the learning of mathematics. Big ideas comprise a network of 'little ideas' or 'micro-content' and teachers who think in terms of them are able to look forwards and backwards from their own year level to identify specific content that a student may not know, and to lay the foundations for what the student needs to know next. Big idea thinking encourages teachers to deconstruct and reconstruct their knowledge. Teachers can actively engage in this by beginning with a mathematical idea such as place value and building a concept map showing the various pieces of 'micro-content' that contribute to the development of the concept and considering how the content is connected. In this chapter, we demonstrate this by considering two big ideas in the mathematics curriculum: the multiplicative array and pattern.

1 Introduction: What are big ideas?

It may be helpful to begin this chapter with some definitions of big ideas starting with the following description by Charles (2005):

> A Big Idea is a statement of an idea that is central to the learning of mathematics, one that links numerous mathematical understandings into a coherent whole. (p. 10)

The key element of Charles' definition is the notion of linking or connecting ideas, a characteristic that features in other definitions and descriptions. However, contemporary discussion around the notion of the big ideas of mathematics is quite diverse. While there is agreement about the need to think in terms of big ideas and about certain elements of them, it is difficult to find a single definition agreed to by most or all educators, for what constitutes a big idea. As well, no particular number, set, or list of big ideas representing any sort of consensus has been generated (Charles, 2005; Clarke, Clarke, & Sullivan, 2012), and it has not been made clear what to actually do with big idea thinking in terms of curriculum development.

Siemon, Bleckly, and Neal (2012) generally concurred with Charles' definition but noted that, in his description of his identified big ideas, there was no discussion of developmental progressions or learning trajectories to facilitate linking them to a curriculum document. Siemon et al. cited the work of Kuntze, Lerman, Murphy, Siller, Kurz-Milcke, and Winbourne (2009) noting four characteristics of big ideas as having

> … high potential for building conceptual understanding, meta-knowledge about mathematics as a science, meaningful communication strategies, and professional reflection.
> (Siemon et al., 2012, p. 22).

To provide further clarity, Siemon et al. cited earlier work by Siemon (2006, 2011) describing a big idea as

> An idea, strategy or way of thinking about some key aspect of mathematics without which, student's progress in mathematics will be seriously impacted; [which] encompasses and connects many other ideas and strategies; [and] provides an organizing structure or a frame of reference that supports further learning and generalizations.
> (Siemon et al., 2012, p. 22).

As with Charles' definition there is reference to connections between ideas but also a sense of structure to support teachers in assisting their students. Similarly, in seeking to define big ideas, Askew (2013) described them in this way:

> [being] big enough to connect together seemingly disparate aspects of mathematics, but not so big that it is unwieldy; [and having] currency across all the years of primary schooling, because then children get to revisit big ideas across the year groups. (Askew, 2013, p. 7-8).

Again, as with the definition by Siemon et al. (2012), there is a clear allusion to structure in that understanding can be built over time if the connections are made explicit for children.

Charles (2005) and Clarke et al. (2012) both noted that it would be highly unlikely for any group of mathematics educators to agree on what the big ideas were or how many of them there should be. Furthermore, both Charles and Clarke et al. stated that this is not the point of big idea thinking and that

> the value of a discussion of important ideas or big ideas is precisely because it stimulates each teacher respondent (or participant in a professional learning program) to deconstruct her/his own conceptual structures, and not in order to assess fidelity to a pre-ordained list. (Clarke, Clarke & Sullivan, 2012, p. 15)

It is this characteristic of big idea thinking that is seen as the most important. That is, enabling or facilitating teachers to break down an important mathematical idea to its component parts and reassembling those parts in a meaningful and connected way. Charles (2005) described it in terms of using big ideas to stimulate conversations about mathematics and this sits well with the point made by Siemon et al. (2012) about 'personal reflection'. However, whatever one identifies as a big idea is characterized by one common element – the connections within a big idea and between it and other big ideas (Australian

Association of Mathematics Teachers, 2009; Hurst, 2015; Hurst & Hurrell, 2014; Siemon, Bleckley, & Neal, 2012).

2 Historical Development of Big Idea Thinking

Big idea thinking is not new but discussion of it has gained momentum since Charles' article was published in 2005. It could be argued that it had its beginnings in the work of John Dewey (1859-1952), in particular about concept formation. In a review of Dewey's work, Laverty (2016) noted that it is essential for teachers to help students develop an understanding of concepts and to know how to use them, implying a sense of connection, as opposed to the acquisition of words and empty symbolisms. Laverty also noted that concepts are never completely developed which sits well with the earlier-mentioned notion of deconstructing and reconstructing knowledge (Clarke et al., 2012). Concept formation also features strongly in the work of Bruner (1960) who stated that the teaching of structure, rather than teaching isolated facts and procedures, promotes transfer of knowledge through the relationships between ideas. The notions of aiding transfer and reducing the cognitive load were also seen by Charles (2005) as beneficial features of big idea thinking.

Clark (2011) discussed concepts rather than big ideas. The common thread between Clark's work and that of others is again the idea of connecting segments of knowledge and Clark argued that it is the connections which are powerful tools enabling humans to construct knowledge. He described concepts as follows:

> My working definition of concept is a big idea that helps us makes sense of, or connect, lots of little ideas. Concepts are like cognitive file folders. They provide us with a framework or structure within which we can file an almost limitless amount of information. One of the unique features of these conceptual files is their capacity for cross-referencing. Because concepts focus on similarities and homologies, they provide powerful linkages between what would otherwise be considered disparate and seemingly incompatible information. (Clark, 2011, p. 33)

3 The Structure of Big Ideas

It has already been noted that both Charles (2005) and Clark et al. (2012) stated how it would be difficult to obtain agreement as to what the big ideas of mathematics should be. However, Askew (2013, p. 6) is of the view that mathematics education researchers "should be able to reach agreement on defining big ideas and then describe what these might be in practice". This is clearly worth pursuing as there has been a wide variety of big ideas put forward by various researchers and it would be helpful to reach some agreement in order to assist teachers to interpret curricula. There is a sense of urgency in how Askew discusses big ideas in that he notes a lack of explicit attention to them in the South African curriculum documents which could lead to "a fragmented knowledge of mathematics and, consequently, be a factor in continued low attainment in mathematics in South Africa" (Askew, 2013, p.6). The same criticism has been levelled at the Australian Curriculum: Mathematics and the Common Core State Standards for Mathematics in the USA (Hurst, 2015).

Charles (2005) proposed 21 big ideas which are shown in Table 1. He did not present any form of learning progression or trajectory but rather gave a series of dot points showing main points within each big idea. For example, one of his big ideas is estimation and he provided the following points about it. In relation to numbers, he said that

- The numbers used to make an estimate determine whether the estimate is over or under the exact answer.
- Division algorithms use numerical estimation and the relationship between division and multiplication to find quotients.
- Benchmark fractions like 1/2 (0.5) and 1/4 (0.25) can be used to estimate calculations involving fractions and decimals.
- Estimation can be used to check the reasonableness of exact answers found by paper/pencil or calculator methods. (Charles, 2005, p. 17).

Table 1
Big ideas according to Charles (2005)

Number	Estimation	Orientation & Location
Base Ten Numeration	Patterns	Transformations
Equivalence	Variables	Measurement
Comparison	Proportionality	Data Collection
Operations, meanings & relationships	Relations & Functions	Data Representation
Properties	Equations & Inequalities	Data Distribution
Basic Facts & Algorithms	Shapes & Solids	Chance

Charles (2005) also provided similar descriptors about estimation in measurement. The points he made are certainly important components of estimation as a big idea but, as noted by Siemon et al. (2012), there is no sequencing or learning progression and it is difficult to see how that could be developed for this type of big idea. Hence, a categorization of big ideas is proposed and this will be discussed following documentation of the big ideas suggested by other researchers.

Siemon et al. (2012) proposed six big ideas of number which should be developed through the primary, middle, and early secondary years – Trusting the Count, Place Value, Multiplicative Thinking, Multiplicative Partitioning, Proportional Reasoning, and Generalizing. Hurst and Hurrell (2014) represented the six big ideas as a series of ellipses, as shown in Figure 1. The ellipses have purposefully been drawn in an overlapping manner to emphasize the connectedness of the ideas and how key component knowledge for one idea develops simultaneously with other ideas. For example, a deep understanding of place value depends on having at least some understanding of multiplicative thinking. As Askew (2013) pointed out, this level of understanding of place value has implications for how it is taught.

> Were learners to meet the big idea of place value as a multiplicative thinking process based on powers of ten, then later difficulties with

ideas such as decimals and standard notation for numbers might be reduced. (Askew, 2013, p. 8).

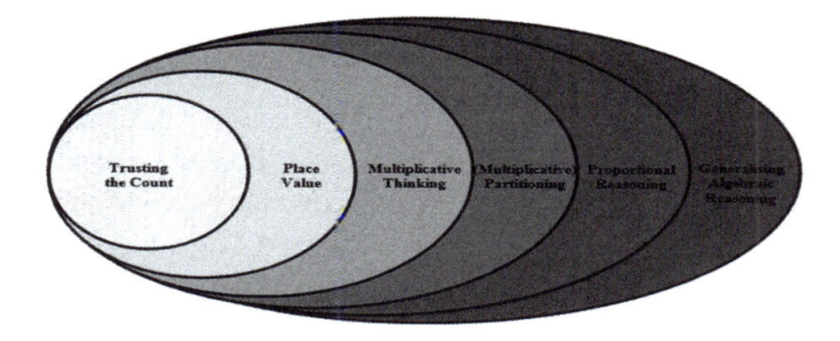

Figure 1. Big Ideas of Number (Hurst & Hurrell, 2014)

It has been emphasized that connections within and between big ideas are the essence of big idea thinking and Siemon et al. (2012) have spoken of this in terms of learning sequences and trajectories. Hence, it is necessary to document what such trajectories or connections of ideas might look like. Hurst and Hurrell (2014) coined the term 'micro-content' to identify the specific content descriptors that contribute to the understanding of big ideas like place value and multiplicative thinking. Micro-content for the first three big ideas of number in Figure 1 is characterized as being concerned with grouping or patterns. Grouping originates from early number experiences of classifying and sorting while pattern originates from early experiences with ordering and building patterns. Some examples of how concept maps and possible sequences of content can be developed are presented later in the section: *Making connections: Playing around with big ideas.*

Ernest (2015) described Charles' list of big ideas as 'topics' and provided comment about what he considers to be the scope of big ideas.

> Mathematics contains many of the deepest, most powerful and exciting ideas created by humankind. These extend our thinking and imagination, as well as providing the scientific equivalent of poetry, offering noble, aesthetic, and even spiritual experiences. (p. 192)

Ernest offered pattern, modelling, symmetry, structure, equivalence, proof, paradox, recursion, randomness, chaos, and infinity as big ideas. However, as with many of Charles' big ideas, it is difficult see how such ideas can be sequenced.

Askew (2013) offered a selection of what he considered to be big ideas, including place value, position on a number line, equivalence, meanings and symbols, and distinguishing between quantities and numerals. As with Charles, he offers an intriguing variety of ideas, some of which could be relatively easily sequenced (e.g., place value) but others not so easily. Similarly, the Australian Association of Mathematics Teachers (AAMT) (2009), in discussing the mathematics needed for the 21st century, differentiated between big ideas (dimension, symmetry, transformation, algorithm, patterns, equivalence, and representation) and concepts (quantity, shape and space, uncertainty, and variables, relationships and change). AAMT added a third section called 'mathematical actions' including reasoning, using procedures, communicating, and solving problems. These are similar in scope and name to the proficiencies of the Australian Curriculum: Mathematics (ACARA, 2018), and the Common Core State Standards for Mathematical Practice (NGA Centre, 2010). Hurst (2015) called such actions 'big process ideas'.

The Department of Education of Western Australia (DEWA), as part of the First Steps in Mathematics Project, identified 'key understandings' for each of the main curriculum areas of Number, Space, Measurement, and Chance and Data. For the outcome of 'Understand Whole and Decimal Numbers', eight key understandings (KU) were provided. For example, KU1 stated "We can count a collection to find out how many are in it", KU5 said "There are patterns in the way we write whole numbers that help us remember their order", and KU7 "We can extend the patterns in the way we write whole numbers to write decimals" (DEWA, 2013a, p. 11). Whether or not these KUs fall into the category of big ideas is debatable, but they are certainly of critical importance. As well, DEWA provided an elaborate set of diagnostic maps showing learning sequences, which seems to be an important element of big ideas, especially according to Siemon et al. (2012).

4 Categories of Big Ideas

There appear to be three main types or categories of big ideas, or at least of the big ideas named here. First, there are the big ideas that have a clear learning sequence or learning trajectory and this group includes the six big ideas of number (Hurst & Hurrell, 2014; Siemon et al., 2012), and DEWA's (2013a; b) 'key understandings'. Other big ideas that could be developed in a sequential manner and with clear trajectories for learning include Operations, Meanings and Relationships, Shapes and Solids, Transformations, orientation and Location, Data Collection, Representation, and Distribution, Chance, and Measurement (Charles, 2005). Indeed, many of these big ideas have well-documented diagnostic maps developed through the First Steps in Mathematics Project (DEWA, 2013a; b).

Second, there are big ideas which will be called 'big umbrella ideas' which sit above and inform the development of aspects of the first set of big ideas. This group includes Patterns, Equations and Inequalities, Properties, Estimation, Variables, Comparison, Equivalence, Relations and Functions (Charles, 2005), Representation (AAMT, 2009), Meanings and Symbols (Askew, 2013), and Symmetry (Ernest, 2015). For example, Patterns is clearly an important idea which informs many of the content areas of a mathematics curriculum. There are patterns in the way we read, write and say numbers, patterns in multiples, patterns in 2D shapes and 3D objects, patterns in calendars, and patterns in tessellating shapes.

Third, there are big ideas which will be called 'big process ideas' (Hurst, 2015) and these include the proficiencies of reasoning, justifying, problem solving (ACARA, 2018), and 'mathematical actions' such as generalizing, comparing, communicating, asking questions, visualizing, and explaining (AAMT, 2009). This third category provides the conduit for developing other big ideas and for realizing the many connections within and between big ideas. Figure 2 provides a graphical representation of the emerging model of big ideas.

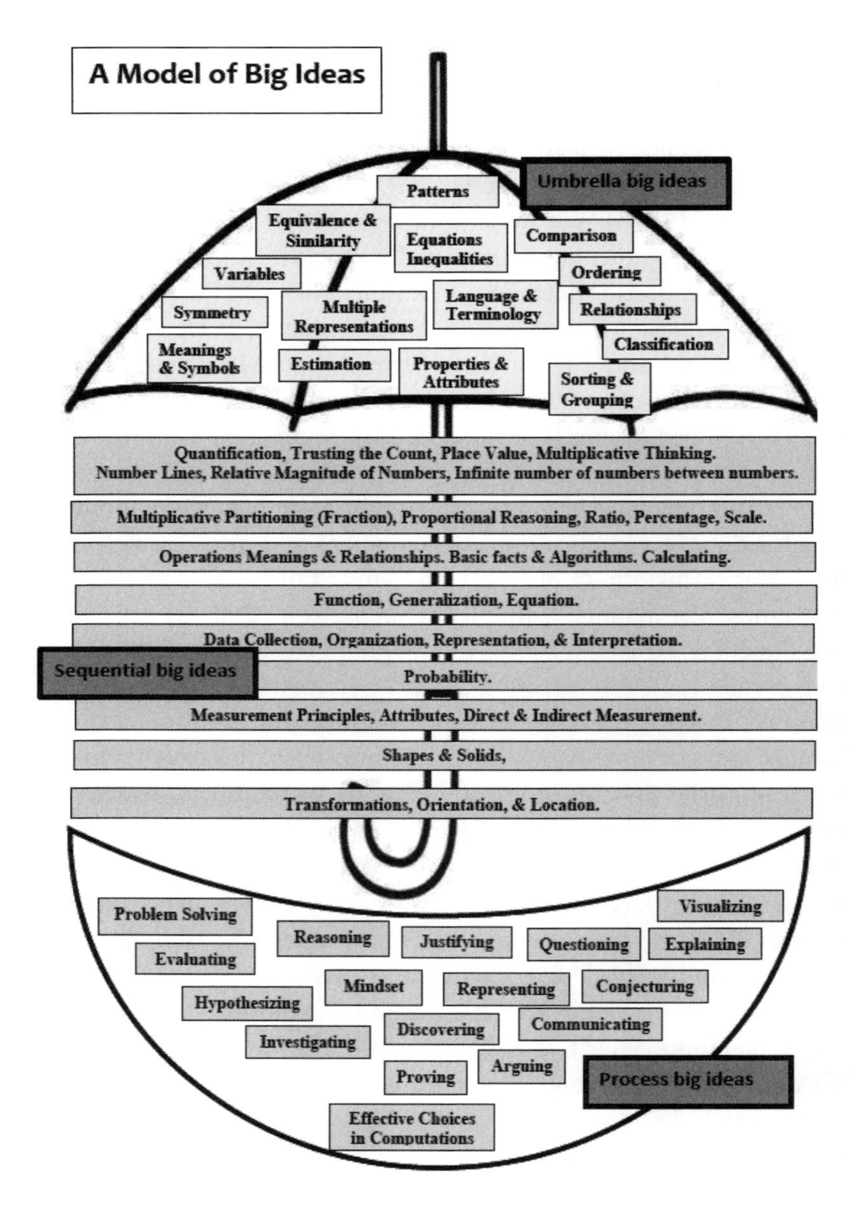

Figure 2. An emerging model for big ideas

5 Making Connections: Playing Around with Big Ideas

The main idea to come from this chapter is the engagement of teachers in big idea thinking through deconstructing a key idea to its component parts and reconstructing those parts in a meaningful and connected way. Initially, a useful way to become familiar with big idea thinking is to literally 'play around' with the concept. This can begin as an individual or group brainstorm on a contextual theme, such as 'how to build a playground' and identifying the mathematical content needed. This can then be mapped against the curriculum and involve looking for links and connections between aspects of the mathematics (Corovic, 2017). Alternatively, it can begin with a particular mathematical idea and compiling a concept map or mind map as connections are visualized. Figure 3 is an example of a concept map brainstormed by a graduate teacher who began with the idea of probability.

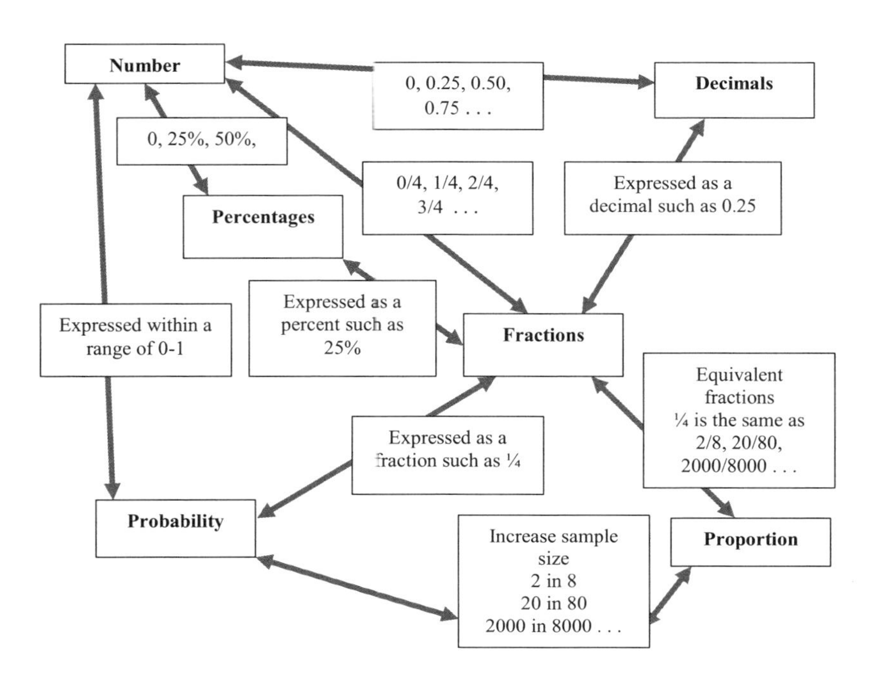

Figure 3. Teacher's concept map showing connections between ideas

It needs to be remembered that if a group of teachers engaged in the same exercise with probability as the starting point, it is likely that there would be as many different responses as there were teachers. It is the process and deconstructing and reconstructing knowledge that is important. Figure 4 shows another attempt at compiling a concept map based on 'division'. While it might not be generally considered as a big idea, it is the process of connecting ideas and attempting to find a sequence for development that are important.

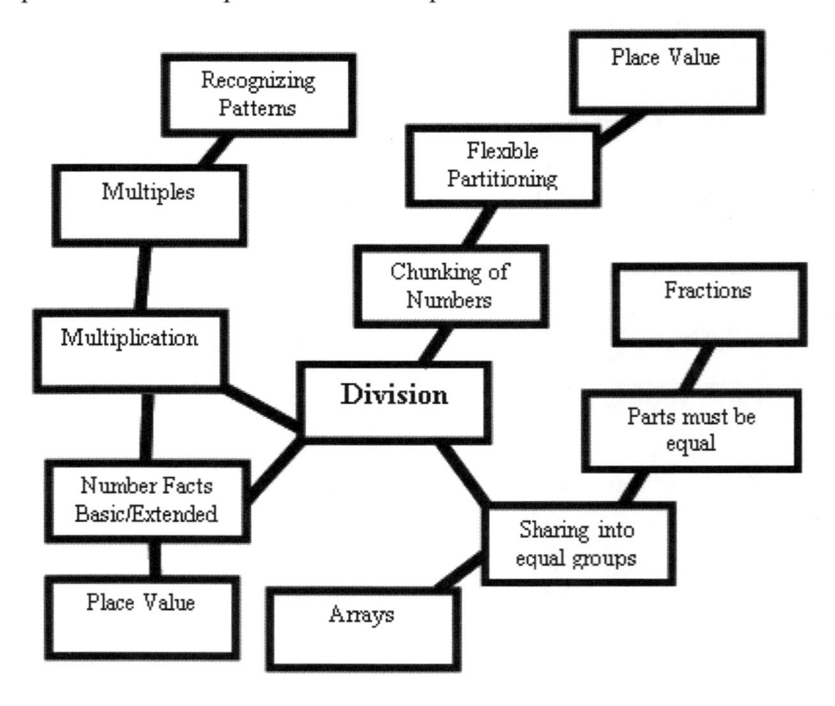

Figure 4. Teacher's concept map based on division

While both Figures 3 and 4 do not present a clear learning sequence or trajectory, they represent initial efforts of two young teachers coming to grips with big idea thinking. The next step for them would be to attempt to build such a sequence.

Hence, a useful way of facilitating big idea thinking is to ask teachers to begin by compiling a concept map about a big idea, such as place value. Figure 5 is one such example of one teacher's concept map based on the idea of Grouping and Patterns and how those ideas underpin aspects of place value. Teachers can share their concept maps and look for common aspects and then work in pairs or small groups to attempt to build a learning sequence based on the micro-content they identified in their concept maps. Table 2 represents a possible sequence of micro-content tied to the notion of grouping, but does not represent the only way to depict the items of micro-content. As has been mentioned before, the strength of big idea thinking is not arriving at exactly the same point in exactly the same way as others, but in the process of deconstructing and reconstructing one's knowledge (Clarke et al., 2012).

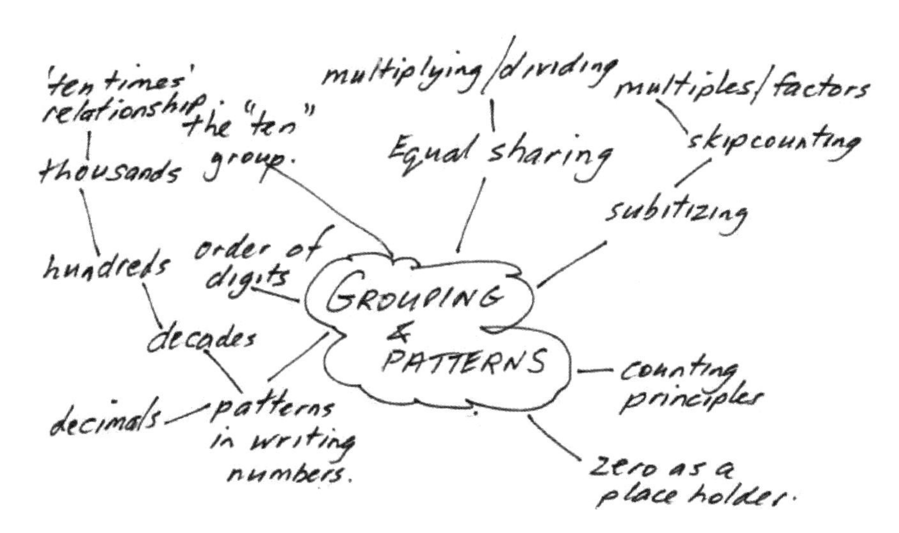

Figure 5. One teacher's concept map about grouping and patterns

In a similar way of developing Table 2, teachers could use their concept maps as the basis for discussion and arrive at a possible sequence of micro-content related to the notion of pattern. This is shown in Table 3.

Table 2

Possible Sequence of Micro-content for Counting, Place Value, and Multiplicative Thinking Related to Grouping

Specific Items of Micro-content
Objects can be classified and sorted into groups.
Groups of objects can be split or shared equally.
Numbers can be split or shared into equal groups.
Subitizing is a way of quantifying.
Skip counting by any amount gives the same result as one to one counting.
Multiples can be derived through skip counting.
Factors divide evenly into other numbers which are multiples of those factors.
Objects can be shared equally and described as the number of groups of a certain size.
When objects are shared, the more portions there are, the smaller are the portions.
Multiplication and division (multiplicative situation) are the inverse of each other.
Quotition division involves finding the number of groups (factor).
Partition division involves finding the size of each group (factor).

Table 3

Possible Sequence of Micro-content for Counting, Place Value, and Multiplicative Thinking Related to Pattern

Specific Items of Micro-content
The principles of counting.
The order of digits makes a difference.
There are patterns in reading, writing, and saying numbers.
Zero can hold a place and means 'nothing of something'.
The 0-9 sequence repeats in the decades and within each decade.
Values increase by a power of ten each time we move from right to left.
The 0-9 sequence repeats in the hundreds and thousands.
The cyclic pattern of naming numbers continues to the thousands and beyond.
A ten times multiplicative relationship exists between the places.
The multiplicative relationship extends to the right of the units for numbers less than one.
Symmetry in the number system is around the ones place.
The decimal point separates whole numbers from parts of wholes.
Digits to the right of the decimal point have decreasing values in powers of ten and represent tenths, hundredths, thousandths, and so on.

6 The Multiplicative Array: A Big Idea to Explore

Much has been written about the power of the multiplicative array (Hurst, 2014; Jacob & Mulligan, 2014; Young-Loveridge, 2005) in helping children understand multiplicative concepts. It might well be considered as a big idea as it is a conduit for connecting mathematical content and it is possible to build learning sequences around the array. The array can be used to depict many connected ideas such as the commutative property of multiplication, numbers of equal groups in the multiplicative situation, equivalent fractions, and area. The relationship between division, multiplication, fractions, and the 'times as many' notion can be shown with arrays as in Figure 6.

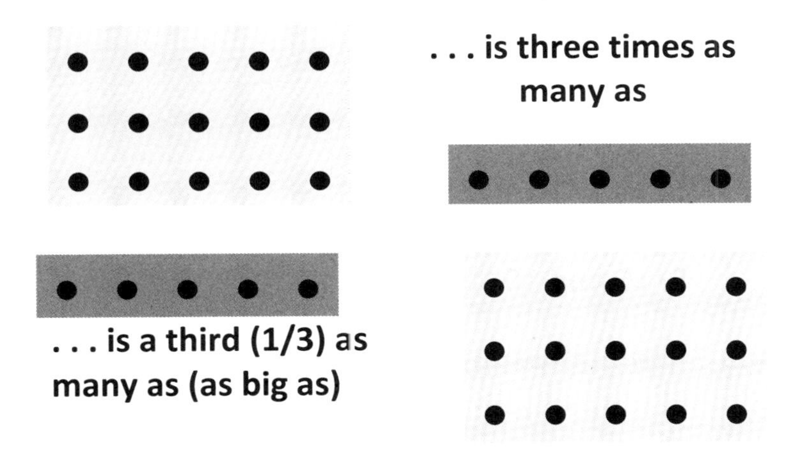

Figure 6. Connecting ideas with an array

Similarly, the relationship between fractions and ratio can be clearly shown, as in Figure 7.

Another example of the importance of the array in helping teachers make connections explicit for their students is in the idea of 'six facts for the price of one'. In Figure 8, a six by four array is used to explicitly connect multiplication, division and fraction facts. As it stands, the array depicts the multiplication fact $6 \times 4 = 24$ (that is, six rows of four). It can be rotated by 90 degrees to show the commutated fact of $4 \times 6 = 24$. The division facts can be derived by considering the array in terms of 'If I have 24 tiles in six rows, how many are there in each row?' or 'If I have 24 tiles in rows of six, how many are there in each row?' The division facts of $24 \div 6 = 4$ and $24 \div 4 = 6$ are thus derived. Finally, in considering the idea presented in Figure 6 that each row is one sixth of the array and in the rotated array, each row is one fourth (or quarter) of the array, we then have the associated fraction facts: $1/6 \times 24 = 4$ and $1/4 \times 24 = 6$.

Ratio of part: whole is 1:3 or 5:15.
This is 1/3 or 5/15

Ratio of part: part is 1:2
That is 1/3 and 2/3
or
5/15 and 10/15

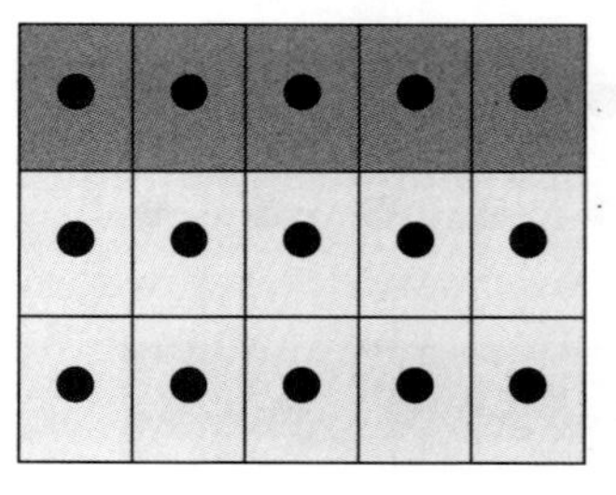

Figure 7. Connecting fraction and ratio using an array

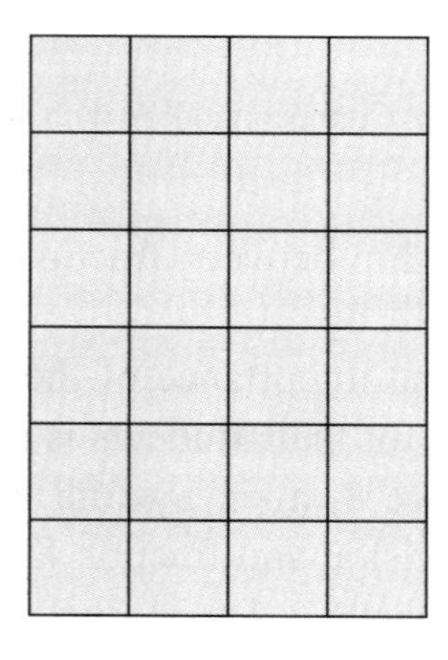

Figure 8. A six-by-four array

The array plays an important role in the development of conceptual understanding of the mathematical structure that underpins written algorithms. The latter has been suggested as a big idea (AAMT, 2009) and is also a component of the Understand Operations strand of the First Steps in Mathematics Project (DEWA, 2013). It is an appropriate topic to consider in terms of connections between ideas and also learning sequences. A number of mathematical ideas have an impact on the understanding of written algorithms, including the properties of multiplication, and the inverse relationship with division, which can be depicted with an array. Other key ideas are the base ten property of place value, flexible partitioning, and basic and extended number facts. The relationships between these ideas are presented in Figure 9. A possible learning trajectory could be that the properties of place value inform the development of partitioning, which in turn informs the distributive property of multiplication. This then manifests itself as grid multiplication, which is the basis for the written algorithm. Alongside this sequence, the base ten property of place value and the idea of zero as a place holder combine with basic number facts to inform the development of extended number facts, which are required if a student is to understand the algorithm. These points, along with others, are shown as Figure 9, a learning sequence for algorithms (Hurst & Hurrell, 2018).

7 Pattern: An 'Umbrella Big Idea' to Explore

Pattern may well be the biggest of the big ideas in primary mathematics. Indeed, mathematics has been described as the science of patterns and Devlin (1998, p. 2) stated that "The patterns of mathematics are found all around us". Patterns certainly abound throughout the content strands of mathematics curricula. As noted earlier, pattern is an umbrella big idea and is generally not sequential in nature, although there are aspects of it that can be developed in a sequential way, particularly in the number strand. The idea of pattern could be explored by teachers in the same way as suggested earlier through brainstorming, and use of concept maps. The following ideas are examples of patterns in numbers.

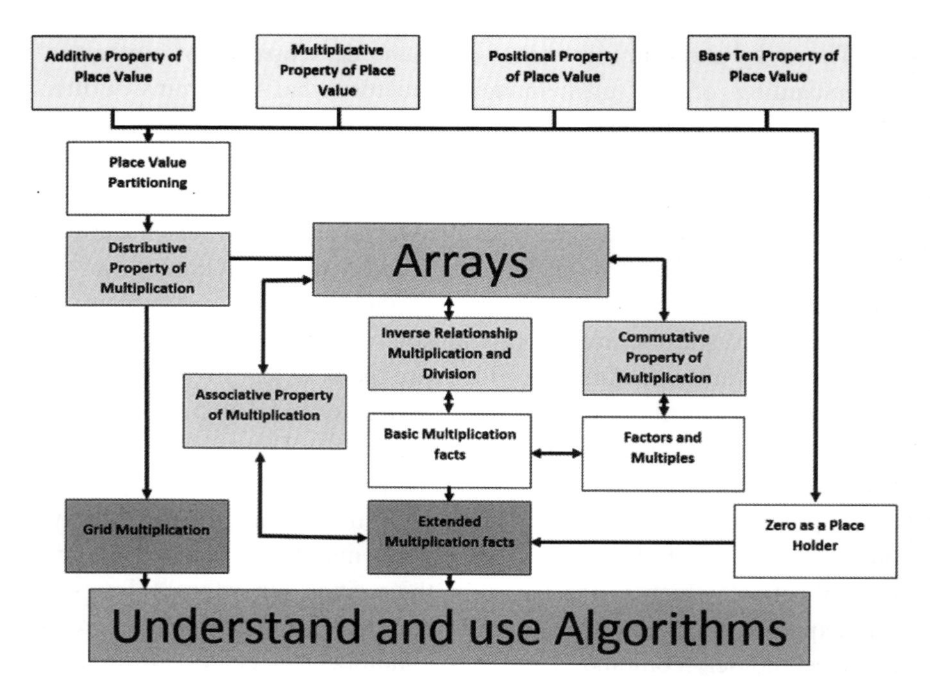

Figure 9. Arrays as an integral part of learning about algorithms

- Counting follows a pattern in which the 0-9 sequence is repeated in and within the decades, hundreds, thousands, and so on, to form an infinite sequence of numbers.
- There is a ten times multiplicative relationship and pattern between places in the number system.
- Skip counting is based on a pattern which equates to multiples of the counting unit/group (e.g., 4, 8, 12 . . .). This is also represented by the pattern in basic number facts and divisibility rules.
- The number system contains an infinite pattern of consecutive odd and even numbers.
- There are patterns in the way we read, write, and say numbers.
- Patterns can be observed when using mental computation strategies such as doubling and halving, and compensating.

Also, the First Steps in Mathematics materials contain six key understandings in 'reasoning about number patterns' including using pattern to infer from one thing to another, recognizing, describing, and investigating patterns, and using patterns to facilitate working with numbers (DEWA, 2013b).

With regard to other strands, there are patterns based on the naming of measurement units and geometric shapes and objects, with the prefixes derived from Greek and Latin. For instance, two dimensional (2D) shapes are named according to the number of sides they have and there are also patterns in the sizes and sums of internal angles of 2D shapes. Three dimensional (3D) objects are similarly named according to their attributes and there are patterns in the numbers of faces, edges, and vertices of pyramids and prisms.

Patterns are also valuable for inferring about solutions to investigations and problems. That is, if a pattern is evident in the solution/s generated, it is reasonable to assume that the solution/s is/are appropriate and that all possible solutions have been found. The following problem provides a simple example.

AB + BA = CC

In this problem, the letters A, B, and C represent different single digits which when combined, form two digit numbers. AB and BA are the reverse of one another and CC has the same digit in the tens and ones places. The question can be posed – "How many solutions are there?" Upon investigation, it is seen that the possible values for CC range from 33, 44 to 99. One solution is possible for each of 33 and 44 (12 + 21 and 13 + 31). Two solutions are possible for each of 55 and 66 (14 + 41, 23 + 32, and 15 + 51, 24 + 42). Three solutions are possible for each of 77 and 88 (16 +61, 25 + 52, 34+ 43, and 17 +71, 26 + 62, 35 + 53) and four solutions are possible for 99 (18 + 81, 27 + 72, 36 + 63, 45 + 54). The existence of this 1, 1, 2, 2, 3, 3, 4 pattern suggests that all possible solutions have been found.

Pattern is clearly a big idea and one that can be found across all strands of mathematics curricula. Other umbrella big ideas such as equivalence, comparison, and estimation similarly underpin traditional content strands such as number and algebra, geometry and measurement, and statistics and probability.

8 Conclusion

Big idea thinking has the capacity to transform the way mathematics is viewed and taught in primary classrooms, and perhaps more importantly, the way in which mathematics curricula are interpreted. An oft made claim is that contemporary curricula are 'overcrowded' and in need of 'thinning out' (Clarke et al., 2012; Siemon et al., 2012) and that a focus on the connections found within and between big ideas is the way to achieve that. Whether or not contemporary mathematics curricula are indeed overcrowded is debatable and the real issue may be the need for better ways of interpreting curricula on the basis of mathematical structure, not compartmentalized content presented in year levels. That is certainly the view taken here.

Big Idea thinking is not something that is presented in a particular format. Rather, its value lies in teachers exploring, 'unpacking', and 'playing with' big ideas along the lines suggested earlier. The aim of this practice is to enable teachers to reflect on what they know and ask themselves questions such as, 'How are these ideas connected?' As they engage with this process, they deconstruct their knowledge and reconstruct it in a more connected way. As they become more aware of the connectedness of the mathematical structure and how key concepts are built over time, they develop the ability to 'look backwards' and 'think forward' about the specific mathematics their students need to have in order to understand a particular concept. As well, they are better equipped to ensure that they are putting in place the underpinning mathematics their students will need to make further progress in their understanding. In reality, curricula need to be seen as massive concept maps with clear learning progressions or pathways made explicit. In that way, the sequential big ideas discussed earlier can bring about change in planning, teaching, and learning.

To conclude this discussion and provide another example of the power of big idea thinking, the notion of 'targeted teaching' is considered. Siemon (2017, p. 4) discussed 'targeted teaching' as a means of implementing big idea thinking, noting that targeted teaching is characterized by an unrelenting focus on big ideas. Specifically Siemon said that targeted teaching requires, among other things,

> assessment tools and techniques that expose students' thinking and provide valid and reliable information about where students are 'at' in relation to an important big idea [and] a grounded knowledge of underlying learning progressions, key steps in the development and application of big ideas and how to scaffold these (p. 1).

Siemon provided research based evidence showing that considerably larger effect sizes were attained in schools where students were subjected to targeted teaching about multiplicative concepts compared to those from reference schools.

The essence of targeted teaching is the content knowledge of teachers informed by big idea thinking and knowing deeply about the myriad connections between mathematical ideas. Armed with that sort of knowledge and thinking, teachers are well placed to identify specific needs and explicitly target the particular mathematics needed by each student. Alongside the sequential big ideas, the umbrella big ideas such as pattern, equivalence, estimation, comparison, and ordering have the power to inform the development of the sequential big ideas through explicit teaching, while the big process ideas of reasoning, justifying, comparing, questioning, and visualizing can become the conduit for developing a connected understanding of mathematics concepts.

Acknowledgement

The author thanks the anonymous students whose work has provided part of the content for this chapter.

References

Askew, M. (2013). Big ideas in primary mathematics: Issues and directions. *Perspectives in Education, 31*(3), 5-18.

Australian Association of Mathematics Teachers (AAMT). (2009). *Discussion paper: School mathematics for the 21st century*. Adelaide: AAMT.

Australian Curriculum Assessment and Reporting Authority (ACARA). (2018). *Australian curriculum: Mathematics, v 8.3*. Retrieved from https://www.australiancurriculum.edu.au/f-10-curriculum/mathematics

Bruner, J. (1960). *The process of education*, Cambridge, MA: Harvard University Press.

Burns, M. (2004). Writing in math. *Educational Leadership*, *62*(2), 30-33.

Charles, R. I. (2005). Big ideas and understandings as the foundations for elementary and middle school mathematics. *Journal of Mathematics Education Leadership, 7*(3), 9-24.

Clark, E. (2011). Concepts as organising frameworks. *Encounter: Education for Meaning and Social Justice, 24*(3), 32-44.

Clarke, D. M., Clarke, D. J., & Sullivan, P. (2012). Important ideas in mathematics: What are they and where do you get them? *Australian Primary Mathematics Classroom, 17*(3), 13-18.

Corovic, E. (2017). Big ideas in mathematics. *The Common Denominator, 3*(17), 1-7.

Department of Education, Western Australia (DEWA). (2013a). *First steps in mathematics: Overview*. Retrieved from http://det.wa.edu.au/stepsresources/detcms/portal

Department of Education, Western Australia (DEWA). (2013b). *First steps in mathematics: Number book 2*. Retrieved from http://det.wa.edu.au/stepsresources/detcms/portal

Devlin, K. (1998). *Life by the numbers*. New York: John Wiley & Sons.

Ernest, P. (2015). The social outcomes of learning mathematics: Standard, unintended or visionary? *International Journal for Education in Mathematics, Science, and technology, 3*(3), 187-192.

Hurst, C. (2014). The multiplicative situation. *Australian Primary Mathematics Classroom, 20*(3), 10-16.

Hurst, C. (2015). New curricula and missed opportunities: Crowded curricula, connections, and big ideas. *International Journal for Mathematics Teaching and Learning*. Retrieved from https://www.cimt.org.uk/ijmtl/index.php/IJMTL

Hurst, C. (2016). Towards a framework for making effective computational choices. *Australian Primary Mathematics Classroom, 21*(4), 16-21.

Hurst, C., & Hurrell, D. P. (2014). Developing the big ideas of number. *International Journal of Educational Studies in Mathematics, 1*(2), 1-18. Retrieved from http://www.ijesim.com

Hurst, C., & Hurrell, D. P. (2018). Algorithms are great: What about the mathematics that underpins them? *Australian Primary Mathematics Classroom, 23*(3), 22-26.

Jacob, L., & Mulligan, J. (2014). Using arrays to build towards multiplicative thinking in the early years. *Australian Primary Mathematics Classroom, 19*(1), 35-40.

Laverty, M. J. (2016) Thinking my way back to you: John Dewey on the communication and formation of concepts. *Educational Philosophy and Theory, 48*(10), 1029-1045, doi:10.1080/00131857.2016.1185001

National Governors Association Center for Best Practices, Council of Chief State School Officers (NGA Center). (2010). *Common core state standards for mathematics.* Retrieved from http://www.corestandards.org/the-standards

Siemon, D. (2006). *Assessment for common misunderstandings materials.* Prepared for and published electronically by the Victorian Department of Education and Early Childhood Development. Retrieved from http://www.education.vic.gov.au/student learning/teachingresources/maths/common/default.htm

Siemon, D. (2011). *Developmental maps for number P-10.* Materials commissioned by the Victorian Department of Education and Early Childhood Development, Melbourne.

Siemon, D. (2017, February 27). Targeting 'big ideas' in mathematics. *Teacher.* Retrieved from https://www.teachermagazine.com.au/articles/targeting-big-ideas-in-mathematics

Siemon, D., Bleckly, J., & Neal, D. (2012). Working with the Big Ideas in Number and the Australian Curriculum: Mathematics. In B. Atweh, M. Goos, R. Jorgensen & D. Siemon (Eds.), *Engaging the Australian National Curriculum: Mathematics – Perspectives from the Field* (pp. 19-45). Online Publication: Mathematics Education Research Group of Australasia.

Young-Loveridge, J. (2005). Fostering multiplicative thinking using array-based materials. *Australian Mathematics Teacher, 61*(3), 34-40.

Chapter 6

Teaching towards Big Ideas: Challenges and Opportunities

CHOY Ban Heng

The inclusion of *Big Ideas* in the 2020 Secondary Mathematics syllabus in Singapore signals a shift towards teaching mathematics as a coherent and inter-connected discipline. This move can potentially foster a better understanding of and a deeper appreciation for the subject. However, preparing to teach towards these big ideas can be pedagogically challenging for teachers, and at times, poses tensions between the need for mathematical rigor and making important ideas accessible to students. In this chapter, I will first present a notion of teaching towards big ideas, before I discuss the opportunities for teaching towards big ideas, the challenges involved when doing so, and suggestions for teaching by using examples from the Singapore curriculum.

1 Introduction

The introduction of *big ideas* in the coming 2020 Secondary Mathematics syllabus in Singapore signals important shifts in how mathematics is perceived as a subject and how it is taught in schools. As National Council of Teachers of Mathematics (2000) has emphatically stated:

> Teachers need to understand the big ideas of mathematics and be able to represent mathematics as a coherent and connected enterprise. (p. 17)

This is a move towards teaching mathematics as a coherent interconnected discipline to counter the perception, held by many students, that mathematics is a set of rules and facts to be remembered (Charles, 2005). In addition, effective teachers organize their teaching around key mathematical ideas, and focus their efforts on supporting students to make connections between these ideas (Charles, 2005), which is a hallmark of students' understanding (Hiebert & Carpenter, 1992). Following similar arguments, other researchers have proposed to frame a curriculum around key disciplinary ideas (Mansilla & Gardner, 1999; Ritchhart, 1999), and have highlighted big ideas as possible ways to enhance teaching and learning (Mitchell, Keast, Panizzon, & Mitchell, 2016).

However, simply articulating big ideas within the curriculum is not likely to result in enhanced teaching and learning. Instead, teachers have to make sense of these ideas and think about the pedagogical approaches to adopt for students to access these ideas. Doing this is challenging (Mitchell et al., 2016) and teachers may require support in identifying, describing, and connecting these big ideas (Clarke, Clarke, & Sullivan, 2012; Tout & Spithill, 2015). How teachers can be supported to do this remains unclear. Moreover, there is an even more fundamental question—what does it mean to teach towards big ideas? In this chapter, I will describe what teaching towards big ideas entails, highlight some of the challenges and issues involved, and illustrate these ideas using some of the big ideas identified in the syllabus.

2 Teaching towards Big Ideas

2.1 *What is a big idea?*

One of the main problems of teaching towards big ideas is the lack of a clear definition of big idea (Askew, 2013). Despite the different definitions offered (Askew, 2013; Charles, 2005; Clarke et al., 2012; Mitchell et al., 2016), there are at least three important characteristics of a big idea expressed in the various definitions. First, a big idea is a statement, or a phrase, or a word that expresses something for students and teachers to focus on. In fact, the origin of the word *idea* comes from the Greek word *idein*, which means "to see". Following the root word of idea, I argue that the purpose of a big idea is for teachers to articulate something important for students to see. Second, a big idea is big. In other words, it should be central to the discipline of mathematics (Ritchhart, 1999), expressing the essence of what mathematics is, and how one goes about doing mathematics (Charles, 2005). This is different from expressing a rule or an instructional objective, which narrowly focuses on a specific technique or result or concept. Last but not least, a big idea offers a way to make connections in mathematics (Tout & Spithill, 2015), or to generate new perspectives or extensions of a mathematical concept (Ritchhart, 1999). One definition that encompasses these three characteristics is offered by Charles (2005):

> A big idea is a statement of an idea that is central to the learning of mathematics, one that links numerous mathematical understandings into a coherent whole. (p. 10)

In Singapore, the articulation of big ideas in the 2020 Mathematics syllabus document (Ministry of Education, 2018) is largely aligned to Charles' (2005) definition, and is centred about the following four recurring themes, which frame the study of mathematics as a coherent whole (Ministry of Education, 2018, pp. S2-2 to S2-3):

1. Properties and Relationships (What are the properties of mathematical objects and how are they related?);
2. Operations and Algorithms (What meaningful actions can we perform on the mathematical objects and how do we carry them out?);
3. Representations and Communications (How can the mathematical objects and concepts be represented and communicated within and beyond the discipline?); and
4. Abstractions and Applications (How can the mathematical objects be further abstracted and where can they be applied?).

The eight big ideas, and the themes associated with them, are listed in the syllabus document (pp. S2-4 to S2-5) as follows:

1. Big Idea about Equivalence (Properties and Relationships, Operations and Algorithms);
2. Big Idea about Proportionality (Properties and Relationships);
3. Big Idea about Invariance (Properties and Relationships);
4. Big Idea about Measures (Abstractions and Applications);
5. Big Idea about Functions (Properties and Relationships, Abstractions and Applications);
6. Big Idea about Notations (Representations and Communications);
7. Big Idea about Diagrams (Representations and Communications); and
8. Big Idea about Models (Abstractions and Applications).

The syllabus document highlights that the preceding ideas are to be perceived as clusters of big ideas, and not to be taken as authoritative nor comprehensive. Each cluster of big idea is given a label (e.g., Functions), followed by a brief description of some possible big ideas that fall within the category. For example, in Figure 1, we see how the cluster of big idea about functions is described. A more elaborate description is also given for teachers to consider as they think about the various opportunities to bring in these ideas for students (see Figure 2).

Big Idea about Functions

Main Themes: Properties and Relationships, Abstractions and Applications

A function is a relationship between two sets of objects that expresses how an element from the first set (input) uniquely determines (relates to) an element from the second set (output) according to a rule or operation It can be represented in multiple ways, e.g. as a table, algebraically, or graphically. Functional relationships undergird many of the applications of mathematics and are used for modelling real-world phenomena. Functions are pervasive in mathematics and undergird many of the applications of mathematics and modelling of real-world phenomena.

Figure 1. Big Idea about functions (Ministry of Education, 2018, p. S2-5)

The given descriptions provide a good starting point for teachers to make sense of these big ideas. Furthermore, the document provides some suggestions for teachers as they prepare to teach towards big ideas. These suggestions highlight possible teaching actions or learning activities for students that might provide the opportunities for students to encounter these big ideas (see Figure 3). However, not much details are given, and the question still remains. What do we mean when we teach towards big ideas in Mathematics? How does such a lesson look like?

The idea of functions, is implicit in many mathematical situations.

- When students use a graph of y against x to determine the value of y (output) for a given value of x (input), they are implicitly treating the graph as the rule in a function.

- When students learn about finite sequences, they can think of them as functions that give, for every position term position, n (input), the term T_n in the sequence (output) following a general rule. Such functions may be presented in a form of a table.

- When students learn about trigonometric functions, they learn that the sine, cosine and tangent functions give for any angle (input), the sine, cosine or tangent ratio (output) obtained as a ratio of pairs of sides from its related right-angled triangle (rule). These functions can be presented in graphical forms.

- When students learn to compute the distance of a point (x, y) from the origin, they can think of it as a function that gives for every co-ordinate pair (input), the distance of the point from the origin (output) using the formula $\sqrt{x^2 + y^2}$ (rule).

In the last case, the input is not a single real number. In fact, many operations on a set of mathematical objects can be considered as functions. The elements in the set of mathematical objects serve as the input, the operation provides the rule, and the elements in the set of outcomes of the operations are the output.

Figure 2. Elaboration of big ideas (Ministry of Education, 2018, p. S2-13)

How do we teach towards the big ideas about Functions?

To help students build a deeper understanding and appreciation of the big ideas about functions,

- Introduce the concept of input, rule and output and the different forms in which a function may be represented, including non-examples of function.
- Develop the concept of functions of real values in the topics of linear, quadratic, logarithmic, exponential and trigonometric functions and their application.
- Provide students with a lens to look at functions – what input does it take, what output does it produce, and what is the rule that uniquely determines the output for a given input.
- Provide students with opportunities to write codes.

Figure 3. Suggestions for teaching big ideas (Ministry of Education, 2018, p. S2-14)

2.2 *How do we teach towards big ideas?*

An important consideration when planning to teach towards big ideas is to recognize that big ideas are generative—in that they provide teachers with a sense of *what*, *why*, and *how* important ideas can be made explicit to the students. Ritchhart (1999), for example, suggests that big ideas should be phrased in ways that point students to understand important (and possibly difficult) concepts, apply their understanding to solve problems in real world contexts, and to make connections within and beyond mathematics. This, according to him, ensures that the inclusion of big ideas goes beyond a cosmetic change to fundamentally changing how mathematics is taught in classrooms. Hence, big ideas are "best phrased as a sentence with a verb" (Mitchell et al., 2016, p. 600) to "explain and inform teaching practices" (p. 601). Consequently, a single word or short phrase to label a big idea is not useful. For example, it may not be instructive to tell students that we are now teaching the big idea of function when we introduce the exponential function in class. Instead, a big idea, when phrased as a sentence such as "*Exponential functions can be represented using an equation, a table of values, and a graph to show us different perspectives of the same growth phenomenon*" is potentially more pedagogically powerful for several reasons.

Firstly, the sentence expresses one of the motivations behind learning exponential functions: to describe growth phenomenon. Secondly, it highlights why different representations of functions are useful. Thirdly, the phrasing suggests certain connections to the themes identified in the syllabus. For example, we see the themes of Properties and Relationships, Representations and Communications, and Abstractions and Applications, featured in the big idea. Lastly, the big idea, when phrased in a sentence, provides a clearer direction for teachers to emphasize the key ideas in the topic (in this case, exponential functions), and incorporate certain features into the tasks they design to teach towards these big ideas. It becomes apparent that teachers need to *provide students opportunities* to model growth phenomena using exponential functions and represent them using an equation, a table, and a graph. This ties in nicely with the idea of *learning experiences*, as stipulated in the current syllabus.

According to the current syllabus, learning experiences are intended to influence teaching and learning of mathematics. They are expressed in the form "*students should have opportunities to ...*" and these statements highlight the student-centricity of these experiences. In a nutshell, learning experiences "describe actions that students will perform and activities that students will go through, with the opportunities created and guidance rendered by teachers" (Ministry of Education, 2012, p. 20). Hence, the intent of learning experiences is to provide opportunities for students to experience the essence of mathematics, which lies at the heart of big ideas. In some ways, these learning experiences are similar to the notion of performances of understanding proposed by Mansilla and Gardner (1999). Performances of understanding are tasks designed for students to demonstrate their understanding by engaging in practices central to the discipline. In other words, teachers design tasks and orchestrate discussions to bring about learning experiences in which students have opportunities to work on mathematical problems, rediscover mathematical results through conjecturing and proving, or present their solutions using mathematical language (Ministry of Education, 2012).

Although designing high-quality tasks to bring about learning experiences is one feasible means to teach towards big ideas, how the tasks are implemented will be crucial if we want to bring students to see the big ideas. As Henningsen and Stein (1997) have pointed out, it is possible for a high-quality task to be implemented without tapping its potential to bring about students' engagement with rich mathematical ideas and processes. With the aim of supporting teachers to use tasks more effectively, Smith and Stein (2011) propose five practices—anticipating, monitoring, selecting, sequencing, and connecting—for teachers to orchestrate their mathematics discussions. In particular, the practice of connecting students' ideas is aimed at directing students' attention to the connections between different strategies they use to solve the task. This enables students to shift their focus from solutions to mathematical ideas, and teachers can begin to support students' efforts in understanding the concepts, and hence the big ideas, targeted in the lesson (Smith & Stein, 2011; Stein, Engle, Smith, & Hughes, 2008).

There are at three types of connections that help to bring out the big ideas during mathematics lessons. First, teachers can try to make connections within the topic by highlighting how different solutions are related to the key ideas of the topic. For example, in a study by Choy and Dindyal (2017), they share how an experienced teacher Alice connected students' different solutions to the same problem with the concept of matrix multiplication. Next, teachers can provide opportunities for students to see important ideas across different topics in their mathematics curriculum. For example, the Additional Mathematics syllabus is centered about the study of different functions—quadratic, cubic, trigonometric, exponential, logarithmic, and gradient functions. These different topics are instantiations of the important idea of functions, as a way to model real-world context problems using relationships between variables. Last but not least, teachers can draw connections between what students study in mathematics to other related ideas in other subjects. For example, teachers can provide opportunities for students to see how the notion of gradient is used to model rates of reactions in Chemistry, kinematic problems in Physics, and bacteria growth in Biology or population growth in Geography. Making these connections can help to strengthen students' development of key mathematical processes central to mathematics and foster their interest in the discipline by highlighting one of the main purposes of learning mathematics—problem solving.

Mathematical problem solving has been at the heart of the Singapore mathematics curriculum since the 1990s (Kaur et al., 2015). Teaching towards big ideas is *not* a shift away from mathematical problem solving. Instead, it offers a means to enhance students' problem solving skills by emphasizing connections, and seeing generalizations from students' solutions to problems. Particularly in the looking back phase of the problem solving approach (Polya, 1945), teachers can create opportunities for students to learn about big ideas in mathematics from the process of problem solving. For example, teachers can use students' solutions to word problems that suggest a matrix solution to highlight the connections between a matrix solution and an arithmetic solution (Choy

& Dindyal, 2017), and bring out the big idea that matrices are used to represent huge amount of information in ways that can be operated upon easily. Similarly, there are opportunities to emphasize big ideas when students are engaged in mathematical modelling activities, or when students are engaged to solve problems set in real world contexts. For example, when students work on the Biggest Box Problem (see Figure 4), they have opportunities to see how *functions* can be used to *model* real world situations, use *diagrams* to represent the real world situation, and work with different *notations* when solving mathematical problems.

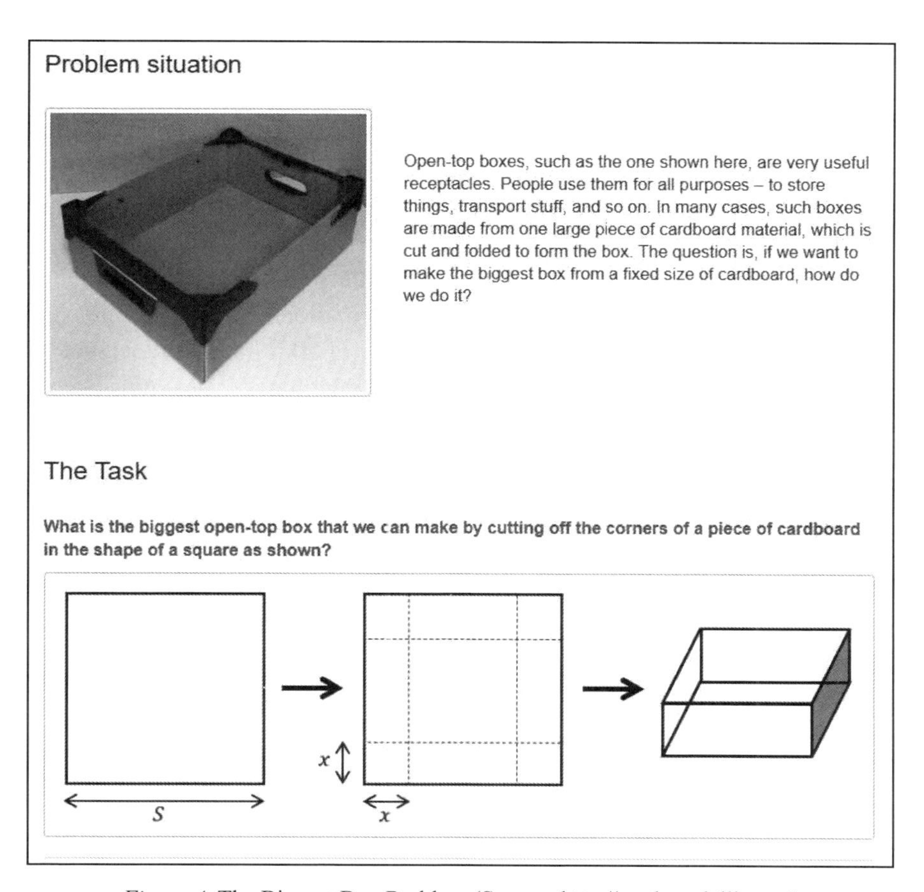

Figure 4. The Biggest Box Problem (Source: http://mathmodelling.sg)

Hence, teaching towards big ideas is not entirely a new approach to teaching mathematics in Singapore; rather, teaching towards big ideas can be seen as deepening students' appreciation of mathematics as a discipline by teaching mathematics through problem solving (and modelling).

3 Teaching towards Big Ideas: Issues and Challenges

Although the notion of teaching towards big ideas is built upon the foundations of our existing mathematics curriculum framework, it may not be easy to teach towards big ideas in the classrooms. Primarily, understanding the big ideas and what they entail may not be straight-forward, and hence, teachers may need to think more deeply about big ideas and how they can be brought across to students. For example, there may be some confusion regarding the big ideas about invariance and equivalence. According to the new syllabus document, invariance refers to a property of a mathematical object, which remains unchanged when the object undergoes some form of transformation. As an example, the document suggests that the commutative law of addition is an instance of invariance. However, this may be problematic as not every mathematician considers a rearrangement as a transformation, and thus there may be issues as to whether the commutative law is considered as an example of invariance. Furthermore, it is unclear whether the commutative law counts as an example of equivalence. While some may discount this discussion as purely academic, teachers may find it challenging to grapple with these nuances during lessons.

Another important consideration is that big ideas, such as invariance, are useful for designing problem solving tasks that engage students in mathematical inquiry (Libeskind, Stupel, & Oxman, 2018). In their article, they suggest a number of instances that highlight the concept of invariance (see p. 113). Here are a few examples Libeskind et al. (2018) have suggested:

1. Difference of two natural numbers is invariant under the same translation. That is, $a-b=(a+x)-(b+x)$ for any x.

2. For any a, the sum of roots for the equation $ax^2 + ax + c = 0$ is invariant.
3. Four congruent circles are cut from a square sheet as shown below:

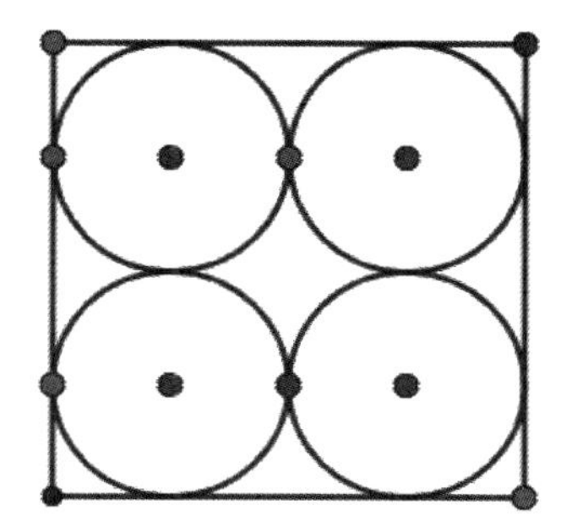

Find the percentage of the square sheet that is wasted. It turned out that this percentage is invariant when n^2 circles are cut from the same square sheet!

We see that these examples of invariance provide some directions for teachers to design investigative tasks for students to "look out for the changes" when there is repetition (Libeskind et al., 2018, p. 107). Following their argument, it seems that what matters is not so much about the property being invariant under some changes; rather, what is important is whether the invariant property of interest can potentially lead to a generative task—a task which has the affordance for teachers to make connections, produce new concepts, and understandings (Ritchhart, 1999). However, there remains a gap between understanding the big idea and transforming it into a generative task.

Moreover, the concept encapsulated in the cluster of big ideas may also be problematic. For instance, as highlighted, there may be issues thinking about the differences between invariance and equivalence. This issue is confounded by a somewhat less rigorous definition of equivalence in the syllabus document—Equivalence is a relationship that expresses the 'equality' of two mathematical objects that may be represented in two different forms. This definition emphasizes the equality of two mathematical objects, and it is possible that some students will focus solely on the equality and view all relations involving an equal sign as instances of equivalence. This is an issue because the notion of equivalence is related to the concept of equivalence relation, which is a binary relation that is reflexive, symmetric, and transitive. For example, the idea of equivalence can be exemplified in the notion of equivalent fractions as follow:

1. $\frac{1}{2}$ is equivalent to $\frac{1}{2}$ (Reflexive);
2. $\frac{1}{2}$ is equivalent to $\frac{2}{4}$ if and only if $\frac{2}{4}$ is equivalent $\frac{1}{2}$ (Symmetric); and
3. If $\frac{1}{2}$ is equivalent to $\frac{2}{4}$, and $\frac{2}{4}$ is equivalent to $\frac{3}{6}$, then $\frac{1}{2}$ is equivalent to $\frac{3}{6}$ (Transitive).

The emphasis in the fraction example is not solely about the idea that $\frac{1}{2} = \frac{2}{4} = \frac{3}{6}$, but more importantly, the idea of equivalence is that these fractions are related in a particular way and they belong to the same equivalence class. Furthermore, students may not see similarity as an example of equivalence because there is no equal sign in the relation!

Although teachers are familiar with the concept of an equivalence relation, they may hesitate to bring these ideas into their lessons. In particular, teachers may struggle with the tensions between the rigor of a big idea and the accessibility of the big idea. Furthermore, this tension may be exacerbated by the constraints of the syllabus. For example, in the case of functions, teachers will find it challenging to bring across the need to specify the input, rule, and process to determine whether a given relation is a function. This is so because the concept of function is not formally dealt with in the syllabus. Instead, functions are introduced using the idea of a function machine in terms of an input, a rule, and a unique output. However, without introducing concepts of domain, codomain, and range, it will be difficult to determine whether a given relation is a function, which is one of the suggested ways to bring in the big idea of functions. As an illustration, consider the following logarithmic relation, $f:\mathbb{R}\to\mathbb{R}, x\mapsto \ln x$. This is not a function because not every element in the set of real number has an image under *f*. Students will not be able to appreciate this as it is not explicitly taught that each and every element in the domain must have an image. Even if we restrict ourselves to consider a function in the contexts of its largest possible domain (without using the term domain), there may be issues when students cannot access concepts of domain and codomain. For example, it may not be immediately clear to the students that the following two functions are different:

$$f(x) = x,$$
$$g(x) = \frac{x(x-1)}{(x-1)}$$

These two functions are different because their domains are different and not just because they look different! Without access to the conventions and definitions that are essential for understanding functions, students may be able to work with functions but not fully appreciate the mathematical idea of functions. So, should teachers introduce these "missing" concepts when teaching towards the big ideas of functions? Or should teachers discuss functions subjected to the constraints of the syllabus at the expense of mathematical rigor? How can teachers introduce these ideas in a more accessible manner without sacrificing rigor? There are no straightforward answers to these questions. Instead, teachers have to think more deeply about these big ideas, and find a balance between rigor and accessibility for themselves in the context of their students.

Besides grappling with the definitions of big ideas, teachers also have to deal with the challenges of transforming their understanding of the big ideas into tasks or lessons. More specifically, they would need to examine the curriculum for opportunities to bring in the big ideas, and select, design, or adapt tasks that students can work with to learn about these big ideas. This will require them to comprehend and transform their understanding of the big ideas in mathematics into "forms that are pedagogically powerful and yet adaptive to the variations in ability and background presented by the students" (Shulman, 1987, p. 15). This involves teachers thinking and selecting instructional tasks and examples that can provide opportunities for teachers to connect students with these big ideas. Therefore, teaching towards big ideas places an increased emphasis on orchestrating classroom discussions because teachers have to draw on their own understanding and appreciation of mathematics during lessons to connect different topics in the school curriculum. This aspect of instruction is challenging, and it puts enormous demands on teachers' instructional repertoire and skills.

4 What can teachers do?

There is little argument against introducing big ideas into the Singapore curriculum. Teaching towards big ideas is a step in the right direction although the path ahead may not be easy. In this section, I will suggest what teachers can do to mitigate some of the difficulties surrounding the worthwhile task of teaching towards big ideas.

4.1 *Unpacking the curriculum*

With the aim of understanding the big ideas embedded in the curriculum, it is critical that teachers spend time making sense of the school mathematics curriculum. From our research, we have found that competent teachers in our studies have developed a strong grasp of what they need to teach by carefully examining school mathematics content (Choy & Dindyal, 2017, 2018a). More specifically, it is helpful for teachers to have a systematic way of distilling the essence of the mathematics curriculum from curriculum materials. For example, teachers could try to unpack the ideas encapsulated in a topic by breaking them into *concepts*, *conventions*, *results*, *techniques*, and *processes* (Backhouse, Haggarty, Pirie, & Stratton, 1992). The aim of this unpacking is to support teachers in using what they know, and guiding them to see new connections between different aspects of the mathematics that they are teaching. What matters is that teachers see both *the forest and the trees* (Choy & Dindyal, 2018b) by zooming in and out of the curriculum, and noticing systematically the details of the mathematics they are teaching. Doing so will get teachers delve deeper into school mathematics, and identify opportunities during the unit to introduce and develop the eight big ideas.

4.2 *Thinking about connections*

A key aspect of teaching towards big ideas is making connections—within the same topic, across different topics, and between different subjects. To do this, teachers have to learn how to attend to the whole curriculum, discern the details of the big idea, and perceive affordances in tasks for bringing out the big ideas. Hence, it will be useful for teachers to think beyond a single lesson by thinking about a lesson as part of a sequence of lessons within the same unit, and conceptualising a lesson as a sequence of tasks. Thinking beyond a single lesson provides teachers a way to focus their attention on the connections between lessons and can potentially heighten their awareness of the big ideas encapsulated within the tasks. Another important way to work out these connections in big ideas is to have conversations around big ideas in mathematics. These conversations could be part of a larger professional learning team activities, or informal conversations with other colleagues, where teachers can discuss the different notions of big ideas and explore ways in which the idea can be exploited within the curriculum.

4.3 *Paying attention to students' ideas in classroom discourses*

The role of classroom discussions cannot be overemphasized when we discuss the notion of teaching towards big ideas. Discussions are integral for teachers to make connections with students around mathematical ideas. Besides engaging in the five productive practices, proposed by Smith and Stein (2011), teachers will also need to adopt a more listening stance during these discussions. Here, the ideas of *listening for*, *listening to*, and *listening with* will be useful for teachers as they reflect upon their classroom practices (Davis, 1997). Do we *listen for* certain ideas we are anticipating so that we can forward the agenda of our lesson? Or do we try to *listen to* our students in order to *listen with* them to make sense of their mathematical ideas? I suspect the latter listening stance has more potential in extending students' ideas about mathematics beyond giving the correct answers to seeing mathematics as a form of sense making. This will be a key pedagogical shift if we were to enhance our mathematics teaching with the introduction of big ideas.

5 Concluding Remarks

In conclusion, teaching towards big ideas is not entirely a new endeavour. It is how competent teachers have always been teaching—highlighting connections and making mathematics meaningful and coherent for students. Articulating the eight big ideas in the Singapore curriculum thus offers an opportunity for teachers to think more deeply about the content they teach, the classroom interactions they orchestrate, and the connections they make between tasks and mathematics. If teachers see big ideas as generative in terms of content and pedagogy, they can begin to see them as ways to sharpen their pedagogical reasoning by unpacking the content and focusing on making connections in class by orchestrating mathematically productive discussions with students. Teaching towards big ideas is hard work, but it is also exciting and certainly worth teachers spending their time to bring our students to experience the beauty of mathematics!

> Mathematics has beauty and romance. It's not a boring place to be, the mathematical world. It's an extraordinary place; it's worth spending time there.—Marcus du Sautoy

References

Askew, M. (2013). Big ideas in primary mathematics: issues and directions. *Perspectives in Education, 31*(3), 5-18.

Backhouse, J., Haggarty, L., Pirie, S., & Stratton, J. (1992). *Improving the learning of mathematics*. London, England: Cassell.

Charles, R. I. (2005). Big ideas and understandings as the foundation for elementary and middle school mathematics. *Journal of Mathematics Education Leadership, 7*(3), 9-24.

Choy, B. H., & Dindyal, J. (2017). Snapshots of productive noticing: orchestrating learning experiences using typical problems. In A. Downton, S. Livy, & J. Hall (Eds.), *Proceedings of the 40th Annual Conference of the Mathematics Education Research Group of Australasia: 40 years on: We are still learning!* (pp. 157-164). Melbourne: MERGA.

Choy, B. H., & Dindyal, J. (2018a). An approach to teach with variations: using typical problems. *Avances de Investigación en Educación Matemática,* (13), 21-38.

Choy, B. H., & Dindyal, J. (2018b). Orchestrating mathematics lessons: beyond the use of a single rich task. In J. Hunter, P. Perger, & L. Darragh (Eds.), *Proceedings of the 41st Annual Conference of the Mathematics Education Research Group of Australasia: Making waves, opening spaces* (pp. 234-241). Auckland, New Zealand: MERGA.

Clarke, D. M., Clarke, D. J., & Sullivan, P. (2012). Important ideas in mathematics: What are they and where do you get them? *Australian Primary Mathematics Classroom, 17*(3), 13-18.

Davis, B. (1997). Listening for differences: an evolving conception of mathematics teaching. *Journal for Research in Mathematics Education, 28*(3), 355-376.

Henningsen, M., & Stein, M. K. (1997). Mathematics tasks and student cognition: Classroom-based factors that support and inhibit high-level mathematical thinking and reasoning. *Journal for Research in Mathematics Education, 28*(5), 524-549.

Hiebert, J., & Carpenter, T. P. (1992). Learning and teaching with understanding. In D. A. Grouws (Ed.), *Handbook of research on mathematics teaching and learning* (pp. 65-97). New York: Macmillan Publishing Company.

Kaur, B., Soh, C. K., Wong, K. Y., Tay, E. G., Toh, T. L., Lee, N. H., . . . Tan, L. C. (2015). Mathematics education in Singapore. In S. J. Cho (Ed.), *Proceedings of the 12th International Congress on Mathematical Education* (pp. 311-316). Cham: Springer.

Libeskind, S., Stupel, M., & Oxman, V. (2018). The concept of invariance in school mathematics. *International Journal of Mathematical Education in Science and Technology, 49*(1), 107-120. doi:10.1080/0020739X.2017.1355992

Mansilla, V. B., & Gardner, H. (1999). What are the qualities of understanding? In S. Wiske (Ed.), *Teaching for Understanding: Linking Research with Practice*. San Francisco, CA: Jossey-Bass.

Ministry of Education. (2012). *O- and N(A)-level mathematics teaching and learning syllabus.* Singapore: Curriculum Planning and Development Division.

Ministry of Education. (2018). *2020 secondary mathematics syllabuses (draft).* Singapore: Curriculum Planning and Development Division.

Mitchell, I., Keast, S., Panizzon, D., & Mitchell, J. (2016). Using 'big ideas' to enhance teaching and student learning. *Teachers and Teaching, 23*(5), 596-610. doi:10.1080/13540602.2016.1218328

National Council of Teachers of Mathematics. (2000). *Principles and standards for school mathematics*. Reston, VA: National Council of Teachers of Mathematics.

Polya, G. (1945). *How to solve it.* Princeton, NJ.: Princeton University Press.

Ritchhart, R. (1999). Generative topics: building a curriculum around big ideas. *Teaching Children Mathematics, 5*(8), 462-468.

Shulman, L. S. (1987). Knowledge and teaching: Foundations of the new reform. *Harvard Educational Review, 57*(1), 1-22.

Smith, M. S., & Stein, M. K. (2011). *5 practices for orchestrating productive mathematics discussions*. Reston, VA: National Council of Teachers of Mathematics.

Stein, M. K., Engle, R. A., Smith, M. S., & Hughes, E. K. (2008). Orchestrating productive mathematical discussions: Five practices for helping teachers move beyond show and tell. *Mathematical Thinking and Learning, 10*(4), 313-340. doi:10.1080/ 10986060802229675

Tout, D., & Spithill, J. (2015). Big ideas in mathematics teaching. *The Research Digest,* (11). Retrieved from http://www.qct.edu.au

Chapter 7

Big Ideas about Equivalence in the Primary Mathematics Classroom

YEO Kai Kow Joseph

The content of primary school mathematics has always been subjected to some forms of classification. In recent years, the Singapore primary mathematics curriculum is organised in terms of three content strands - Number and Algebra, Measurement and Geometry as well as Statistics. While some of these organisations (e.g., Number and Algebra) might be regarded as really big ideas in school mathematics, they are too broad to inform mathematics teachers' teaching practice. This chapter therefore reviews the big ideas in mathematics, big ideas for teaching and learning, and big ideas about equivalence. This chapter also describes three activities on big ideas about equivalence that can be infused in the teaching and learning of mathematics at the primary level. Three activities are highlighted that teachers might trial in their mathematics lessons to provide opportunities for pupils to learn about the big ideas of equivalence.

1 Introduction

It is critical for mathematics teachers to understand the big ideas of mathematics and be able to exemplify mathematics as a coherent and connected enterprise (National Council of Teachers of Mathematics (NCTM), 2000, p. 17). In recent years, mathematics teachers are encouraged to teach towards big ideas of mathematics. Yet if you ask a group of teachers or any group of mathematics educators for examples of big ideas, you will receive a range of responses. Some will offer a topic, like equations, others will suggest a strand, like algebra, others will offer a benchmark, such as those found in Principles and Standards for School Mathematics (NCTM, 2000) and some will even offer an aim, such as those found in the mathematics syllabus. Although all of these are essential, none seems to be eligible as a big idea in mathematics. There is an agreement that designing meaningful learning experiences in the mathematics lesson can be reinforced by an emphasis on big ideas related to mathematics and mathematics teaching. The prerequisite for big ideas require professional knowledge of mathematics teachers, particularly in the areas of content knowledge (CK) and pedagogical content knowledge (PCK). While we are concerned about the amount of mathematical knowledge primary teachers have, it may be more pertinent to deliberate how the knowledge is held (Hill & Ball, 2004). Therefore, it is more appropriate to examine the mathematical knowledge needed by primary school teachers in terms of the numerous connections and relations that are presented within and between mathematical ideas. If primary mathematics teachers can be educated to grasp these connections and relations to emphasise the big ideas of mathematics, there is the potential to change the approach in how they think about mathematics and their lesson planning. In addition, Clarke, Clarke and Sullivan (2012) and Kuntze et al. (2011) established that pre-service and in-service mathematics teachers should be given opportunities to identify, describe or link important ideas in mathematics. The interpretation here is that teaching mathematical content knowledge using big ideas as central themes is the way to deepen the understanding of teachers and to have a positive influence on their pedagogies. Gojak (2013) also supported that it is time to change the way in which mathematics education is presented and that pupils need to be taught by teachers who understand mathematical concepts intensely.

The latest secondary mathematics syllabus in Singapore which was released in 2018 and which will be implemented in 2020, indicated that big ideas "bring coherence and connect ideas from different strands and levels" (Ministry of Education, 2018 p. S2-3). This notion of big ideas of mathematics is desirable as it enables teachers to meet the requirements of the curriculum. However, the 2013 primary mathematics syllabus in Singapore, which was released in 2012 and was implemented in 2013, continued to present content in a familiar linear way. Although big ideas have existed for some time, they are not part of Singapore primary mathematics curriculum, teaching and learning as well as assessment. Moreover, Siemon, Bleckly and Neal (2012) reinforced that “a focus on the big ideas is needed to ‘thin out’ the over-crowded curriculum” (p. 20). The increasing evidence of understanding the big ideas and its importance is timely to start these discussions. This chapter therefore reviews the big ideas in mathematics, incorporating mathematics big ideas in teaching and big ideas about equivalence. This chapter also describes three activities on big ideas about equivalence that can be infused in the teaching and learning of mathematics at the primary level.

2 Review of Literature

This section explains what constitutes big ideas in Mathematics and what is teaching towards big ideas. In addition, reviews on pupils’ understanding of the big ideas about equivalence are also discussed.

2.1 *Big ideas in mathematics*

Big ideas in mathematics are not foreign to many mathematics educators but recently it has drawn some attention among the mathematics educators and researchers (Charles, 2005; Clarke et al., 2012; Siemon et al., 2012). Prior to a discussion on big ideas in mathematics, the view of big ideas can be found in the work of Bruner (1960) with respect to concept attainment and the spiral curriculum. Clark (1997) cited Bruner’s work by deliberating the significance of concepts as follows:

> My working definition of “concept” is a big idea that helps us makes sense of, or connect, lots of little ideas. Concepts are like cognitive file folders. They provide us with a framework or structure within which we can file an almost limitless amount of information. One of the unique features of these conceptual files is their capacity for cross-referencing. (p. 94)

Many researchers and mathematics educators have defined big ideas differently. Explanations of each of the definitions by some of the researchers are provided as follows. In the 1990s, Schifter and Fosnot (1993) pointed out big ideas within the discipline of mathematics. They defined them as "the central, organizing ideas of mathematics – principles that define mathematical order" (p. 35). The seminal work of Ma (1999) did use the label big ideas but termed knowledge packages as being the way in which ideas are structured and developed. She also labelled concept knots that symbolise the vehicles for connecting and linking ideas that are related to one another. Similarly, the National Council of Teachers of Mathematics (NCTM) expressed that "teachers need to understand the big ideas of mathematics and be able to represent mathematics as a coherent and connected enterprise" (NCTM, 2000, p.17). Later, Charles (2005) explains a big idea as "a statement of an idea that is central to the learning of mathematics, one that links numerous mathematical understandings into a coherent whole" (p. 10). He emphasised that big ideas are crucial because they allow the learners to view mathematics as a coherent set of ideas that bring a deep understanding of mathematics, promote transfer of learning, enhance memory and reduce the amount of information to be recalled. He also documented 21 such ideas including equivalence, the base ten number system, estimation, patterns, proportionality as well as orientation and location.

Siemon (2006) contended that without developing big ideas, pupils' mathematical learning might be limited. She reasoned that big ideas provide unifying structures that support further learning and generalisations. For this reason, Siemon referred big idea as an idea, approach, or way of thinking about key structures of mathematics and it includes and connects many other ideas and approaches. She also indicated that without big ideas, pupils' progress in mathematics would be delayed. In addition, she mentioned that big ideas may not be clearly defined but can be linked through activity. In 2012, Siemon and her team introduced the big number ideas and established counting, place value, multiplicative thinking, multiplicative partitioning, proportional reasoning, and generalising (Siemon et al., 2012). Later, Hurst and Hurrell (2014) continued the work of Siemon et al. (2012) and suggested micro content or basic parts that encompass big number ideas or concepts. However,

Charles (2005) and Siemon et al. (2012) believed that it would be a challenge to have a unique definition of big idea from mathematicians or teachers. There appears to be differing perspectives to consider when searching for a sufficient and workable definition of big ideas in mathematics. The different definitions of big ideas in mathematics, as described in this section, are not mutually exclusive. The different definitions that were discussed, highlighted the overarching concepts that are fundamental to our pupils' learning and to connect various mathematical ideas into a coherent whole.

2.2 *Incorporating mathematics big ideas in teaching*

As teaching towards big ideas is a critical part of the teaching and learning process in classrooms, mathematics teachers need to keep abreast of new developments in big ideas and be equipped with the necessary knowledge and skills of teaching towards big ideas. The big ideas concept contends for an integrated, coherent and connected way to learning and teaching, which supports pupils' development of a deep understanding of mathematics. In addition, Charles (2005) reported that a focus on big ideas and the links between them is necessary as it builds up pupils' understanding and enhance teachers' mathematical knowledge and confidence for teaching mathematics. This was supported by many researchers who studied the characteristics of effective teachers of mathematics advocated the importance of teaching towards big ideas (Askew, 1999; Charles, 2005; Hattie, 2003; Ma, 1999). In fact, Siemon et al. (2012) stated that:

> Effective teachers recognise the connections between different aspects and representations of mathematics. They ask timely and appropriate questions, facilitate and maintain high-level conversations about important mathematics, evaluate and respond to student thinking during instruction, promote understanding, help students make connections, and target teaching to ensure key ideas and strategies are understood. (p. 21)

Providing rich learning experiences in the mathematics classroom is essential for instructional quality and the development of competencies of pupils. Many researchers have emphasised the role of overarching concepts or fundamental ideas in mathematics and its teaching and learning for creating conceptually rich learning opportunities (Bishop, 1988; Schweiger, 2006).

The concept of big ideas is notable as it helps teachers in developing a coherent structure of mathematics. However, Clarke et al. (2012) reported that some teachers found that it is a challenge to articulate the big ideas that inform their teaching. Later, in their research studies on extensive analysis of primary teachers' unit plans, Roche, Clarke, Clarke, and Sullivan (2014) also reported that teachers have a challenge to articulate the big ideas. Their research showed a high level of variation in the identification and phrasing of key ideas for units of work. The implication of the research is that teachers' understanding of key mathematical ideas will also influence their selection and use of appropriate activities.

3 Big Ideas about Equivalence

The brief review of literature in the previous section has provided a sense of how big ideas are imperative to mathematics teaching and learning. With the introduction of big ideas in the 2020 secondary Mathematics syllabus in Singapore, it is also timely to introduce big ideas in primary mathematics curriculum to bring coherence and connected ideas from different mathematical ideas. The big idea about equivalence is a very important concept for primary school pupils and should be developed from a young age. The operational definition of equivalence in this chapter is referred to as "a relationship that expresses the 'equality' of two mathematical objects that may be represented in two different forms" (Ministry of Education, 2018, p.S2-4). Mathematically, equivalence is a big idea because mathematical objects that appear different can be connected to the same basic idea. For instance, $\frac{1}{10}$, 0.1 and 10% are all equivalent symbolic representations of the idea of a tenth. When pupils have made a robust connection between their concept of fractions and decimals, the topic of percentage can be introduced. Instead of teaching percentage as a new and separate idea, pupils should be provided a pictorial representation that shows a connecting link between fractions, decimals and percentages.

Theoretically, equivalence is a big idea for two main reasons. Firstly, numbers and measures can be stated in many equivalent forms by different ways of partitioning and factorising, with different symbolic representations highlighting different properties (Askew, 2013). For example, the whole number 81 can be expressed as: 9×9, $80 + 1$, 3^4, $90 - 9$, $3 \times 3 \times 3 \times 3$, $181 - 100$, $243 \div 3$, $100 - 19$. These different symbolic representations highlight different properties of numbers. From $3 \times 3 \times 3 \times 3$, we know that 2 is not a factor of 81, while 9×9 shows that 81 is a perfect square. Writing different equivalent expressions for 80 and 82 can show just how different these are from 81. This will make these three numbers more interesting than simply state it as three consecutive whole numbers in the eighties. Secondly, in every explanation about equivalence, there is a mathematical object (e.g. a number, an expression, an equation or statement) and a condition for 'equality' or the equivalence condition (e.g. value(s) and solution sets) (Ministry of Education, 2018). For instance, a number equation, a fraction or ratio, or an algebraic expression can be written in different but equivalent forms. Equivalence is also conceptually important especially when it is easier to compute an equivalent calculation than the one actually given. For example, find the value of 98×5, one could reason that 'ninety multiplied by five is 450, eight multiplied by five is 40, so the answer is 450 plus 40 which is equal to 490. In another way of computation, 100 multiplied by 5 is 500 and the answer must be 10 less, so that the answer is 490. The equivalent number equation could be written as: $98 \times 5 = (100 - 2) \times 5 = (100 \times 5) - (2 \times 5) = 500 - 10 = 490$. The change or conversion from one form to another equivalent form is the beginning of many manipulations for analysing and comparing them, as well as algorithms for finding solutions (Ministry of Education, 2018).

For these reasons, big ideas about equivalence with the following three activities is discussed in this chapter. Among others, primary mathematics teachers should consider implementing the following activities in their instructional process: Equivalence of two number sentences, Equivalence of two fractions and Equivalence of two ratios. All these activities, which can be easily carried out in the primary mathematics classrooms, are described below.

3.1 *Equivalence of two number sentences*

Anecdotal evidence from many experienced lower primary mathematics teachers has indicated that lower primary pupils do not accept the following as correct number equations: $4 + 2 = 3 + 3 = 6$. One possible way of laying the foundation for a deeper understanding of equivalence is to provide pupils with opportunities to discover the equivalence of two number sentences, such as the Activity 1 outlined in Figure 1.

Activity 1: Equivalence of two number sentences (Lower Primary)

Manipulative: Mathematics balance

Procedure: Place some weights on either side of the mathematics balance to make it balance.

Task 1

Write the addition equation that is shown on the mathematics balance.

(a) 10 balances with 5 and 5
(b) 3 and 4 balance with 1 and 6
(c) 2, 3 and 4 balance with 9

Task 2

Write the addition equation that is shown on the mathematics balance.

(a) 3 and 2 balance with 2 and 3
(b) 7 and 1 balance with 1 and 7

Task 3

Show 3 and 5 on the right side of the balance.

Next, make a balancing load by putting weights on two numbers on the other side.

(a) How many different ways can you do this?
(b) Write the addition equation that is shown on the mathematics balance.

Task 4

Put five weights on hook 2 on the left-hand side.

Put two weights on hook 5 on the right-hand side.

Figure 1. An activity to focus on equivalence of two number sentences.

Activity 1 aims to show how a mathematics balance could be used to reinforce the equivalent ways that the answer might be written. A mathematics balance is a worthy manipulative to use in exploring equality statements (Cathcart, Pothier, Vance, & Bezuk, 2006). Using a mathematics balance allows lower primary pupils to visually explore the concepts of equivalence. In addition, using a mathematics balance helps pupils discover misunderstandings in how they interpret numerical representations in equations and expressions. Lower primary pupils typically encounter the equal sign in number sentences that have operations on the left side of the equal sign and an answer blank on the right side. Pupils may not think about the equal sign as a symbol of equivalence and they may only perform the calculations on the left side of the equal sign to get an answer. As a result, pupils associate the equal sign with the arithmetic operations performed to get a final answer. In Activity 1, the mathematics balance is to strengthen understanding and computation of numerical expressions and equality. In understanding equality, pupils must realise that equal sign is a relational symbol of equivalence, not an operation. The commutative property of addition and the commutative property of multiplication are the easiest properties for pupils to explore with the mathematics balance concretely. One illustration is to use the mathematics balance to show the visual representation of the commutative property of multiplication. For example, in task 4, the numerical expression 5×2 (five weights hanging on the hook 2) and 2×5 (2 weights hanging on the hook 5) appears different; yet they are equivalent. Pupils can be asked what they notice on the balance when they put the weights on the mathematics balance. Teachers should encourage their pupils to verbalise that two numerical expressions are equivalent because the value of both the numerical expressions is the same.

3.2 *Equivalence of two fractions*

The beginning of developing fraction sense is understanding of and proficiency with equivalence. This includes, primarily, creating equivalent fractions and extends to understanding decimals and common percentages. As pupils develop their understanding of equivalence, which actually begins at the primary school level, they should be able to represent equivalent fractions using regions and manipulatives as well as paper folding. Although pupils appear to have opportunities to develop an understanding of equivalence, teachers tend to focus on the procedures of listing equivalent fractions. It is important for teachers to realise that equivalence is connected to and builds on ability and experience with a variety of representations, such as the Activity 2 outlined in Figure 2.

Figure 2 shows a paper-folding activity that illustrate the concept of equivalent fractions. Activity 2 provides opportunities for pupils and teachers to discuss how their representations are equivalent. This activity also allows pupils to establish the idea of equivalence between the concrete and symbolic representations, with the unfold paper providing a visual image of why the fractions are equivalent. At the first glance of Activity 2, it appears pupils may not be able to identify symbolic representations of $\frac{1}{2}$, $\frac{2}{4}$ and $\frac{4}{8}$ are equivalent. In class discussions, teachers could draw attention to equivalence relationships by helping pupils understand that different partitions result in equivalent amounts. Pupils would realise that half of a coloured region can be halved and that $\frac{1}{2}=\frac{2}{4}$. In other words, pupils should be able to articulate the coloured parts of the area in the rectangular strip of paper, representing respective fractions, depict the mathematical idea of equality. The mathematics teacher should also provide opportunities for pupils to explain their thinking so that other pupils can pose different sets of equivalent fractions. With more exposures and practices with similar concrete representations, the pupils might eventually make sense of the concept of equivalence between and among fractions.

<u>Activity 2: Equivalent fractions (Primary 3)</u>

Manipulative: A long strip of paper and a colour pencil.

Procedure:

1. Fold a piece of paper into two equal parts. Unfold the paper and colour one part.
   (a) How many equal parts are coloured?
   (b) What fraction of the paper is coloured?

2. Fold the same paper again to get 4 equal parts. Unfold the paper.
   (a) How many equal parts are coloured?
   (b) What fraction of the paper is coloured?

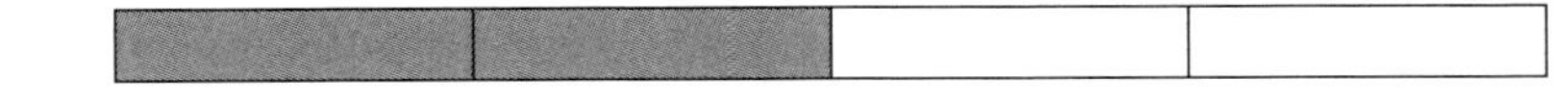

3. Fold the same paper again to get 8 equal parts. Unfold the paper.
   (a) How many equal parts are coloured?
   (b) What fraction of the paper is coloured?
   (c) Has the amount of paper coloured changed?
   (d) What can you say about the fractions $\frac{1}{2}$, $\frac{2}{4}$ and $\frac{4}{8}$?

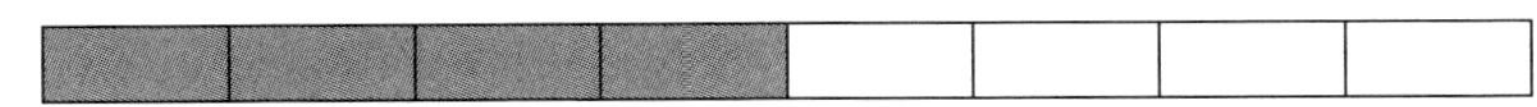

4. Fold the same paper again to get 16 equal parts. Unfold the paper.
   (a) How many equal parts are coloured?
   (b) What fraction of the paper is coloured?
   (c) Has the amount of paper coloured changed?
   (d) What can you say about the fractions $\frac{1}{2}$, $\frac{2}{4}$, $\frac{4}{8}$ and $\frac{8}{16}$?

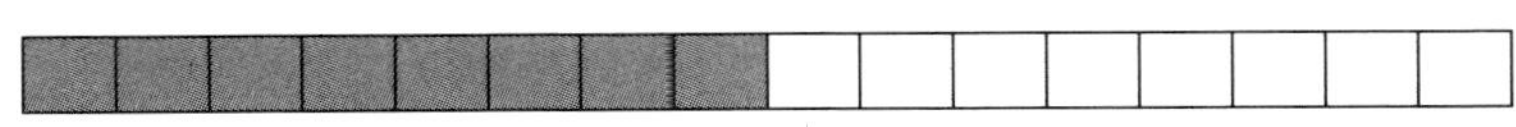

Figure 2. An activity tc focus on equivalence of fractions

3.2 *Equivalence of two ratios*

Two ratios that are equivalent possess the same within or unit relation (Noelting, 1980). Nonetheless, this idea of equivalence of two ratios is rather abstract for primary school pupils because pupils often base their intuitive thinking about ratio on counting, adding and combining. Pupils' understanding of the concept of equivalent ratios may not be linked to a multiplicative relationship. For example, the ratio 1 : 3 expresses the relationship between 4 squares and 12 circles in a box of 16 shapes. The word multiplicative is crucial here because we could also say the relationship between the squares and circles in additive terms, that is, as a difference, '8 more circles'. Therefore, pupils should be given the opportunity to construct their own representation of mathematical concepts, rules, and relationships related to the equivalence of two ratios. However, we should expect them to have an understanding that goes beyond concrete cases. For instance, teachers may begin with concrete representations or physical manipulatives to encourage pupils to use their own strategies for making sense of the equivalence of two ratios. In other words, pupils learn to find equivalent ratios by regrouping the objects into equal groups of another size. They also learn that equivalent ratios show the same comparison of numbers or quantities.

Activity 3 helps pupils understand the process of finding equivalent ratios. Teachers could lead pupils to the concept of equivalent ratios and ratio in its simplest form by using the guiding questions shown in Activity 3. It is necessary for pupils to note the following: (1) the stickers may be grouped in twos and the ratios is 2 : 6, (2) the stickers may be regrouped in fours again and the ratio is 1 : 3. Pupils should be encouraged to discover and verbalise the procedures for generating equivalent ratios. In Activity 3, pupils need to note that the simplest form of the ratio 1 : 3 does not mean that there are only 1 square sticker and 3 cicular stickers. The actual numbers of stickers remain unchanged.

Activity 3: Equivalent ratios (Primary 5)

Manipulatives: Four square stickers and twelve circular stickers.

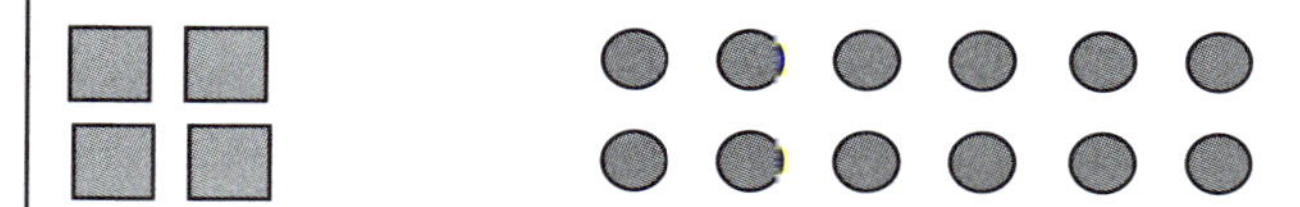

1. What is the ratio of the number of square stickers to the number of circular stickers?
2. How many different ways can you group these stickers?
3. If you group the stickers in groups of 2, what is the ratio of the number of square stickers to the number of circular stickers?
4. If you group the stickers in groups of 4, what is the ratio of the number of square stickers to the number of circular stickers?
5. What do you call these three ratios?
6. Is there any pattern or relationship that you have gathered about the numbers in the equivalent ratios?
7. What is the rule to determine whether two ratios are equivalent?
8. Explain the rule.
9. Which ratio is in its simplest form?

Figure 3. An activity to focus on equivalence of two ratios

These three activities exemplify equivalence as a relationship that expresses the 'equality' of two mathematical objects that could be represented in two different forms. The three activities have pertinent key oral questions to lead pupils to connect and identify the big ideas about equivalence. Opportunities such as these can be implemented so that they take little time away from an already crowded curriculum, yet still allow teachers to emphasise the big ideas about equivalence. The three activities are just a first step towards teaching big ideas about equivalence in the classroom. The emphasis is on explaining the connections and guiding pupils to uncover these connections rather than teaching the content in isolation.

4. Conclusion

The aim of teaching towards big ideas about equivalence is to deepen pupils' understanding and make connections. Equivalence is certainly a big idea in mathematics, which is stated in the Singapore secondary mathematics curriculum as being essential. Mathematics teachers need to have an in-depth understanding of big ideas themselves and pedagogical practices that provide learning experiences for pupils to explore and connect big ideas from different strands and levels. This approach is predominantly useful for pupils who have difficulty recognising the big ideas about equivalence or are easily confused by procedures of obtaining different equivalent forms.

If teachers were to focus on big ideas in mathematics, it could enable them to consider the mathematics curriculum in a different way. They would not be constrained by the linear structure of the content but could observe how mathematical ideas are established across different grade levels. Perhaps most meaningfully, they could see that planning to teach mathematics in this way would have numerous benefits for their pupils. Finally, teaching towards big ideas that engages all pupils is a gradual process, and both the teacher and pupils must have a paradigm shift regarding teaching and learning of big ideas.

References

Askew, M. (1999). It ain't (just) what you do: Effective teachers of numeracy. In I. Thompson (Ed.), *Issues in teaching numeracy in primary schools* (pp. 91-102). Buckingham, England: Open University Press.

Askew, M. (2013). Big ideas in primary mathematics: Issues and directions. *Perspectives in Education, 31*(3), 5-18.

Bishop, A. J. (1988). Mathematics education in its cultural context. *Educational Studies in Mathematics, 19,* 179-191.

Bruner, J. S. (1960). *The process of education.* Cambridge, MA: Harvard University Press.

Cathcart, W., Pothier, Y., Vance, J., & Bezuk, N. (2006). *Learning mathematics in elementary and middle schools: A learner-centred approach* (4th ed.). Upper Saddle River, NJ: Pearson Prentice Hall.

Charles, R. I. (2005). Big ideas and understandings as the foundations for elementary and middle school mathematics. *Journal of Mathematics Education Leadership, 7*(3), 9-24.

Clark, E. (1997). *Designing and implementing an integrated curriculum: A student-centred approach*. Brandon, Vermont: Holistic Education Press.

Clarke, D. M., Clarke, D. J., & Sullivan, P. (2012). Important ideas in mathematics: What are they and where do you get them? *Australian Primary Mathematics Classroom, 17*(3), 13-18.

Gojak, L. M. (2013). *It's elementary! Rethinking the role of the elementary classroom teacher*. Retrieved from http://www.nctm.org/about/content.aspx?id=37329.

Hattie, J. (2003, October). *Teachers make a difference: What is the research evidence?* Paper presented at the annual research conference of the Australia Council for Education Research, Melbourne: ACER. Retrieved from http://www.acer.edu.au/documents/RC2003_Proceedings.pdf.

Hill, H. C., & Ball, D. L. (2004). Learning mathematics for teaching: Results from California's mathematics professional development institutes. *Journal for Research in Mathematics Education, 35*(5), 330-351.

Hurst, C., & Hurrell, D. (2014). Developing the big ideas of number. *International Journal of Educational Studies in Mathematics, 1*(2), 1-18. Retrieved from http://www.ijesim.com/arsiv/volume-1-issue-2-december-2014.

Kuntze, S., Lerman, S., Murphy, B., Siller, H.-S., Kurz-Milcke, E., Winbourne, P., & Fuchs, K.-J. (2011). *Awareness of big ideas in mathematics classrooms: Final report.* Ludwigsburg: Pädagogische Hochschule.

Ma, L. (1999). *Knowing and teaching elementary mathematics: Teachers' understanding of fundamental mathematics in China and the United States.* Mahwah, NJ: Erlbaum.

Ministry of Education (2018). *2020 Secondary mathematics syllabuses (draft).* Singapore: Curriculum Planning and Development Division.

National Council of Teachers of Mathematics (NCTM). (2000). *Principles and standards for school mathematics*. Reston, VA: NCTM.

Noelting, G. (1980). The development of proportional reasoning and the ratio concept. Part II: Problem-structure at successive stages; problem-solving strategies, and the mechanism of adaptive restructuring. *Educational Studies in Mathematics, 11*(3), 331-363.

Roche, A., Clarke, D. M., Clarke, D. J., & Sullivan, P. (2014). Primary teachers' written unit plans in mathematics and their perceptions of essential elements of these. *Mathematics Education Research Journal, 26*(4), 853-870.

Schifter, D., & Fosnot, C. T. (1993). *Reconstructing mathematics education: Stories of teachers meeting the challenge of reform*. New York: Teachers College Press.

Schweiger, F. (2006). Fundamental ideas: A bridge between mathematics and mathematical education. In J. Maaß & W. Schlöglmann (Eds.), *New mathematics educational research and practice* (pp. 63-73). Rotterdam/Taipei: Sense Publishers.

Siemon, D. (2006). *Assessment for common misunderstandings materials. Prepared for and published electronically by the Victorian Department of Education and Early Childhood Development.* Retrieved from http://www.education.vic.gov.au/student learning/teachingresources/maths/common/default.htm.

Siemon, D., Bleckly, J., & Neal, D. (2012). Working with the Big Ideas in number and the Australian curriculum: Mathematics. In B. Atweh, M. Goos, R. Jorgenson., & D. Siemon (Eds.), *Engaging the Australian national curriculum: Mathematics – Perspectives from the field* (pp. 19-45). Online Publication: Mathematics Education Research Group of Australasia.

Chapter 8

Teaching Mathematics: A Modest Call to Consider Big Ideas

TAY Eng Guan

Big Ideas are defined here as overarching concepts that occur in various mathematical topics in a syllabus. Knowing these will guide teachers to help students develop a better understanding of mathematics, by making visible the central ideas, the coherence and connection across topics and the continuity across levels. In this chapter, we shall give a few examples of Big Ideas across the Pre-university mathematics syllabuses and how to exploit them for better pedagogy. A deeper consideration of what mathematics is will guide our discussion.

1 Introduction

Charles (2005) defines a Big Idea in mathematics as "a statement of an idea that is central to the learning of mathematics, one that links numerous mathematical understandings into a coherent whole." (p. 10) He breaks the definition down into: (i) A statement, such as "Mathematical expressions can be expressed in more than one way", which then for ease of discussion may be given a title such as Equivalence; and (ii) the centrality and connectivity of the idea, i.e., its importance in connecting many important concepts into a coherent whole. It follows then that understanding Big Ideas in mathematics develops a deeper and more robust understanding of individual topics in mathematics.

It would seem obvious that to recognize a Big Idea in mathematics, one ought to be able to pin down what mathematics is in the first place. There is however no universally accepted 'definition' of mathematics and avoiding it, in our opinion, makes us all the poorer – in appreciating good mathematics and in the teaching of mathematics. To this end, we offer a simultaneous description of both music and mathematics, with the purpose of understanding both better through comparison and reflection.

Music is the *organization* of sound. Mathematics is *reasoning* with and about (well-defined) abstract concepts/objects, often involving the concept of number. Thus, the din of traffic is not music because it is random and there is no agency, human or otherwise, organizing the output. Arguing like the ancient Greeks that the result of cutting a piece of iron again and again until it is indivisible results in an atom (άτομο - cannot be cut) of iron is not mathematics because the concepts of 'cutting' and 'indivisible' are not well-defined concepts. Instead, in proper mathematics, we may argue that it is possible to continue to divide a positive real number by 2 an infinite number of times. Big Ideas about mathematical reasoning such as proofs, axioms, definitions and theorems need to be understood to obtain a deeper appreciation of individual topics in mathematics.

Musicians organize sound (and its complement, silence) using various instruments (including the human voice) to inspire, entertain, evoke and communicate. Mathematicians reason with and about abstract concepts using various techniques, representations and instruments (including the calculator and computer) to solve problems, create, entertain, inspire and communicate. Thus, the techniques, representations and instruments for mathematical reasoning are also important accoutrements that a student of mathematics ought to understand. These accoutrements themselves can also be coherently linked and understood by Big Ideas about them.

Charles (2005) goes on to argue that teachers can increase their effectiveness when they teach by consistently connecting new ideas to Big Ideas and connecting topics across grades through relevant Big Ideas. In a way, Big Ideas help to achieve the following adage in teaching: "Make the new familiar and make the familiar new."

This chapter defines Big Ideas as overarching concepts that occur in various mathematical topics in a syllabus. We focus only on Big Ideas relevant to the Singapore Pre-University mathematics. Guided by our 'definition' of mathematics, we shall consider Big Ideas in methods of techniques, representation and reasoning in mathematics.

2 Some Big Ideas in Pre-University Mathematics

2.1 *Solving equations*

Information about one or more variables may be presented as equation(s). Solving equation(s) uses techniques that result in a 'clearer' expression for the variable(s).

This is a small Big Idea with regard to techniques in pre-university mathematics. In the syllabus, students encounter solving linear, quadratic, simultaneous, trigonometric and differential equations. Students should learn that each often involves isolating the variable and expressing it as a range of numbers or as an expression in terms of another variable. A 'target expression' approach will then involve suitable algebraic manipulations, e.g. completing the square, partial fraction decomposition.

The following is an example of how trigonometric equations can be taught with reference to the Big Idea of Solving Equations.

- Solve $2\sin^2 2\theta + 3\sin 2\theta + 1 = 0$.
- Recall that in solving simple linear equations such as $2x+1=0$, the technique involves 'isolating the unknown' as follows:

$$2x = -1$$

$$x = -\tfrac{1}{2}.$$

- We can see that the original equation is a quadratic equation in $\sin 2\theta$. To be efficient, we obtain $\sin 2\theta$ by using the formula as obtained by completing the square (a technique for isolating the unknown $\sin 2\theta$) as follows:

$$2\sin^2 2\theta + 3\sin 2\theta + 1 = 0$$

$$\sin 2\theta = \frac{-3 \pm \sqrt{9-8}}{4}$$

$$\sin 2\theta = -1 \text{ or } -\tfrac{1}{2}.$$

- Finally, we isolate θ by taking the inverse relation $\sin^{-1}$ and then taking half of 2θ as follows:

$$\begin{array}{lcl} \sin 2\theta = -1 & \text{or} & \sin 2\theta = -\tfrac{1}{2} \\ 2\theta = n\pi + (-1)^n \sin^{-1}(-1) & \text{or} & 2\theta = n\pi + (-1)^n \sin^{-1}(-\tfrac{1}{2}),\ n \in \mathbb{Z} \\ 2\theta = n\pi + (-1)^n(-\pi) & \text{or} & 2\theta = n\pi + (-1)^n(-\tfrac{\pi}{6}),\ n \in \mathbb{Z} \\ 2\theta = n\pi + (-1)^{n+1}\pi & \text{or} & 2\theta = n\pi + (-1)^{n+1}\tfrac{\pi}{6},\ n \in \mathbb{Z} \\ \theta = \tfrac{\pi}{2}(n + (-1)^{n+1}) & \text{or} & \theta = \tfrac{\pi}{6}(6n + (-1)^{n+1}),\ n \in \mathbb{Z} \end{array}$$

Throughout, we use the technique of 'isolating the unknown' to move across different types of equations, thus making links between linear, quadratic and, in this hypothetical lesson, the new trigonometric equation.

For good measure, we provide another example below.

- Solve $3\sin\theta - \cos 2\theta + 1 = 0$.
- This looks like an equation with two unknowns θ and 2θ.
- We look for a trigonometric identity that makes it an equation with only one unknown. We choose:

$$\cos 2\theta = 1 - 2\sin^2\theta$$

- Then we proceed to obtain:

$$3\sin\theta - \cos 2\theta + 2 = 0$$

$$3\sin\theta - (1 - 2\sin^2\theta) + 2 = 0$$

$$2\sin^2\theta + 3\sin\theta + 1 = 0$$

which is the equation as before.

2.2 *Equivalence*

Any number, measure, numerical expression, algebraic expression, or equation can be represented in an infinite number of ways that have the same value (Charles, 2005).

Students may wonder why one should bother to work out the Maclaurin series for sin x. The enlightened teacher may explain that this is an equivalent expression for sin x that is computable. Mathematicians exploit or create different equivalent expression forms so that among them, they may choose a form that is most useful for a particular purpose.

Again, we surmise a teaching situation as follows. The teacher may ask her students, "How would your calculator compute sin 2?" To be humorous, she may proceed to suggest that when one presses sin 2 on the calculator, a little man inside the calculator wakes up, uses a ruler and protractor to draw a right-angled triangle with angle 2. He then measures the opposite side and the hypotenuse and obtains sin 2 by dividing the opposite with the hypotenuse!

Instead, the teacher adds, the Maclaurin series for sin x, i.e.,

$$\sin x = x - \frac{x^3}{3!} + \frac{x^5}{5!} - \frac{x^7}{7!} + \ldots + \frac{x^{2n-1}}{(2n-1)!} + \ldots$$

is an equivalent expression that can be computed, manually, and by a computer/calculator. Thus

$$\sin 2 = 2 - \frac{2^3}{3!} + \frac{2^5}{5!} - \frac{2^7}{7!} + \frac{2^9}{9!} - \frac{2^{11}}{11!} + \ldots = 0.9093 \text{ (correct to 4 d.p.).}$$

She can also elaborate on computational thinking in mathematics by coding a simple VBA program in Microsoft Excel as shown in Figure 1.

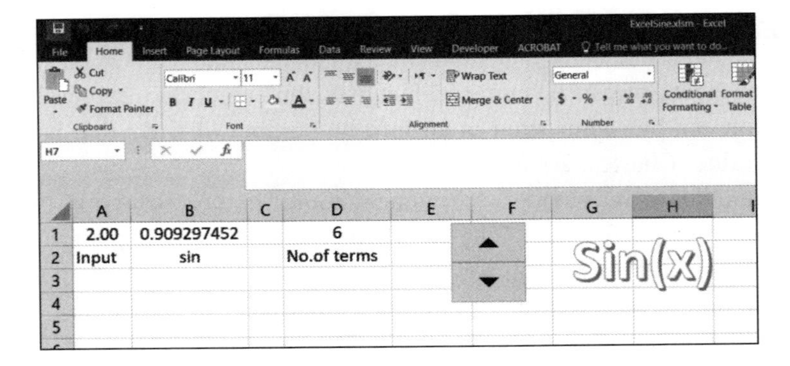

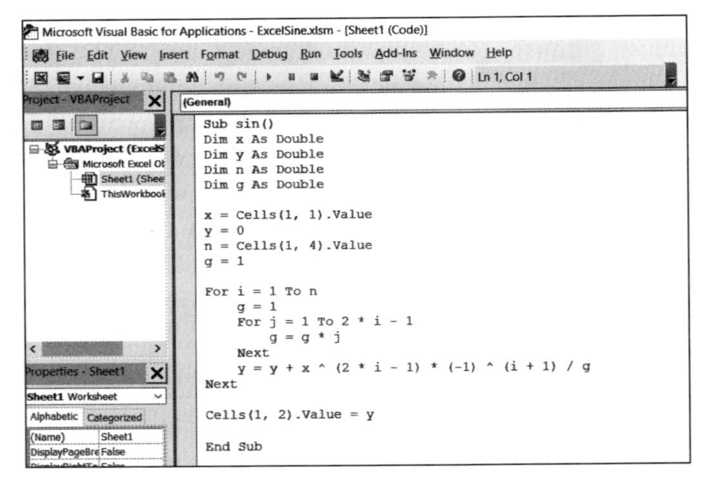

Figure 1. Excel program for Maclaurin series of sin x

Anecdotally, many students do not understand why they ought to learn Maclaurin series. By explaining the Big Idea of Equivalence and the purpose of having different representations of the same thing to exploit the most appropriate expression, the teacher would have one means to motivate the student. In the example given above, students will also be able to understand the underlying computability consideration when looking for equivalent expressions.

To be exact, we would like to add in a disclaimer that calculators typically do not use Maclaurin series to compute trigonometric functions. In a Question and Answer forum (Texas Instruments, 2018), Texas

Instruments states very clearly that they do not use polynomial approximations. Also, in those cases that do use polynomial approximation, the preferred polynomials are not the familiar ones like the Taylor series but those more related to Chebyshev polynomials.

2.3 *Definition and notation*

Definitions formalize concepts so that they can be reasoned on. Notations are definitions written as shorthand symbols for mathematics communication.

In the earlier example of Greek philosophers musing on the essence of a material and postulating that an atom of that material would result from dividing until one can divide no more, we see that the flaw in that argument was the non-rigorous definition of 'dividing'. We know now through science that all material when 'divided' breaks into the same fundamental particles and that it is the structure that determines the material. We are still not sure if these fundamental particles can be divided further.

Mathematics skirts these 'real-life' constraints by imposing definitions a priori so that the reasoning that follows is perfectly logical. If the end-result conforms to something in real-life as it often does, we are pleased. If not, the reasoning remains true and what results is some abstraction that could still be useful. Indeed, the mathematician physicist Freeman Dyson (1996) commented on the permanence of mathematics:

> [M]athematical papers, provided that they are correct and not trivial, have permanent value, whereas most papers in physics journals are ephemeral. For this reason, it is customary to publish complete collected works of mathematicians but only selected works of physicists. (p. 1)

Tay and Leong (2012) give three examples on how to explain the notations a^0, a^n and 0! in relation to the development of mathematical concepts and their communication through building definitions and theorems. Indeed, a particularly important type of definition is notation.

When students learn the Big Idea of Notation, they would be able to first appreciate the simplification the notation in a topic affords and then to spend time to carefully understand each symbol of the notation.

Across different topics, whenever new notation is introduced, the teacher should explain how the notation encapsulates a concept in short form – thus, visually, a good notation should summarize all key components of the concept with as few symbols as possible. For example, the notation for summation in the given example

$$\sum_{r=1}^{n} r^2$$

gives visually the Greek capital letter sigma which alliterates with 'sum', the range from 'low' at the bottom to 'high' at the top, and the general term in prime position.

As a further example, the Singapore A-Level Further Mathematics syllabus document (Singapore Examinations and Assessment Board, 2018) has the following notations for the derivative.

1. $\frac{dy}{dx}, \frac{d^2y}{dx^2}, \ldots, \frac{d^ny}{dx^n}$ the first, second, …, nth derivatives of y with respect to x
2. $f'(x), f''(x), \ldots, f^{(n)}(x)$ the first, second, …, nth derivatives of $f(x)$ with respect to x
3. $\dot{x}, \ddot{x}, \ldots$ the first, second, … derivatives of x with respect to time

Now, students may be confused as to why three different notations need to be used for the derivative. The teacher can explain that the usefulness of each notation varies with the context. The first notation is useful when the equation $y = f(x)$ is regarded as a functional relationship between the dependent variable y and the independent variable x. This notation is by Leibniz and in it, the relationship between the variables is made explicit. It has the added advantage of making the chain rule easy to remember:

$$\frac{dy}{dx} = \frac{dy}{du} \cdot \frac{du}{dx} \ .$$

The second notation is due to Lagrange. It is useful when the derivative evaluated at a particular value is needed. For example, the Maclaurin series expansion for a real or complex valued function $f(x)$ that is infinitely differentiable at 0 is the power series nicely expressed in terms of $f^{n}(0)$:

$$f(x) = f(0) + \frac{f'(0)}{1!}x + \frac{f''(0)}{2!}x^2 + \ldots + \frac{f^{(n)}(0)}{n!}x^n + \ldots \ .$$

See how bad it would look if we use the first notation:

$$y = y\Big|_{x=0} + \frac{\left.\frac{dy}{dx}\right|_{x=0}}{1!}x + \frac{\left.\frac{d^2y}{dx^2}\right|_{x=0}}{2!}x^2 + \ldots + \frac{\left.\frac{d^ny}{dx^n}\right|_{x=0}}{n!}x^n + \ldots \ .$$

The third notation is due to Newton. This notation is popular in areas of mathematics connected with physics such as differential equations. Perhaps, for the reason of future familiarity, it has been included in the Further Mathematics syllabus. This may however not be necessary if students learn throughout their school mathematics what the purpose of notation is, thus anticipating the links across topics – *make the new familiar*.

3 Conclusion

In this chapter, we give three examples of Big Ideas that are relevant to the Singapore Pre-University mathematics syllabuses. Certainly, there are more Big Ideas with links stretching back through secondary to primary school and forward to university mathematics. These Big Ideas lend an important layer to the tapestry of mathematical knowledge that the student accumulates in his studies. This layer provides familiarity when new material is encountered and motivation when old material is revised. We hope that teachers will themselves seek more Big Ideas to tie their teaching together in a coherent and inspirational manner for the learning of their students.

References

Charles, R. I. (2005). Big ideas and understandings as the foundation for elementary and middle school mathematics. *Journal of Mathematics Education Leadership*, *7*(3), 9-24.

Dyson, F. J. (1996). *Selected papers of Freeman Dyson with commentary*. MA: American Mathematical Society.

Singapore Examinations and Assessment Board (2018). *9649 Further Mathematics GCE Advanced Level H2 Syllabus (2019).* Retrieved from https://www.seab.gov.sg/docs/default-source/national-examinations/syllabus/alevel/2019Syllabus/9649_2019.pdf

Tay, E. G., & Leong, Y. H. (2012). Explaining definitions in secondary mathematics: a^0, a^n, 0! *Australian Senior Mathematics Journal*, *26*(2), 28-37.

Texas Instruments (2018). *Solution 11693: Algorithm for solving trigonometric functions (sine, cosine and tangent) on Texas Instruments' graphing calculators.* Retrieved from https://epsstore.ti.com/OA_HTML/csksxvm.jsp?nSetId=74414

Chapter 9

Invariance as a Big Idea

TOH Pee Choon

Understanding big ideas in mathematics allows us to see the subject as a coherent and connected discipline. In this chapter, we shall focus on the big idea of invariance and explicate how this big idea threads a number of topics horizontally across content strands and vertically across levels.

1 Introduction

According to Charles (2005),

> [A] Big Idea is a statement of an idea that is central to the learning of mathematics, one that links numerous mathematical understandings into a coherent whole. (p. 10)

He asserts that big ideas should be the "foundation for one's mathematics content knowledge, for one's teaching practices, and for the mathematics curriculum", and understanding big ideas allows one to view mathematics as a "coherent set of ideas" and not a collection of "disconnected concepts, skills, and facts" (Charles, 2005, p. 10). Charles provides a list of 21 big ideas but states unequivocally that it is impossible to have a list that all mathematicians and mathematics educators agree on. In his conclusion, he urges teachers to develop their own lists of big ideas while keeping in mind that "Big Ideas need to remain BIG and they need to be the anchors for most everything we do" (Charles, 2005, p. 12). In this chapter, we shall focus on a single big idea, namely invariance, which threads a number of topics horizontally across content strands and vertically across levels. Our statement about invariance is as follows.

> Big Idea of Invariance: Valuable information can be obtained by studying the *invariant set* of a general class of *transformations*.

We will illustrate with several examples of invariance that occur in school mathematics, ranging from geometry at the Primary/Secondary Level to number theory at the Pre-University Level. In each example, we shall try to identify explicitly the *invariant set* and the *transformation*.

2 Invariance in Geometry

The concept of invariance is of particular importance in geometry. Many theorems in geometry are formulated as statements about certain invariants. For example, virtually every student of mathematics will know that the angle sum (*invariant set*) of a triangle in the Euclidean plane equals 180°, regardless of how the triangle is drawn (*transformation*). They will also know that the area (*invariant set*) of the triangle ABC, illustrated in Figure 1, will remain constant when the vertex C is translated (*transformation*) to C_1, C_2, or any point along the line C_1C_2 that is parallel to AB.

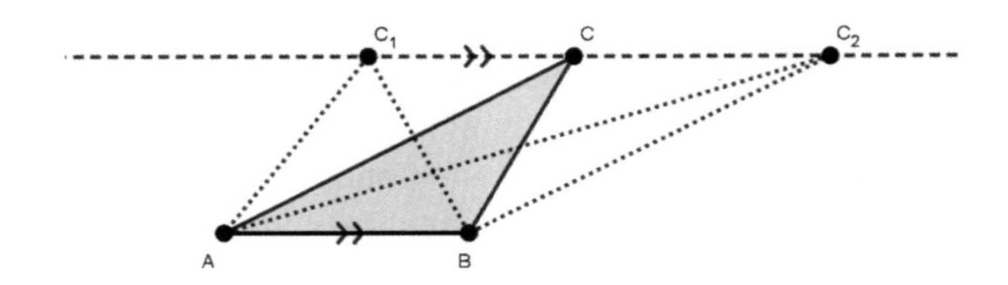

Figure 1. Area is invariant

Building on our previous example, the first square from the left in Figure 2 is typically found in puzzles in geometry. Here one is supposed to calculate what fraction of the square is shaded. The standard algebraic solution is to note that if x is the length of a side of the square and h the height of one triangle, then the height of the other triangle must be $x - h$. Thus the shaded region must have area

$$\frac{1}{2}(h + x - h)x = \frac{1}{2}x^2.$$

The expression is clearly independent of the variable h and so the area of the two triangles (*invariant set*) remains the same when the point of the intersection of the two triangles is translated (*transformation*) to any other point within the square. A student who understands this fact can quickly work out the required answer by translating the point of intersection to the extreme right or even to the top right corner as seen in the two diagrams from the right in Figure 2. An interesting follow-up question for students is to identify precisely when the invariance breaks down if the intersection is translated to a point outside the square.

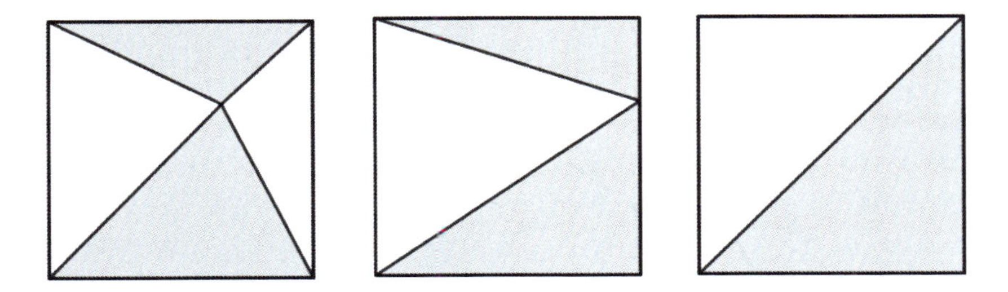

Figure 2. Finding the proportion of a shaded figure

Let us consider another example of a fixed equilateral triangle ABC and any point P in its interior. If we drop a perpendicular to each of the three sides (Figure 3), the sum of the lengths of the three perpendiculars remain constant irrespective of which interior point P was initially chosen. In this example, the transformation refers to any translation of the point P as long as it remains in the interior of the triangle ABC, while the invariant set is the sum of the lengths of the three perpendiculars.

Readers can verify that indeed this sum equals the length of the height of the triangle.

There are many more interesting examples of invariance in geometry. In their book written for teachers, Sinclair, Pimm and Skelin (2012) articulated four big ideas in geometry. The second of which is "geometry is about working with variance and invariance, despite appearing to be about theorems." Libeskind, Stupel and Oxman (2018) give several other geometric examples of invariance in school mathematics, including two examples on conics which might be of interest to A-Level Further Mathematics teachers. They advocate the approach of providing students with the opportunities to investigate invariance through dynamic geometry software such as GeoGebra.

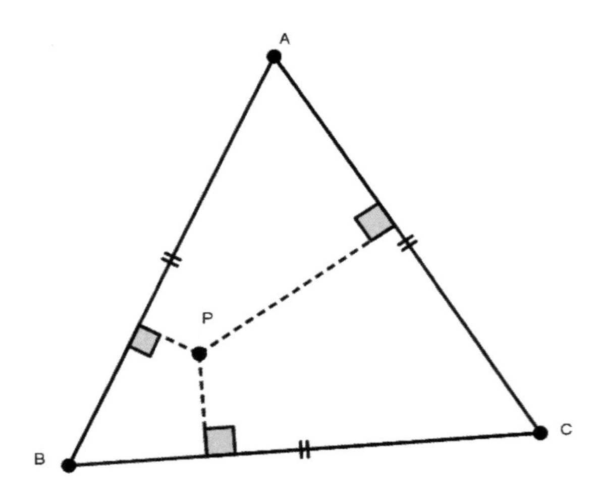

Figure 3. Sum of the lengths of the three perpendicular is invariant

3 Invariance in Numerical Iterations

The following is a simple activity to investigate invariance in numerical iterations. Using a calculator, or preferably a computer, compute (in radian measure) cos(0.5). Next, continue to compute cos(x) where x is the output obtained from the previous computation. This is an example of an iteration scheme: $x_0 = 0.5$; $x_{n+1} = \cos(x_n)$, under the topic of numerical methods in Further Mathematics at the Advanced level. Figure 4 shows two screenshots from a graphing calculator. The left screenshot shows the first four iterations while the right screenshot shows that x_n appear to converge to an approximate value of 0.7390851332 after 60 iterations. The actual value, which we shall denote as c, is called a fixed point of $\cos(x)$ and satisfies $\cos(c) = c$. In other words, the point c is invariant under the transformation $\cos(x)$.

Left screenshot		Right screenshot	
cos(0.5)		cos(Ans)	
	.8775825619		.7390851333
cos(Ans)		cos(Ans)	
	.6390124942		.7390851332
cos(Ans)		cos(Ans)	
	.8026851007		.7390851333
cos(Ans)		cos(Ans)	
	.6947780268		.7390851332

Figure 4. Iterating $\cos(x_n)$

In our example, the initial value of x_0 can be replaced by any value in the interval (0, 1) and the iteration process would still appear to converge to the same fixed point after a sufficient number of iterations. It is also possible to replace the function $f(x) = \cos(x)$. Iterations using

$$f(x) = e^{-x} \quad \text{or} \quad f(x) = 1 + \frac{2x - x^2}{2}$$

would also converge to the respective fixed points of these functions. It would take us too far afield to describe the conditions for such iterations to converge, but it is comparatively simple to show the existence of fixed points.

The set of functions that are continuous on the closed interval [0, 1] contains most of the functions that students have been introduced to in school mathematics. These are, for example, the polynomial functions, the absolute value function, the trigonometric sine and cosine functions, and the exponential function. It is thus somewhat surprising that one can extract useful information from such a general class of functions.

Theorem 1. If $f(x)$ is continuous on $[0, 1]$, $f(0) > 0$ and $f(1) < 1$, then $f(x)$ has a fixed point c in $(0, 1)$. That is, $f(c) = c$.

A rigorous proof of the above theorem is beyond the scope of this chapter but the following explanation is usually sufficient to convince most students. Let us consider the graph of $f(x)$ and $y = x$ on the same interval $[0, 1]$. Since $f(0) > 0$ and $f(1) < 1$, the left and right endpoints of the graph of $f(x)$ are lying respectively 'above' and 'below' the graph of $y = x$. Although we are unable to actually sketch the graph of $f(x)$ as we do not have any additional information about this function other than it being continuous, we can argue that any continuous graph joining the left and right endpoints must intersect the graph of $y = x$. If we use c to denote the x-coordinate of this intersection point, we must have $f(c) = c$.

There are a number of fixed point theorems in mathematics and what we have sketched above is a special case of what is usually known as Brouwer's fixed point theorem (Munkres, 2000). Fixed points theorems illustrate very well the big idea of invariance where useful information can be extracted from very general situations. Munkres' book contains a proof of how Brouwer's theorem can be used to prove that any 3×3 matrix with positive real entries has a positive real eigenvalue. It is no coincidence that the concept of eigenvectors in linear algebra is yet another instance of invariance. In this case, the space spanned by an eigenvector of a linear transformation is a one-dimensional invariant subspace. Finally, we note that fixed point theorems are more than mere intellectual curiosities. These theorems have found applications in economics and game theory (Border, 1985).

4 Invariance in Number Theory

We focus on two examples of invariance in number theory in this section. The first involves simple arithmetic and can be viewed as an instance of a fixed point theorem analogous to those discussed in the previous section. We describe it in the form of a 'magic trick' that can be used in the classroom. The teacher first asks a student to think of a three-digit positive integer with distinct digits. He then pretends to read the student's mind, writes a number down on a piece of paper and folds it in half to prevent anyone from seeing the number. Next, he hands the folded piece of paper to another student for safekeeping. The teacher now asks the first student to rearrange the digits of his chosen number to form the largest possible number and write it on the board. For example, if the student chose 197, the digits would be rearranged to form 971. Next, he asks the student to rearrange the same three digits to form the smallest possible number and subtract this smallest number from the largest. Continuing with our numerical example, the student would compute 971 – 179 = 792. The teacher then asks the student to repeat the calculations with this new number. That is, subtracting the smallest possible number from the largest possible number, both of which are formed by rearranging the digits of the new number. The calculations are repeated until the student arrives at the number 495, whereupon the teacher asks the second student to reveal to the class that the number previously written on the piece of paper is also 495.

The 'magic trick' that we just described can be used in classrooms ranging from Upper Primary to Pre-University. The transformation is the operation of subtracting the smallest possible number from the largest possible number formed by rearranging the digits, while the invariant set or fixed point is the number 495. In other words, one can begin with any three-digit number (with distinct digits) and after at most six iterations of the operation arrive at 495. For more advanced students, the same exercise can be carried out with four-digit numbers, in which case, the fixed point is the number 6174. This is known as Kaprekar's constant (Kaprekar, 1955). Let us see why this works in the three-digit case, where we may assume $a > b > c$, are the three distinct digits. Carrying out the required operation will result in

$$100a + 10b + c - (100c + 10b + a) = 99(a - c).$$

We obtain one of eight possible multiples of 99, because $a - c$ can range from 2 to 9 inclusive. It is then not difficult to observe that every subsequent iteration will result in another multiple of 99 where the difference between the largest digit (which is always 9) and the smallest digit converges to 5, giving 495 When used in appropriate situations, the 'magic trick' just described creates an element of surprise and is an interesting and accessible piece of mathematics for most students. More advanced students may be tasked to investigate the situation of numbers with more digits or to find a different transformation that has an analogous invariant property.

Our second example is more advanced and requires background knowledge of number theory. Number theory is a recent addition to the Higher 3 (H3) Mathematics at A-Level. While not explicitly stated in the syllabus document, we can infer from the specimen examination paper and subsequent actual examination papers that students are expected to know one particular proof of Fermat's Little Theorem. This proof is an exemplar of the big idea of invariance and we will describe it below. We first define the concept of congruence in number theory. If n is a positive integer, we say that a is congruent to b modulo n and write $a \equiv b \pmod{n}$ if $a - b$ is an integer multiple of n. In other words, both a and b have the same remainder when divided by n. A useful way to understand the congruence $a \equiv b \pmod{n}$ is to think of it as a generalization of $a = b$. (More precisely, it is an equivalence relation.) We can now state Fermat's Little Theorem.

> **Theorem 2**. Suppose p is prime number and a is not a multiple of p, then $a^{p-1} \equiv 1 \pmod{p}$.

This means that if we take any integer and raise it to the exponent $p - 1$ for some prime number p, the resulting integer always leaves a remainder of 1 when divided by p. The only exception is when the original integer was already a multiple of p, in which case the remainder is clearly 0. This is a remarkable result because it works for any prime number and any integer that is not a multiple of that prime number. Consider this example: $495^{12} = 216, 402, 556, 571, 320, 625, 160, 840, 087, 890, 625$ has remainder 1 when divided by 13. Equally surprising is the fact that we can actually prove that the theorem holds. To understand the proof, we require the following result.

> **Theorem 3**. Suppose p is prime number and a is not a multiple of p, then the congruence $ax \equiv 1 \pmod{p}$ has a solution.

Theorem 3 is a fundamental result in number theory because it recovers the concept of a multiplicative inverse, which is missing in the integers. For example, the equation $7x = 1$ does not have any solutions in integers but the congruence $7x \equiv 1 \pmod{p}$ has a solution for every prime number p other than 7. Taking $p = 17$, a solution for $7x \equiv 1 \pmod{17}$ is $x = 5$.

All the pieces are now in place for us to use invariance to prove Fermat's Little Theorem. Given a fixed prime number p, the invariant set S, is the set of all possible non-zero remainders after dividing by p. In other words, $S = \{1, 2, \ldots, p - 1\}$. The set of transformations is that of multiplying an element by a, with the condition that a is not a multiple of p, followed by taking the remainder after dividing by p. For example, when $a = 4$ and $p = 7$, then the transformation sends 6 to 3 because $4 \times 6 = 24 \equiv 3 \pmod{7}$. The action of the transformation on all the elements of S are given below:

$$1 \mapsto 4,\ 2 \mapsto 1, 3 \mapsto 5, 4 \mapsto 2, 5 \mapsto 6, \text{and } 6 \mapsto 3.$$

We can also represent the above as

$$\{4 \times 1, 4 \times 2, 4 \times 3, 4 \times 4, 4 \times 5, 4 \times 6\} \equiv \{4, 1, 5, 2, 6, 3\} \pmod{7}.$$

Let us now prove that the set S is invariant under the general transformation by showing that

$$\{a \times 1, a \times 2, \ldots, a \times (p - 1)\} \equiv \{1, 2, \ldots, p - 1\} \pmod{p}. \qquad (**)$$

Consider any pair of distinct elements from the set on the left of (**), namely $a \times j$ and $a \times k$ for some distinct j, k between 1 and p - 1. Suppose we have $a \times j \equiv a \times k \pmod{p}$, then we can multiply to both sides the solution x that is guaranteed to exist by Theorem 3. Replacing xa by 1 would mean $j \equiv k \pmod{p}$, or simply $j = k$ which contradicts our assumption that they are distinct. We have just shown that none of the elements from the left side of (**) are congruent to each other. Since there are exactly $p - 1$ elements, this can only mean that these elements are simply $\{1, 2, \ldots, p-1\}$ in some order. This completes the proof of (**). We now multiply all the elements from both sides of (**) to obtain

$$\begin{aligned}(a \times 1)(a \times 2) \ldots (a \times (p-1)) &\equiv (1)(2)\ldots(p-1) \pmod{p} \\ \Rightarrow a^{p-1}(p-1)! &\equiv (p-1)! \pmod{p} \\ \Rightarrow \quad a^{p-1} &\equiv 1 \pmod{p}.\end{aligned}$$

In the above, we again used Theorem 3 to cancel the factor $(p-1)!$ from both sides. Although the proof that we have presented is slightly technical, it illustrates perfectly the big idea of obtaining useful information by studying the invariant set of general transformations. The proof can be further adapted to prove a number of important results in number theory, namely Wilson's Theorem, Euler's Criterion for quadratic residues, and Gauss' Lemma. Fermat's Little Theorem itself is commonly used as a test to see if a large integer is a prime number, and thus has important applications to the area of modern cryptography.

5 Conclusion

Understanding big ideas in mathematics and translating them into one's teaching practice can appear to be a daunting challenge. It requires both the breadth of content knowledge and the depth of insight to identify the underlying principles behind mathematical topics or results. In this chapter, we focused on the big idea of invariance and illustrated its connections with examples from the curriculum. We hope that teachers can use these examples as a starting point to develop their own understanding of invariance and other big ideas in mathematics.

References

Border, K. C. (1985). *Fixed point theorems with applications to economics and game theory.* Cambridge: Cambridge University Press.

Charles, R. I. (2005). Big ideas and understandings as the foundation for elementary and middle school mathematics. *Journal of Mathematics Education Leadership, 7*(3), 9-24.

Kaprekar, D. R. (1955). An interesting property of the number 6174. *Scripta Mathematica, 15*, 244-245.

Libeskind, S., Stupel, M., & Oxman, V. (2018). The concept of invariance in school mathematics. *International Journal of Mathematics Education in Science and Technology, 49*(1), 107-120. doi:10.1080/0020739X.2017.1355992

Munkres, J. (2000). *Topology* (2nd ed.). New Jersey: Prentice Hall.

Sinclair, N., Pimm, D., & Skelin, M. (2012). *Developing essential understanding of geometry for teaching mathematics in grades 9-12.* Reston, VA: NCTM.

Chapter 10

Empirical Motivation for Teaching Functions and Mathematical Modelling

YAP Von Bing

What makes the two ideas of Functions and Mathematical Modelling important for secondary school mathematics? This article aims to bring out some compelling features related to empirical considerations of these two ideas, such as perfect predictability for functions and measurement errors for modelling, in an elementary narrative that can serve as a possible teaching approach.

1 Introduction

Big mathematical ideas tend to be abstract, so convincing examples are indispensable for teaching these ideas to any type of students, in particular, secondary school students. This is even more important in situations where the formal definitions of functions and mathematical models are not revealed fully in the secondary mathematics classroom.

Among the Big Ideas recently unveiled by the Ministry of Education, Functions and Mathematical Modelling appear most fascinating because of their close historical and current connections with science, by which I mean the vast enterprise of systematically observing and learning from the real world. Hence, infusing some empirical flavor in the teaching of the topics, like collecting data or making measurements, is pedagogically sound and likely useful.

This chapter presents teaching approaches to the two topics that explicitly incorporate empirical connections, though hands-on activities may not be involved. Rather, empirical considerations are meant to help students grasp the central features of the two ideas, as will be demonstrated. However, the extent of implementation must be calibrated by the desired amount of depth students are expected to attain in the respective topics. It might seem surprising that this last issue is more straightforward in Mathematical Modelling than Functions. But Functions have been a familiar topic in the syllabus, and above all to teachers, and we sometimes need to distance ourselves from a well-known subject to consider how much our students should learn.

2 Functions

How much should secondary school students know about functions? To shed light on this question, it is worthwhile to study the historical origin of the word "function". Useful resources include the Wikipedia entry "History of the function concept", Historical Modules for the Teaching and Learning of Mathematics (Katz & Michalowicz, 2005), and (Eves, 1990). What follows is found on pages 234-235 of (Eves, 1990). Leibniz used it in 1694, and in 1718, Johann Bernoulli thought of a function as "any expression made up of a variable and some constants". Euler thought of it as "any equation of formula involving variables and constants". After Fourier's work on trigonometric series in the early 1800's, Dirichlet defined: "A variable is a symbol that represents any one of a set of numbers; if two variables x and y are so related that whenever a value is assigned to x there is automatically assigned, by some rule or correspondence, a value to y, then we say y is a (single-valued) function of x."

What can we glean from this historical glance? First, the best mathematicians in the world took nearly 200 years to get close to the modern notion of a function, of assigning a unique y value to every x value. Second, during the early part of this period, differentiation and integration techniques were fully established for the important basic analytic functions, which are taught in secondary school mathematics. Based on these observations, if it is deemed necessary to teach the uniqueness idea, we need to do so quite carefully. Here is an approach which is likely to be widely used. State some polynomials as examples of functions; effectively, a function is defined via concrete examples. Use the graphs of the examples to highlight the uniqueness property, in order to cement the abstract idea. The domain and codomain should be kept to subintervals of the real numbers initially, before moving on to other sets, such as the list of student names in a classroom. My reason for stating this obvious approach is to show that it essentially recapitulates the historical development.

What is a non-example of a function? With the approach above, i.e., "definition by concrete examples", a meaningful answer is difficult to find initially. Of course, anything not in the example set is a non-example, but this is hardly convincing. When the uniqueness property is learnt, students may be asked to suggest one, like the circle. The practical benefits of distinguishing functions from non-functions of this type are, however, less clear to me, and contributes to my reservation about teaching the uniqueness property to secondary school students. In the university, it is common to teach relation first, then define function as a special case. But relations are even more abstract; this is like teaching General Relativity first, then derive the Special Relativity. I think the practice is not sound even for university students.

Another angle of attack on the question of a non-example is presented next. It has no historical precedent that I know of, so may be challenging to use in the classroom, though may be of interest as a commentary on the abstract nature of Euclidean geometry.

2.1 *The measured cosine*

The seeds of Euclid's glorious work *The Elements* were likely sown by land surveyors who made large numbers of measurements. When they worked away from the fields, they probably engaged in something like the following. On a piece of paper draw an equilateral triangle ABC, and drop a perpendicular from B to AC at D, as shown in Figure 1. By construction, the angle θ is 60° and its cosine is the ratio of AD (the adjacent side) to AB (the hypotenuse), which is ½.

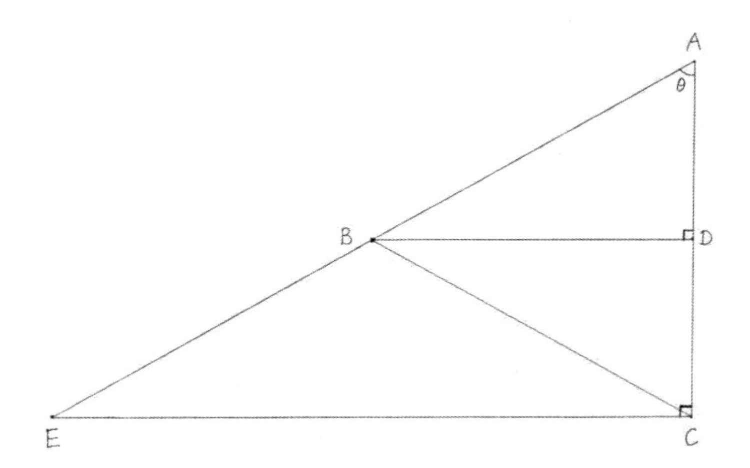

Figure 1. ABC is an equilateral triangle, AE is a straight line, and both BD and EC are perpendicular to AC

"By construction" is noteworthy, as it refers to an ideal world. In real drawings, the quantities are inexact. For simplicity, let us agree that θ is exactly 60°, i.e., a third of the angle on one side of a straight line. Because real lines are not straight, the drawing surface is not flat, and measurement contains errors, the measured lengths of AD and AC will be unlikely to give the ratio ½. Now extend AB to E, so that EC is parallel to BD. Then for the same reasons, the ratio AC/AE should be equal to ½, but not exactly. In particular, it is unlikely to be equal to AD/AB. Hence, a series of measured cosines of 60°, say through further extensions of Figure 1, tend to vary around ½. Thus there is no unique value for a measured cos 60°. By the same argument, the same applies to any other acute angle.

2.2 *The ideal cosine*

Extensive experience showed that the variation in the measured cos 60° is small. It makes sense to assign cos 60° a definite value, so as to facilitate algebraic manipulation of such quantities, and to make predictions. Euclidean geometry comes to the rescue, enabling the "by construction" argument in the last subsection, to derive the value ½ for cos 60°. Therefore, we make the definition of the ideal cosine:

$$\cos 60^\circ = \tfrac{1}{2}\,.$$

The ideal cosine and sine can be similarly derived and defined for some other angles, like 30° or 45°. Or one can make use of the addition formulae to define the ideal cosine for all acute angles which are rational multiples of 60°, which are dense in the set of all acute angles. For example, from $\cos 60^\circ = 2\cos^2 30^\circ - 1$, we get $\cos 30^\circ = \sqrt{3}/2$. Then it is intuitive to interpolate smoothly to all acute angles, so that the ideal cosine is really a function of the acute angles, assigning a unique value to each angle. This function can be written as a formula, where it is more convenient to measure angles in radians, i.e., the arc length of a circle of radius one subtended by the angle. For example, a right angle corresponds to a quarter of the circle, so $90^\circ = \pi/2$ radian. By a similar reasoning, $60^\circ = \pi/3$ radian. Now let x be an acute angle given in radian, i.e., $0 < x < \pi/2$. Then the cosine is given by the infinite series

$$\cos x = 1 - x^2/2! + x^4/4! - x^6/6! + \dots$$

where $2! = 1\text{x}2$, $4! = 1\text{x}2\text{x}3\text{x}4$, etc. It takes some advanced mathematics to see that for any acute angle x, the infinite series converges, i.e., evaluates to a real number, and that substituting $x = \pi/3$ gives exactly ½. Furthermore, the series converges for any real number x, so that we may speak precisely of the cosine of any real number. Recall that the geometric series

$$1 - x + x^2 - x^3 + \dots$$

only converges for $-1 < x < 1$. So writing down a formula does not guarantee a function is defined on all real numbers.

The explicit discussion of real angles and lengths helps students appreciate the beautiful role mathematics plays in the sciences. Although the fuzziness in the measured cosine is excised in the ideal cosine, the latter has proved tremendously useful in practice, producing predictions that are highly consistent with real measurements. The trigonometric functions played a significant role in the invention of a bewildering array of engineering marvels, from bridges to smartphones, which we take for granted.

2.3 *Dependence and uniqueness*

Plausibly an important motivation for functions is the idea of dependence. The Wikipedia entry "Function (mathematics)" has "formalisation of how a variable y depends on a variable x". However, the modern definition focuses on uniqueness, relegating dependence to a mere motivation. This section makes two remarks in this connection.

First, the mathematical pioneers might not recognise the constant function as a function. To find out more about modern intuition, I asked 14 colleagues who engage in quantitative work, and had at least a university-level calculus class, to consider a game console with five buttons labelled 1, 2, 3, 4 and 5. Upon pressing button 1, one star will appear on the screen. Upon pressing button 2, one star appears, and the same thing happens upon pressing the other button. Let x take value 1, 2, 3, 4 or 5, and y be the number of stars. Two questions were asked:

(1) Does y depend on x?
(2) Imagine you have never learnt mathematical functions. Does y depend on x?

Only two colleagues indicated "yes" to both questions. Three said "yes" to (1) and "no" to (2). All others said "no" to both questions. The responses of these 14 colleagues suggest that it is not intuitive to regard a constant as an example of dependence.

Secondly, the notion of dependence may be broader than functions. Imagine we draw a series of figures like in Figure 1, where θ varies from 10°, 20°, to 80°, and for each figure, we plot two measured cosine values against the angle, to get a graph as in Figure 2. It would seem reasonable to view the measured cosine as dependent on the angle: it decreases steadily from near 1 to near 0, as the angle increases. However, it is more difficult to use the points in Figure 2 to predict y from x, than the ideal curve. This might explain why historically, deterministic dependence was assumed, with uniqueness built in. Empirical graphs of non-functions like Figure 2 are often the starting point of an inquiry into whether y depends on x, and if so, how.

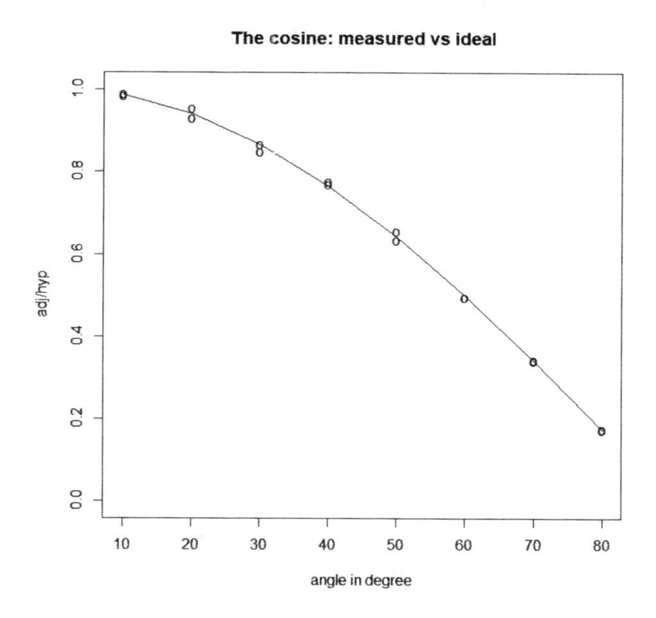

Figure 2. For each angle, two measured cosine values are generated by adding random numbers to the ideal cosine value. The curve interpolates the ideal values.

3 Mathematical Modelling

In the last decade, mathematical modelling has steadily gained prominence in the Singapore secondary school community. Ang (2009) categorises a variety of modelling approaches to three types: Deterministic, Empirical, and Simulation. Additional resource can be found in the website Math Modelling SG (n.d.).

The focus here is on a class of examples that are closest to the Empirical type, where to some extent, the variables' definitions and measurements are intuitively clear, and a real data set is available. I will attempt to illustrate the tension between the mathematical model, which almost always involves a function, and the model-inspiring data, which are typically quite messy and often not functions, if one thinks about that more carefully.

A relevant example is about rate of leaf growth, where time is measured in weeks and leaf size is measured in total area (Ang, 2009). Both variables are intuitive. Then one focuses on finding a suitable function that describes how leaf size depends on time, which can be used to make predictions and may shed light on the nature of the growth process. However, the function will likely be highly dependent on multiple factors, such as the species of plant, the condition of the environment, etc., to the extent that we do not expect to uncover any general pattern, or something like a "universal law".

3.1 *Data and the fitted curve*

To get a handle on the key issue, namely the relationship between the data and the model function, I will instead consider an object falling freely in vacuum, or more realistically, an object with a small surface area falling in the atmosphere, so that the effect of friction is negligible. How does the distance travelled depend on the time since its release? Many of us learn in a physics class that a single quadratic function works remarkably well on almost all of the earth's surface. But this was not obvious until around the 17th century. The ancient Greek philosopher Aristotle thought the object would fall at constant speed, so he would have proposed a linear function to model the distance travelled.

The Wikipedia entry "Free fall" includes a data set from the following experiment. A steel ball is released from a height. An equipment is used to measure the time for it to drop certain fixed distances, as shown in Table 1.

Figure 3 shows that distance increases faster than a constant rate, as time passes. Indeed, the increase is quadratic, as shown by the good fit of the curve given by

$$y = 4.9x^2.$$

Table 1

Data from free fall experiment

Time (second)	Distance (metre)
0.00	0.00
0.22	0.25
0.33	0.50
0.39	0.75
0.45	1.00

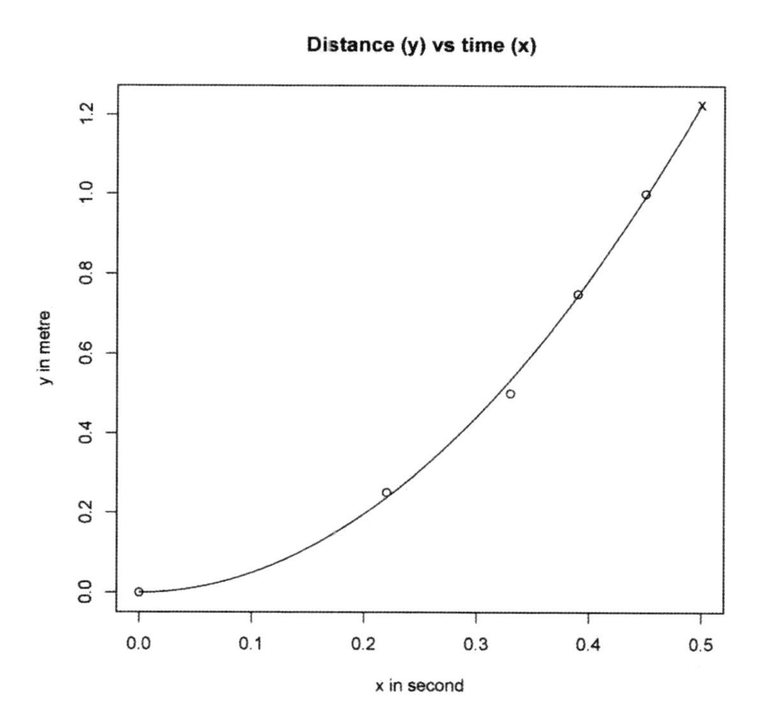

Figure 3. Circles represent measurements from the free fall experiment. The curve is $y = 4.9x^2$. The cross is the predicted distance travelled at 0.5 s, using the curve.

Students of physics may find this equation familiar, for it is none other than an elementary kinetic equation, where the acceleration is due to gravity, taken as 9.8 metres per second squared. But in the mathematics classroom, there is no need to place emphasis on this knowledge; it suffices for students to observe the close fit to the data.

The curve is slightly to the right of the second point, and slightly to the left of the third point. This is mainly due to measurement error. In this case, the measurement of times when the metal reaches the specified distances can be slightly too high or too low, despite the best efforts. After all, the ball is dropping fast. The curve need not pass through any of the five points, but it is still a useful description of the data.

3.2 *Empirical modelling*

In this experiment, time and distance are related, and the relationship is captured very well by a quadratic function. This act of using a function to describe a relationship between quantities is a prototype of mathematical modelling, of the empirical type.

The function can also be used to make predictions. For example, how far would the ball have dropped 0.50 seconds after the drop? Since

$$4.9 \times 0.50^2 \approx 1.23,$$

we predict that it should be 1.23 m. This can be checked against a measurement, to see if it is satisfactory. In this case, many similar experiments have been done to convince us that the prediction is excellent. The remarkable power of mathematical modelling has led to tremendous savings in the labour and cost of making measurements. Of course, one has to gauge whether the prediction works. If the ground is 1.10 meters below the release point, then the prediction above does not make sense. The ball would have hit the ground before then. The model predicts that to happen after $\sqrt{(1.10/4.9)} \approx 0.47$ s.

There is prefect reproducibility in the quadratic function, in that if we repeatedly substitute x with the same value, the same y value will always be obtained. An experiment is almost never perfectly reproducible. However carefully made, repeated measurements tend to be different. Thus, the relationship that holds between the measured quantities, as seen in the plot, is not exact. More explicitly, it is inexact relative to the function used to describe the data. If a function seems to track the data quite closely, the relationship between the quantities is said to be well-modelled by the function. Many data sets on the internet are perfectly modelled by some functions. If the data were measured experimentally, this is almost too good to be true. More likely, the y values are predictions from the function itself.

3.3 *Goodness-of-fit: The least squares*

Now we deal with the issue of goodness-of-fit. Visually, $y = 4.9x^2$ fits the data better than $y = 4.3x^2 + 0.05$ and $y = 5.8x^2 - 0.06$, as shown in Figure 4. The comparison can be made quantitative via a measure of the overall discrepancy between a curve and the data. The most common measure is the sum of squares (SS), i.e., the sum of the squares of the difference between the observed y value and the predicted y value according to the curve. Table 2 calculates the SS for $y = 4.9x^2$ to be 0.001 to 3 decimal places. The SS of the other two curves are 0.011 and 0.028, so a small SS seems to go with a better visual fit.

Table 2

For each x, the predicted value y' is given by 4.9x².

x	y (observed)	y' (predicted)	$(y - y')^2$
0.00	0.00	0.00	0.0000
0.22	0.25	0.24	0.0002
0.33	0.50	0.53	0.0011
0.39	0.75	0.75	0.0000
0.45	1.00	0.99	0.0001

It turns out that among all curves of the form $y = mx^2$, where m is a real number, there is a unique value of m that gives the minimum SS. In this example, that the m value turns out to be 4.89 to 2 decimal places.

The curve with minimum SS is by definition the closest curve to the data, among all curves of the stipulated type. The theoretically correct model (recall 4.9 is half the value of the acceleration due to gravity) is very close to the best-fit curve. But this should not be too surprising, for it is the observation of this close fit that gave the 17th century scientists the idea that objects in free fall are subject to a constant acceleration, ignoring air resistance. This was by no means obvious to Aristotle.

Assume that there is a true function for the experiment: $y = Mx^2$, where M is an unknown constant. Then the SS-minimising value 4.89 can be viewed as an estimate of M. It is called the least-square estimate. This estimation technique, and its more sophisticated cousin called regression, is the workhorse of all empirical sciences, and was first published by Legendre in 1805, though Gauss claimed that he had discovered it 10 years earlier (Stigler, 1990).

In addition to contributing to the deviation of data from the fitted curve, measurement error also accounts for the deviation of real data from the predictions from the true model. In the free fall example, the errors are mainly in the time measurements, but in most applications, the errors are in the y direction, with the x values regarded as known. From this perspective, the least square estimation method works by minimising the overall size of measurement errors. The least-square method is a good gateway towards statistical modelling, where the measurement errors are explicitly assumed to be generated from certain random variables. The advantage of the statistical approach is that it is then possible to calculate approximately the error incurred in the least-square estimate, which is a very important tool for quantifying uncertainty in all kinds of scientific work.

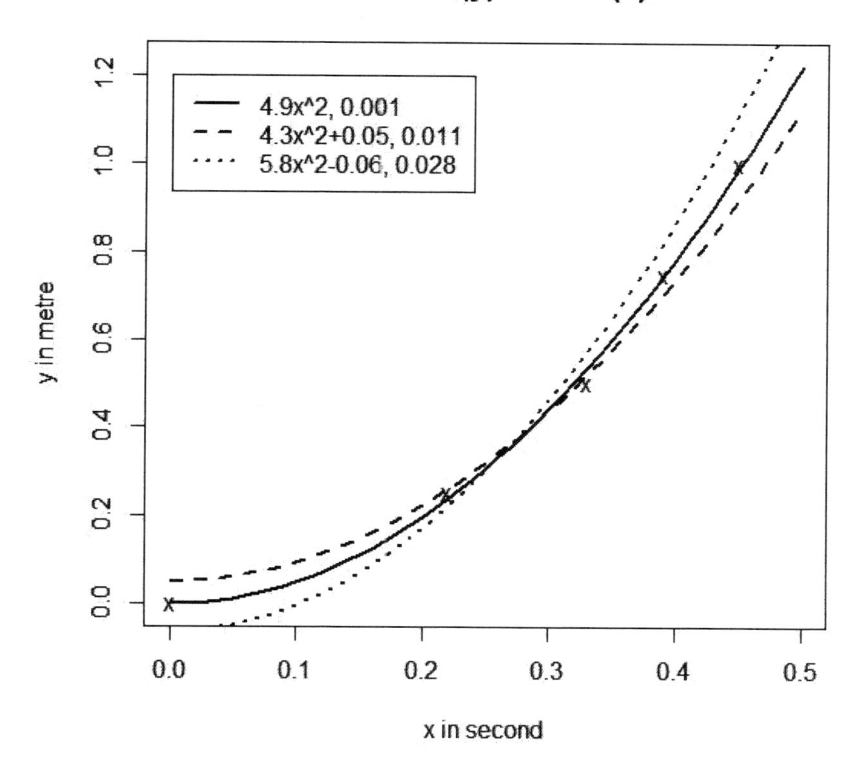

Figure 4. Crosses mark measurements from the free fall experiment. The sums of squares (SS) for the curves are respectively 0.001 ($y = 4.9x^2$, solid curve), 0.011 ($y = 4.3x^2 + 0.05$, broken curve) and 0.028 ($y = 5.8x^2 - 0.06$, dotted curve).

4 Discussion and Conclusion

The two teaching approaches outlined above hinge heavily on empirical data. Ideally, some class time devoted to the actual collection of data should help leave a deeper impression of the intimate connection between mathematics and the sciences. Even without direct experience, exposure to real data or simulated data through graphs like Figures 2 and

3 coupled with careful instruction should still yield much benefit. Hopefully, the teaching approaches outlined above will prompt students to look forward to more advanced topics, namely Calculus and Statistics.

Section 3 shows that there is much to teach if we confine ourselves to only the Empirical type of Mathematical Modelling, and even more so, to only problems where the variables are intuitive to the mind. Even then, there can be surprising obstacles to accurate measurements. In the leaf growth example, the jagged boundary and the unevenness of the blade are real issues. There are a whole class of optimisation problems that are amenable to mathematical modelling, though one has to beware of those where the objective function is challenging to define or measure. Considering the needs of and constraints on secondary school students, some topics should be prioritised. I believe that the empirical aspects of modelling should take precedence over tackling problems where the variables or objective functions are challenging to even imagine.

Figures 2 and 3 look similar. Both contain measured numbers that fluctuate around a smooth curve. Both are grounded in real, empirical measurements, or at least a careful contemplation of making such measurements. Figure 2 attempts to capture the ancient discovery that there is an underlying theoretical cosine that can be sublimated from the mess of measured ratios. Figure 3 represents the more recent discovery of the acceleration of free fall. Imagine a series of graphs like Figure 3, based on measurements on different objects released from different heights at different locations, all nicely fitted with similar quadratic curves. This landscape, analogous to Figure 2, facilitates the next discovery, that the acceleration due to gravity is essentially the same everywhere on earth. In this sense, there is an increase of sophistication in going from Figure 2 to Figure 3. Then the leaf growth example mentioned in the beginning can be put above Figure 3 in this scale, for there, despite the analysis of many data sets, it is still a great challenge to distil any general rules that describe a wide enough class of examples.

Our intellect may have touched the stars and the galaxies beyond, thanks to the revolution started by Galileo and Newton, but numerous wonders on earth still await tidy theorising, perhaps for a very long time.

Acknowledgement

The author thanks Mr Chia Aik Song for numerous fruitful exchanges of ideas.

References

Ang, K. C. (2009). *Mathematical modelling in the secondary and junior college classroom.* Singapore: Pearson Education.

Eves, H. (1990). *Foundations and fundamental concepts of mathematics (3rd ed.).* Boston: PWS-Kent.

Katz, V., and Michalowicz, K (ed.) (2005). *Historical modules for the teaching and learning of mathematics.* Boston The Mathematical Association of America.

Math Modelling SG. (n.d.) Retrieved from http://www.mathmodelling.sg/

Stigler, S. M. (1990). *The history of statistics: The measurement of uncertainty before 1900.* Cambridge: Belknap Press.

Chapter 11

Mathematical Modeling in Problem Solving: A Big Idea across the Curriculum

Padmanabhan SESHAIYER   Jennifer SUH

In this chapter, we consider ways in which introduction to mathematical modeling (MM) as a big idea in teaching mathematics across the curriculum, can promote critical thinking, creativity, collaboration and communication, as well as connect to interdisciplinary topics in STEM and community-based service learning. Moreover, this chapter introduces MM as a big idea to solve real world tasks with the various attributes including Open-endedness; Problem-posing; Creativity and choices and; Iteration and revisions. With a collective understanding of these interrelated attributes of MM in the elementary, middle school and high school grades, this chapter provides an opportunity for teachers to enhance their pedagogical practices to integrate MM into their respective curriculum. This is needed to help students acquire mathematical competencies; to make connections between the real world and mathematics; to maintain the high cognitive demand of the mathematical modeling process and; to provide classroom management that is learner-centered.

1 Introduction

Problem solving is a fundamental competency that can challenge students' curiosity and interest in mathematics. Polya's (1945) seminal book, *How to Solve It*, stated,

> A teacher of mathematics has a great opportunity. If he fills his allotted time with drilling his students in routine operations he kills their interest, hampers their intellectual development and misuses opportunity. But if he challenges the curiosity of his students by setting them problems proportionate to their knowledge, and helps them to solve their problems with stimulating questions, he may give them a taste for, and some means of independent thinking. (p. v)

Polya also suggested four fundamental steps to problem solving, which included:

1) Understanding the problem;
2) Devising a plan;
3) Carrying out the plan and;
4) Looking back.

Problem solving is also one of the important Mathematical Processes in the National Council for Teachers in Mathematics (NCTM) Mathematical Standards along with communication, representations, connections, and reasoning. NCTM (2000) defined *problem solving* as:

> Engaging in a task for which the solution method is not known in advance. In order to find a solution students must draw on their knowledge, and through this process, they will often develop new mathematical understandings. Solving problems is not only a goal of mathematics but also a major means of doing so (p.52).

Polya's Four Steps as well as NCTMs definition of Problem Solving connect well with big ideas in *mathematical modeling*.

1) The first idea involves *observation*, which helps us to recognize the problem, the variables that describe the attributes in the problem as well as any implicit or explicit relationships between the variables.
2) The second idea includes the *formulation* of the mathematical model. In this step, one makes the necessary approximations and

assumptions under which one can try to obtain a reasonable model for the problem. These approximations and assumptions maybe based on experiments or observations and are often necessary to develop, simplify and understand the mathematical model.

3) The third idea addresses *computation* that involves solving the mathematical model for the problem formulated.
4) It is then important as a final step to perform *validation* of the solution obtained that helps to interpret the solution in the context of the physical problem. If the solution is not reasonable, then one can attempt to modify any part of the three steps including observation, formulation and computation and continue to validate the updated solution until a reasonable solution to the given physical problem is obtained.

Mathematical modeling, as a big idea across curriculum may therefore, can be thought of as an unstructured implicit *problem solving* process. A typical mathematical modeling process entails introducing a problem with context, which promotes problem solving and requires a qualitative as well as a quantitative analysis. Explicit problem solving often involves the gathering of information and determining what needs to be solved. However when the problem is unstructured (i.e. mathematical modeling), one is often required to employ implicit problem-solving methods (such as a creative non-routine approach).

2 The Mathematical Modeling Process

Mathematical modeling has a very distinct meaning and involves teachers and students mathematizing authentic situations. Mathematical modeling has been encouraged and is often used in secondary mathematics courses where real world applications are emphasized. However, there is also a need to introduce creative ways to engage teachers and students in the elementary and middle grades to problem solving and mathematical modeling in order to develop their 21st century skills, such as critical thinking, creativity, collaboration and communication, which is the focus of this paper.

We discuss mathematical modeling as a process of connecting mathematics with real-world situations. These real-world problems tend to be messy and require multiple math concepts, a creative approach to math, and involves a cyclical process of revising and analyzing the model. Figure 1 outlines a cycle often used to organize reasoning in mathematical modeling (NCTM, 2009). The steps identified in this four-step modeling cycle illustrates how a real-world problem can be translated to a mathematical model under specific assumptions, which can then be solved using appropriate mathematical tools. The solutions can then be interpreted in terms of the real-world and in turn helps validate the predictive capability of the model proposed.

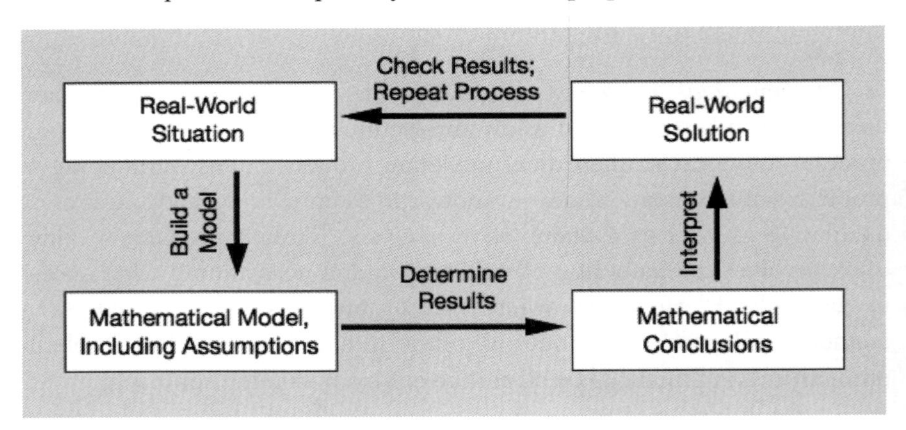

Figure 1. Four-step modeling cycle to organize reasoning about mathematical modeling (NCTM, 2009)

In our work with teachers, we have employed the Mathematical Modeling Process (see Figure 2) that we modified from the steps defined by the Society for Industrial and Applied Mathematics (SIAM) Moody's Mega Math Challenge, which is a national mathematical modeling contest for high school students sponsored by The Moody's Foundation (see MathWorks Math Modeling Challenge, n.d.).

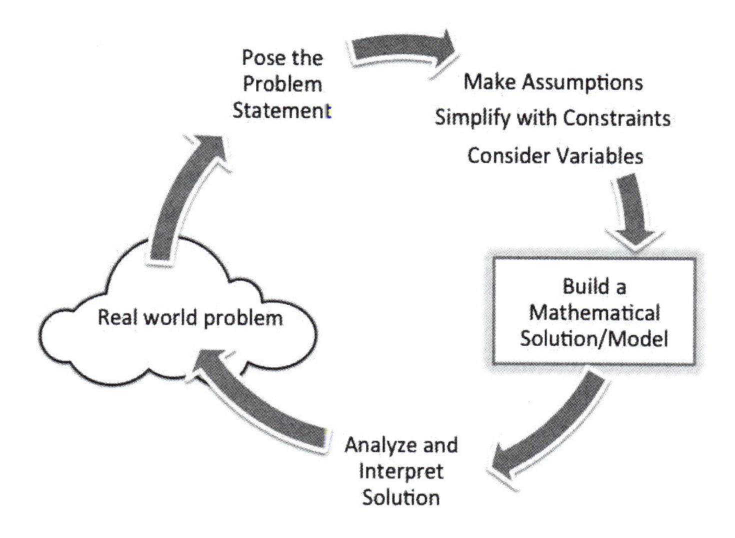

Figure 2. Math Modeling Process

The steps as illustrated in Figure 2 include:

1. **Pose the Problem Statement**: Pose questions. Is it real-world and does it require math modeling? What mathematical questions come to mind?
2. **Make Assumptions, Define, and Simplify**: What assumptions do you make? What are the constraints that help you define and simplify the problem?
3. **Consider the Variables**: What variables will you consider? What data/information is necessary to answer your question?
4. **Build Solutions**: Generate solutions.
5. **Analyze and Validate Conclusions**: Does your solution make sense? Now, take your solution and apply it to the real world scenario. How does it fit? What do you want to revise?
6. **Present and Justify the Reasoning for Your Solution**. (Modified from MathWorks Math Modeling Challenge, n.d.)

These steps also suggest that modeling of mathematical ideas is the process of connecting mathematical reasoning with real-world situations. These types of problems are open-ended and messy and require creativity and persistence. The discovery of this messy nature of the problem not only helps students to engage in an exploration process but also helps teachers to analyze student thinking in the mathematical modeling process. One way to learn how students learn mathematics as they go through this discovery and how teachers develop their teaching practices as they collaborate is through Lesson Study, which we describe next.

3 Lesson Study to Enhance Mathematical Modeling

Lesson Study (Lewis, 2002; Lewis, Perry & Murata, 2006) is a model of professional learning that offers situated learning through collaborative planning, teaching, observing, and debriefing that affords opportunities for teachers to reflect individually and collectively. Teacher educators have embraced Lesson Study, originating from Japan, because it empowers teachers and provides a collaborative structure for eliciting reflection for critical dialogue about pedagogical content knowledge among teachers. We have also seen that Lesson Study affords teachers opportunities to deepen their understanding of the mathematical learning progression through observation and analysis of students' thinking through a situated school-based professional development experience (Suh and Seshaiyer, 2014a).

In our Lesson Study, we focus on some of the essential research-based professional development resources that we use as high leverage practices. These include working with teachers to unpack the mathematics, choose worthwhile tasks through cognitive demand analysis, set goals to enact some of the core teaching practices through research lessons, integrate technology to amplify the mathematics and use formative assessment to assess their mathematical proficiency.

When deciding on a research goal for their Lesson Study, teacher teams decide on a content learning goal as well as select several of the core teaching practices that they want to focus on in their research lesson. This allows the teachers to set a personal teaching goal besides the student-learning goal. In the Principles to Actions (NCTM, 2013), the National Council for Teachers in Mathematics calls for the mathematics education community to focus our attention on the essential core teaching practices that will yield the most effective mathematics teaching and learning for all students.

Authors of the Principles to Actions responded to this challenge by defining eight core teaching practices drawn from research that include:

1) Establishing Mathematics Goals to Focus Learning (Hiebert, Morris, Berk, & Jansen, 2007);
2) Implementing Tasks That Promote Reasoning and Problem Solving (Stein, Engle, Smith, & Hughes, 2008);
3) Using and Connecting Mathematical Representations (Lesh, Post, & Behr, 1987);
4) Facilitating Meaningful Mathematical Discourse (Hufferd-Ackles, Fuson, & Sherin, 2004);
5) Posing Purposeful Questions (Chapin & O'Connor, 2007);
6) Building Procedural Fluency from Conceptual Understanding (Hiebert & Grouws, 2007);
7) Supporting Productive Struggle in Learning Mathematics (Dweck, 2008); and
8) Eliciting and Using Evidence of Student Thinking (Sherin & van Es, 2003).

Since many of these eight practices are related to and support one another, teacher teams would often state several of these practices as the focus of their professional learning goals. The example presented in the next section highlights how teachers implemented some of these eight Mathematics Teaching Practices outlined in the Principles to Actions during our Lesson Study. In our work, we have used the Japanese Lesson Study Professional Development model where we immersed teachers in vertical teams, collaboratively planned, taught the first cycle, observed, debriefed, and taught the second cycle then came to reflect both individually and collectively.

Next, we present an example of an authentic real-world problem that was considered by a group of teachers to engage the students in mathematical modeling through a Lesson Study.

4 The Traffic Jam Problem

The research lesson presented in this chapter focused on mathematical modeling with the process of connecting mathematical reasoning with real-world situations. When students model such situations, they are applying mathematics in a way similar to how they would visualize this in reality. Students therefore, need to make genuine choices about what is essential, decide what specific content to apply and finally decide if the solution is reasonable or useful. It provides an opportunity to develop and practice their 21st century skills including critical thinking, creativity, collaboration and communication.

With a goal to engage students on an open-ended messy unstructured real-world problem, one of the teachers from our summer professional development institute presented the "Traffic Jam Problem". The teachers in this lesson felt that this problem met the higher-level demands (*Doing Mathematics*) in terms of a rich mathematical task (Smith & Stein, 2012). While the problem presented is an elementary task, one can see the parallel in the process a modeler will go through in solving this problem. The host teacher began the lesson showing a photo of a major traffic jam. Then stated the problem statement.

> There is a 7-mile traffic jam caused by an accident. How many cars and trucks are on the road? When the accident was cleared, the vehicles drove away from the front, one car every two seconds. Estimate how long it took before the last car moved.

To help students make assumptions, define and simplify the problem, the teacher asked students to think about all the questions that come to their mind. Students shared the following open-ended questions:

a) Does a four lane highway consist of two lanes going in the same direction or four?
b) What types of vehicles are on the highway? 18 wheelers, Smart Cars, Motorcycles, etc…
c) What is the ratio or percent of each type of vehicle?
d) How much, if any, space is between each vehicle?

To answer these preliminary/foundational questions, the students (and teachers) had to find other resources (i.e. videos and traffic definitions. Also, the teacher reviewed some basic vocabulary including vehicle, traffic jam, two lane highway, mile, estimate, assumptions. A two lane highway was defined as a main road for fast-moving traffic, having limited access, separated for vehicles traveling in opposite directions, with two travel lanes in each direction. For the number of lanes, most of the students in the class were also aware of the highway system and confirmed with the teacher that they needed to work with a two-lane system for this problem after revisiting the photo. Students also took a quick fieldtrip out to the parking lot to measure cars. Having some of this information provided students with a starting point but quickly students realized that they had to make some assumptions to define and simplify this problem. For example, they needed to think about the ratio of trucks to automobiles to motorcycles on a given stretch of highway and the average distance between vehicles during a traffic jam. The host teacher mentioned:

> "I know some questions that all of us have encountered in math class, such as this one, require students to recall and use 'common knowledge' that is not provided (i.e. how many feet are in a mile like knowing 5280 feet in a mile). However, how rare of a problem is it that expects, or requires students to do outside research on 'uncommon knowledge,' (Or even better, allows students to set their own parameters or assumptions), before starting the calculations."

4.1 *Implement tasks that promote reasoning and problem solving*

The teacher team chose this particular task and listed the following reasons for why they thought the problem was worthwhile. The task was open-ended; modeled a real life authentic situation, and many students (or most) have had some personal experience that would help them connect with, understand, or at least feel somewhat familiar with the task. Additionally, there was no specific procedure for solving the task; students would have to apply what they know; there was no answer or absolute solution, and students' answers would be as good as their justification of their procedure and rationale. Ultimately, the teachers chose the task because it met the criteria of being a proportional reasoning problem; higher order thinking and 21st century skills were required to work through the task. The task required students to present how they reached their answer. There were numerous variations in problem solving strategies and modeling the original problem (e.g., considering restrictions, car space, ratio of cars to trucks, number of lanes); and the task links to and aligns with many curriculum standards, key skills, and concepts.

4.2 *Elicit and use evidence of student thinking*

The lesson gave an opportunity for the teachers in the Lesson Study to not only observe student learning but also to help enhance their pedagogical practices. The host teacher commented:

> "This idea of giving students a 'real world' math problem was very inspiring. Not only does it reveal and improve student's common sense and/or math sense, but also it provides opportunities for students to apply what they have learned during instruction to these new problems AND see the different methods other classmates use to solve the same problems. I was so impressed by this that I began to implement similar types of problems the last week of school with my 5th grade students."

This collaborative study also gave a chance to analyze student learning through various factors. For instance, the participating teachers reported that most student groups forgot to take into consideration the space between each vehicle and that there were two lanes of traffic. They also noticed that one of the groups that worked most efficiently and presented very professionally fell into this category. Soon after they realized their mistake by way of another group commenting, they went back to correct part of their answer about the two lanes. They were not able to change their answer to take into account space in between each car, which would have been more challenging than multiply their answer by two (i.e. % increase, proportions). See Figure 3.

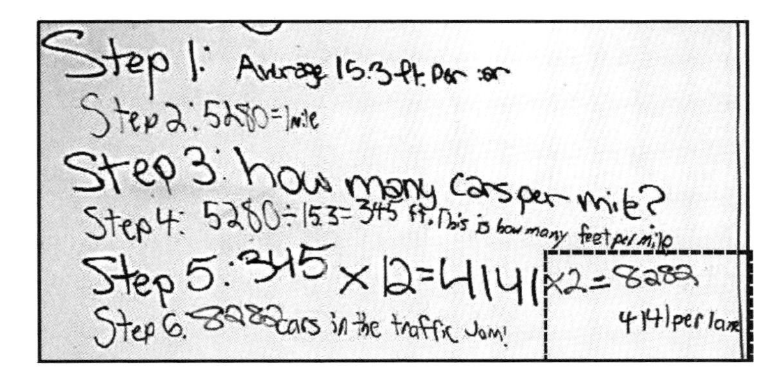

Figure 3. Students making assumptions for number of cars in a mile

4.3 *Supporting productive struggle*

The teachers observed that the student group that was most productive and came up with the most justifiable answer still experienced a significant flaw in their interpretation of their calculations. When asked the question "How long would it take for the cars to clear out of the traffic jam?" the student groups considered the information provided involving a car being cleared every two seconds after.

Students took different approaches and reached solutions that were both different and comparable. Some students took a very simplistic approach, coming up with a single average length of a car and doing appropriate calculations to achieve reasonable numbers consistent with the problem. Nearly 25 percent of the groups did not include space between vehicles in their calculations. Nearly 50 percent based their averages on multiple sized vehicles (including cars, trucks, SUVs, etc.). Some of them employed manipulatives (erasers, dominoes, small or large paper clips or other available classroom items) to conceptualize the problem by doing, providing that the scale assigned to them is the same as the highway. Accounting for the space between the cars, the student group that accounted for 4141 cars in a lane estimated the space to be modeled using additional 731 cars. Assuming 2 seconds to clear a car, the student group found the number of minutes it would take [about 162 minutes]. However, the teachers noticed that the students in the group used long division to attempt to convert their time to hours. Again, their calculation of the hours was correct [2.7 hours]. The flaw was interpreting this decimal as "Two hours and seventy minutes" instead of "two WHOLE hours and seven-TENTHS of an hour." As a result, they converted their interpretation to "Three hours and ten minutes," which they misrepresented again as 3.1 hours. The teachers were curious why they didn't ask, "How many hours are in 162 minutes (two)", and then figure out how many minutes remained (forty-two).

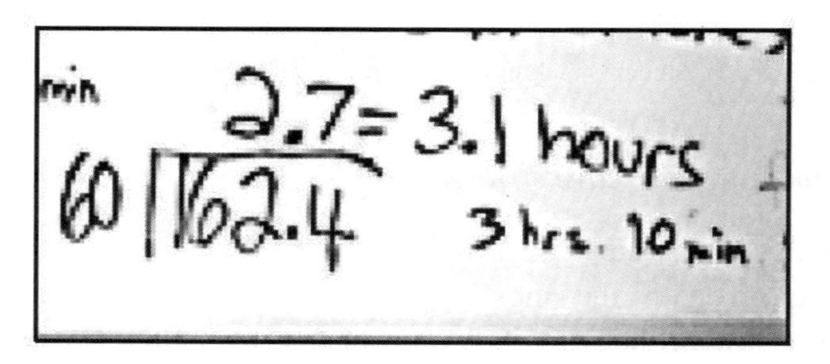

Figure 4. Dividing hours and making sense of decimal

The teachers also observed that one student group struggled to understand the numbers that they were using throughout the entire class, which was ultimately revealed during their presentation. Although this group, with some help, performed the calculations to determine how many vehicles were in the traffic jam [4022.8571 cars], the teachers felt there was one major problem it was a decimal ("How can you have 0.8571 of a car?").

During their presentation, the students were asked if they were able to have 'part' of a car. No one in the group understood what the teachers were talking about. The teachers felt they understood that you could not have 0.8571 of a car driving around the highway, however the teachers realized and discussed during the debrief after the Lesson Study that the students in this group were all 'lost in the numbers,' probably overwhelmed by this 'rich task' that they never fully grasped (This makes the host teacher wonder "how many of my students are like this during math class, and/or while working on math assessments or activities"). During the debrief, the teachers also discussed the possibility of students not being incorrect with the decimal in their answer as they realized there is no reason to expect exact number of cars to be on the 7-mile traffic jam. The host teacher concluded,

> "I feel that this last group's answer reveals some areas of neglect in my math instruction: checking your work, estimating precise numbers, asking questions when you don't understand something, and ultimately using common sense. These skills just mentioned, are what all students need to succeed in the real world (compared with many of the math standards that we are required to teach). I feel these 'rich tasks,' coupled with collaborative learning projects and presentations are an effective way to assess and improve students 'essential math knowledge and skills."

Through this case study and professional development with teachers, we have seen models and modeling of mathematics expressed and developed in interrelated ways. Each of these approaches was shown to contribute towards developing strategic competence in both students and teachers. Through our professional development model, we were able to expand teacher's understanding of modeling math ideas and provide them with practical means to do so. The following reflection from the teacher who led the lesson study for the traffic jam problem captures this:

> "I find it ironic that these 'higher-level demanding' problems are called 'doing mathematics' because you can't start 'doing the math' until you research and/or agree on some assumptions, tasks that you would 'TYPICALLY' find in science and other subjects. The professional development experiences that I take back, are redefining what math problems, math class, and 'doing the math' should encompass: not just plugging numbers into calculators and equations, but exploring and understanding the nature of the tasks, accessing relevant knowledge and experiences and making appropriate use of them while working throughout the task, and applying un-predictable and new approaches that aren't explicitly suggested by the task. All of the descriptions/characteristics of "Higher-Level Demanding Problems (*Doing Mathematics*)" have become relevant because I saw and heard for myself this process with my students while implementing the "Traffic Jam" task."

4.4 *Promoting the 21st century skills*

The Traffic Jam problem also gave the students an opportunity to engage in the process of mathematical modeling with opportunities to develop their 21st century skills (P21) Learning and Innovation Skills, which emphasize the 4 Cs: critical thinking, creativity, collaboration and communication (Table 1).

The problem allowed the students to engage in the actions identified in Table 1 and helped them to begin the work they were assigned without any bias or opinions about what is or is not expected from the task. The misconceptions and misrepresentation that arose in the process helped them to get a better insight. Such student insights often open up new problem-solving strategies or even questions overlooked by the teacher.

Table 1.
21st century skills (Partnership for 21st Century Skills, 2011)

Critical Thinking & Creativity

Reason Effectively/ Use Systems Thinking

- Use various types of reasoning (inductive, deductive, etc.) as appropriate to the situation

Make Judgments and Decisions

- Effectively analyze and evaluate evidence, arguments, claims and beliefs
- Analyze and evaluate major alternative points of view

Solve Problems

- Solve different kinds of non-familiar problems in both conventional and innovative ways
- Identify and ask significant questions that clarify various points of view and lead to better solutions

Work creatively with others

- *Demonstrate* originality and inventiveness in work and understand the real world limits to adopting new ideas

Collaborate with Others

- Demonstrate ability to work effectively and respectfully with diverse teams. Assume shared responsibility for collaborative work, and value the individual contributions made by each team member

Communication

- Articulate thoughts and ideas effectively using oral, written and nonverbal communication skills in a variety of forms and contexts;
- Use communication for a range of purposes (e.g. to inform, instruct, motivate and persuade (Partnership for 21st Century Skills, 2010)

4.5 *Evaluation of the mathematical modeling process*

The Traffic Jam problem also illustrates the importance of evaluation of the mathematical modeling process as a substantial part of the didactical contract established between the student and the teacher. This was seen in many ways including:

Problem Posing

- Are the problems that students posing mathematical in nature?
- Can the problems be posed in a way that allows students to predict, optimize, and make decisions on solutions to the problem?
- Will the students get to the mathematical goals set for the lesson?

Making Assumption and Constraints:

- How do students discuss and determine the assumptions and constraints to define the problem?
- Are their assumptions close to reality? What are the basis for their assumptions?
- How much reasonableness do they demonstrate?

Considering the Variables

- What variables do students define?
- As students consider variables to mathematize or build a solution to the problem, are they considering the appropriate variables?
- Are they quantifiable variables?
- How do they use data and information to build a solution?

Building a solution

- What problem solving strategies are they using as they build their solution?
- Do students have opportunities to use mathematical modeling to make decisions, predict, optimize, predict and determine a solution to a problem?
- Do students have opportunities to model with tools, diagrams, graphs, tables, or equations? Are connections made between multiple representations? What features of the representation(s) do they focus on?

Analyzing and making conclusions:

- How do students learn and practice justifying their solutions and/or explaining their mathematical thinking?

- In what ways are they making sense of the results: Ways in which students are prompted to connect solutions back to the problem situation.

Applying their solution to the real world scenario

- In what ways are students able to connect the solutions back to the problem scenario?
- How does it fit? Do they see ways they may want to revise/refine their thinking?

A structured way for teachers to capture these evaluation points also helped enhance the pedagogical practices of teacher through:

1. **Pre-observation**: What important mathematical ideas and competencies will the task and context afford you as you engage your students in this mathematical modeling process?
2. **Post Observation**: How did students actually engage with mathematical ideas in this lesson? How did mathematical modeling support students use of math ideas, tools and reasoning to answer questions about a contextual situation?
3. **Next Steps**: How will we leverage students' mathematical ideas, misconceptions revealed, and mathematical opportunities that presented itself in this lesson to build on future lessons?

5 Conclusion and Discussion

This chapter offers a snapshot of what mathematical modeling for solving real-world problems looks like in an elementary and middle grades classroom. Launching a unit with a motivating and intriguing mathematical modeling task can help students see how mathematics is an important tool we use to make decisions in our everyday life. The Traffic Jam problem considered here helped to engage students at multiple levels in conceptual learning and helped to enhance their understanding of the application of mathematics in the real world. Such problems provide rich mathematical and inquiry-based tasks that require the use of 21st century skills, namely, critical thinking, creativity, collaboration and communication. In addition, as we worked with teacher co-designers on lessons focused on mathematical modeling, we identified five critical norms (Figure 5) needed in the classroom to ensure success.

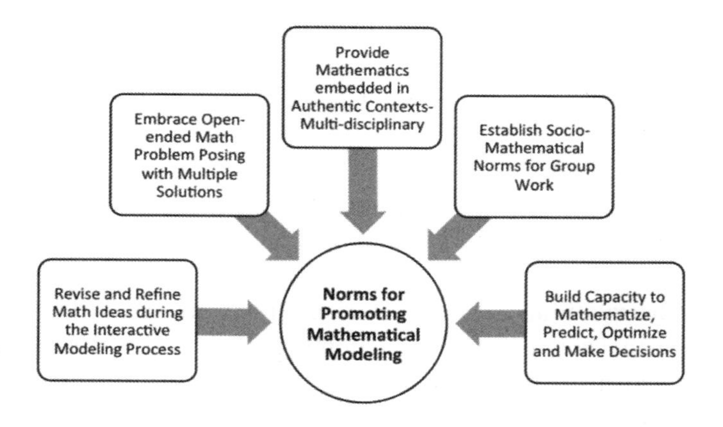

Figure 5. Norms for Promoting Mathematical Modeling

The Traffic Jam problem discussed in this work helped the teachers to recognize that they needed to (a) choose a context that provided an authentic problem in which students can engage in the mathematics; (b) provide appropriate scaffold to help students revise and refine their mathematical ideas during mathematical modeling and; (c) build in opportunities and capacity to mathematize, predict, optimize, and or make decisions for the mathematical modeling process to be worthwhile. It also helped the teachers demonstrate strategic competence through their ability to (a) formulate, represent, and solve problems; (b) model mathematical ideas; and (c) demonstrate representational fluency, that is, the ability to translate and connect within and among multiple representations with accuracy, efficiency, and flexibility (Suh & Seshaiyer, 2014b). Both teachers and students needed to embrace the open-ended nature of the mathematical modeling task that had multiple solutions depending on the assumptions made. Mathematical modeling also helped both teachers and students to establish agreed upon sociomathematical norms such as persevering through complexity and productive struggle.

This study suggests that developing proficiency in mathematical modeling with a focus on eliciting students' strategic competence, required more than the analysis of students' diverse strategies. It also required providing time and space for students to reason by sharing, arguing, and justifying their strategic thinking. Through the experience of working together, teachers helped each other gain an appreciation for multiple strategies. As the teachers in the study relearned mathematics through authentic problem solving, they felt more confident and more strategically competent using mathematical modeling and strategies while posing rich problems and engaging in meaningful mathematics discourse in class with their students. The study also illustrated how students can effectively engage in aspects of mathematical modeling that leads to sense making as well as algebraic, computational and spatial thinking. The study also helped to emphasize that mathematical modeling not only enhances pedagogical practices of teachers but also helps to improve student mathematical learning.

Acknowledgement

The authors thank the National Science Foundation (DRL 1441024) that supported this work. The authors also thank the members of the lesson study team who participated in this activity.

References

Chapin, S. H., & O'Connor, C. (2007). Academically productive talk: Supporting students' learning in mathematics. In W. G. Martin, M. Strutchens, & P. Elliot (Eds.), *The learning of mathematics* (pp. 113-139). Reston, VA: NCTM.

Dweck, C. (2008). *Mindsets and math/science achievement*. New York, NY: Carnegie Corporation of New York Institute for Advanced Study.

Hiebert, J., Morris, A. K., Berk, D., & Jansen, A. (2007). Preparing teachers to learn from teaching. *Journal of Teacher Education*, *58*(1), 47-61.

Hiebert, J., & Grouws, D. A. (2007). The effects of classroom mathematics teaching on students' learning. In F. K. Lester (Ed.), *Second handbook of research on mathematics teaching and learning* (pp. 371-404). Charlotte, NC: Information Age.

Hufferd-Ackles, K., Fuson, K. C., & Sherin, M. G. (2004). Describing levels and components of a math-talk learning community. *Journal for Research in Mathematics Education, 35*(2), 81–116.

Lesh, R., Post, T., & Behr, M. (1987). Representations and translations among representations in mathematics learning and problem solving. In C. Janvier (Ed.), *Problems of representation in the teaching and learning of mathematics* (pp. 33–40). Hillsdale, NJ: Erlbaum.

Lewis, C. (2002). *Lesson study: A handbook of teacher-led instructional change.* Philadelphia, PA: Research for Better Schools.

Lewis, C., Perry, R., & Murata, A. (2006). How should research contribute to instructional improvement? The case of lesson study. *Educational Researcher, 35*(3), 3-14.

MathWorks Math Modeling Challenge. (n.d.). Retrieved from http://m3challenge.siam.org/resources/modeling-handbook

National Council of Teachers of Mathematics (NCTM). (2000). *Principles and standards of school mathematics*. Reston, VA: Author.

National Council of Teachers of Mathematics (NCTM). (2009). *Reasoning and sense making*. Reston, VA: Author.

National Council of Teachers of Mathematics (NCTM). (2013). *Principles to actions.* Reston, VA: Author.

Polya, G. (1945). *How to solve it* (2nd ed.). Princeton University Press.

Smith, M. S., & Stein, M. K. (2012). Selecting and creating mathematical tasks: From research to practice. In G. Lappan (Ed.), *Rich and engaging mathematical tasks: Grades 5-9*. Reston, VA: NCTM.

Stein, M. K., Engle, R. A., Smith, M. S., & Hughes, E. K. (2008). Orchestrating productive mathematical discussions: Five practices for helping teachers move beyond show and tell. *Mathematical Thinking and Learning, 10*(4), 313-40.

Sherin, M. G., & van Es, E. A. (2003). A new lens on teaching: Learning to notice. *Mathematics Teaching in the Middle School*, *9*(2), 92-95.

Suh, J., & Seshaiyer, P. (2014a). Examining teachers' understanding of the mathematical learning progression through vertical articulation during lesson study. *Journal of Mathematics Teacher Education, 17,* 1-23.

Suh, J., & Seshaiyer, P. (2014b). Developing strategic competence by teaching using the common core mathematical practices. In K. Karp (Ed.), *Annual perspectives in mathematics education 2014* (pp 77-87). Reston, VA: NCTM.

Chapter 12

Unpacking the Big Idea of Proportionality: Connecting Ratio, Rate, Proportion and Variation

Joseph B. W. YEO

Proportionality is a foundation of mathematical knowledge and is one of the most commonly applied mathematics in the real world. Yet many students are unable to reason proportionally and to see the connections among the concepts of proportion, ratio, rate and variation. In fact, proportionality is one of the clusters of big ideas proposed for the 2020 secondary school mathematics syllabus in Singapore. Therefore, the purpose of this chapter is to unpack the meanings of these four terminologies so that teachers are better equipped not only to explain to their students the similarities and differences among these concepts but also to connect these ideas into a coherent whole and to appreciate how proportionality is used in real life. The chapter will also end with some other implications for teaching.

1 Introduction

In Singapore, students learn ratio and rate in Primary 5 and direct and inverse proportions in Secondary 2. However, Primary 5 students are expected to use direct proportional reasoning to solve the following kind of word problems:

Mickey saves \$72 in 18 days. He saves the same amount each day. How many days will Mickey take to save \$160? (Lee, Koay, Collars, Ong, & Tan, 2017b, p. 74)

The following shows how the solution is usually presented in local primary school textbooks:

$$\$72 \rightarrow 18 \text{ days}$$
$$\$1 \rightarrow \frac{18}{72} \text{ day}$$
$$\$160 \rightarrow \frac{18}{72} \times 160 = 40 \text{ days}$$

It will take Mickey 40 days to save \$160.

Since the solution involves finding the quantity corresponding to one unit of another quantity first (in this case, how many days to save \$1), the method of solution is called the Unitary Method. But the basis of the Unitary Method is the equality of two ratios or proportional reasoning (in this case, direct proportional reasoning), which I will elaborate on later in this chapter. Therefore, students in Primary 5 are expected to solve word problems using direct proportional reasoning via the Unitary Method without learning the concept of direct proportion.

However, this kind of word problems is found in the chapter on rate (and not in the chapter on ratio) in all the three local primary school textbooks. So the first question to address in this chapter is, "What are the similarities and differences among the three concepts of ratio, rate and proportion?" Anecdotal evidence suggests that many students are not clear about the three concepts: some believe that ratio and proportion mean the same thing, while others use ratio and rate interchangeably.

Moreover, in the 2001 Singapore secondary school mathematics syllabus, students also learn direct and inverse variations in Secondary 3. But in the 2007 and 2013 syllabi, variation was combined with proportion and taught in Secondary 2 as direct and inverse proportions. So the second question to address in this chapter is, "How is the concept of variation similar or different from the concept of proportion?"

But why are these two questions important? For a teacher, content knowledge is as important as pedagogical content knowledge (Ball, Thames, & Phelps, 2008; Shulman 1986). If a teacher does not understand the subject matter, he or she may not be able to teach it properly. Therefore, the main purpose of this chapter is to unpack the concepts of ratio, rate, proportion and variation. In addition, the chapter will end with some other implications for teaching.

2 Literature Review

Proportional reasoning is an essential basis of mathematical knowledge (Lesh, Post, & Behr, 1988) and is one of the most commonly applied mathematics in the real world (Hoffer & Hoffer, 1988; Lanius & Williams, 2003). Heinz and Sterba-Boatwright (2008) explained:

> Proportional reasoning is at the core of so many important concepts in mathematics and science, including similarity, relative growth and size, dilations, scaling, pi, constant rate of change, slope, rates, percent, trig ratios, probability, relative frequency, density, and direct and inverse variations. (p. 528)

Despite its importance in real life, research has found that many students and even adults have difficulties in reasoning proportionally (Behr, Harel, Post, & Lesh, 1992; Lamon, 1999; Staples & Truxaw, 2012). One reason suggested from research studies is that students tend to see comparative situations in absolute (or additive) rather than relative (or multiplicative) terms (Fielding-Wells, Dole, & Makar, 2014). For example, we can compare the number of boys and the number of girls in a class in two ways. Suppose there are 30 boys and 10 girls in the class. One method of comparison is to ask, "How many more boys are there than girls?" The answer is $30 - 10 = 20$ more boys than girls. This is called comparison in absolute (or additive) terms. Another method of comparison is to ask, "How many times are the number of boys more than the number of girls?" The answer is $30 \div 10 = 3$ times as many boys as there are girls. This is called comparison in relative (or multiplicative) terms. For many students, it is more intuitive to compare two quantities in absolute (or additive) terms (Hart, 1981), partly because they may be more familiar with addition and subtraction than with multiplication and division (Lo & Watanabe, 1997) and partly because it is easier to see the actual difference of 20 boys, i.e. in absolute terms. Also, many students fail to distinguish proportional situations from absolute comparison situations, so there is a need for a more targeted focus on recognising proportional situations (Behr et al., 1992; Van Dooren, De Bock, Hessels, Janssens, & Verschaffel, 2005).

Another gap about proportional reasoning that many students have is that they are unable to distinguish "between and among concepts of decimal, percent, ratio, proportion, and proportional reasoning' (Ojose, 2015, p. 111). This may be because of the way topics are sequenced by teachers during instruction without presenting the opportunity for students to see the big picture and the interconnectedness among the various mathematical concepts (ibid.). In fact, proportionality is one of the big ideas proposed for the 2020 secondary school mathematics syllabus in Singapore (Ministry of Education, 2018). Charles (2005) wrote:

> A Big Idea is defined as a statement of an idea that is central to the learning of mathematics, one that links numerous mathematical understandings into a coherent whole. When one understands Big Ideas, mathematics is no longer seen as a set of disconnected concepts, skills, and facts. Rather, mathematics becomes a coherent set of connected ideas. Highly effective teachers make these connections explicit. (p. 9)

Therefore, it is important to unpack the concepts of ratio, rate, proportion and variation so that teachers and students know what they mean and how they are connected to one another.

3 Comparison between Ratio and Rate

3.1 *Concept of ratio*

Let us first look at how the word 'ratio' is used in Singapore primary and secondary school textbooks which have been approved by the Ministry of Education. Local primary school textbooks seldom give a definition for ratio. Instead, the textbooks just give a few examples of ratios in Primary 5, the level at which students first study ratio. The following shows the first example used by each of the three primary school textbooks:

- "The ratio of the number of blue caps to the number of red caps is 4 : 1. We read the ratio 4 : 1 as 4 to 1 … Units of measurement are not shown in ratios." (Chan, 2017a, p. 142)
- "The ratio of the number of pens to the number of pencils is 2 : 3. We read 2 : 3 as 2 to 3." (Fong, Gan, & Ramakrishnan, 2017a, p. 119)
- "The ratio of the number of cups of plain flour to the number of cups of wheat flour is 3 : 2. We read the ratio 3 : 2 as 3 to 2. A ratio does not have units." (Lee, Koay, Collars, Ong, & Tan, 2017a, p. 85)

All the above examples of ratios compare numbers of two objects. In subsequent examples, ratios also compare numbers of two persons or two measurements, such as lengths, and then ratios of three quantities. Only in Primary 6 do students learn how to write ratios as fractions.

On the other hand, local secondary school textbooks usually provide a definition or a more formal description of ratio as shown below:

- "At the primary level, we have learnt to use a ratio to compare two similar quantities a and b. The ratio of a to b is denoted by $a : b$ which can also be represented by $\frac{a}{b}$, where $b \neq 0$." (Chow, 2013, p. 2)

- "A ratio is used to compare two or more quantities of the *same kind* which are measured in the same unit. The ratio of a is to b, where a and b represent two quantities of the *same kind*, and $b \neq 0$, is written as $a : b$." (Yeo et al., 2013, p. 226)

Unlike primary school textbooks, secondary school textbooks stress that the quantities in a ratio must be 'similar quantities' or 'quantities of the same kind'. Let us unpack this concept of ratio further. First, what do we mean by 'similar quantities' or 'quantities of the same kind'? For example, when we say that the ratio of the number of boys to the number of girls in a class is 3 : 2, we are *not* comparing boys with girls *per se*, but the number of boys with the number of girls. In other words, it is not because boys and girls are of the same kind (i.e. children) but because the number of boys and the number of girls are of the same kind since they are just numbers. This is similar to how letters are used to represent numbers, not objects (whether inanimate or animate), in algebra. For example, if there are x boys and y girls in a class, the letters x and y do not represent boys and girls respectively, but the number of boys and the number of girls respectively. Similarly, x and y in the ratio $x : y$ are numbers, not objects, although the numbers refer to the numbers of different objects.

Second, must quantities of the same kind be measured in the same unit? For example, on a map, 1 centimetre represents 100 metres. Can we write the map scale as 1 cm : 100 m? In the real world, some maps may represent the map scale in this manner, but this will *not* be the ratio notation since the latter has no units. It is just like how the real world would use abbreviations such as hrs and mins (e.g. the train will arrive in 5 mins), but these abbreviations are not short forms of SI units (International System of Units) used in school mathematics, e.g. 5 min, because short forms of SI units do not have plural forms. Therefore, in school mathematics, if we want to write the map scale as a ratio, we have to convert 100 metres to 10 000 centimetres and write the scale as the ratio 1 : 10 000. In other words, the length on a map and the actual distance are still quantities of the same kind, i.e. length: although they are measured in different units, the units are of the same type because they can be converted from one to the other. It is unlike comparing a quantity (length) measured in metres and another quantity (mass) measured in kilograms, which are not of the same kind.

Third, we can write the ratio of two quantities in two ways: $a : b$ or $\frac{a}{b}$. But we cannot write the ratio of three or more quantities as a fraction. Instead, we have to write it using colons, e.g. $a : b : c$.

Fourth, it is often stated that $b \neq 0$ in the ratio $a : b$ or $\frac{a}{b}$. But can $a = 0$? For example, if the numbers of boys and girls in a class are 0 and 40 respectively, the ratio of the number of boys to the number of girls is 0 : 40. But what if we want to compare the number of girls to the number of boys? Then the ratio would be 40 : 0, but we say that $b \neq 0$. Moreover, since 0 : 40 is equivalent to 0 : 20 or 0 : 1 or 0 : 100, do all these ratios mean the same thing as they are equivalent? In the real world, we cannot prevent a quantity from being zero. But if one quantity is zero, it is not so meaningful to compare it with another quantity by using ratio as a basis of comparison: instead, we just say that the quantity is zero. So I propose that the ratio notation $a : b$ be used for both non-zero a and b so that we will not encounter the above problems.

Fifth, all the quantities for ratio given in local primary and secondary textbooks are always positive. This is understandable because in the real world, most quantities are positive. Although temperatures measured in degree Celsius can be negative, in laws involving temperatures such as the Ideal Gas Law, the temperature is always measured in Kelvin, which cannot be negative. So, for most purposes in school mathematics, we can just treat a and b in the ratio notation $a : b$ as positive.

Sixth, a ratio can also be expressed as a percentage because a fraction can be converted to a percentage. For example, the Cambridge Business English Dictionary (2018) gives this example: "The ratio of exports to imports also improved from 70.4% to 83.2%."

Lastly and more importantly, the idea of ratio is to compare two or more quantities in relative (multiplicative) terms, instead of absolute (additive) terms. As explained in Section 2, many students tend to compare two quantities in absolute (additive) terms partly because they may be more familiar with addition and subtraction than with multiplication and division and partly because it is easier to see the actual difference in absolute terms.

To summarise, ratio is a way of comparing two or more quantities of the same kind. The quantities can be numbers (of objects) or measurements. The ratio notation $a : b$, where $a, b > 0$, has no units. A ratio comparing two quantities can also be expressed as a fraction or as a percentage. If two quantities have different units which can be converted from one to the other, we can still use ratio to compare the two quantities but the ratio notation must still have no units. For three or more quantities, their ratio cannot be written as a fraction.

3.2 *Concept of rate*

Unlike ratio, Singapore primary school textbooks do give a definition of rate, after giving some examples, as shown below:

- "A rate is a comparison of two quantities and is expressed as one quantity per unit of another quantity." (Chan, 2017b, p. 34)
- "Rate is the amount of a quantity per unit of another quantity." (Fong, Gan, & Ramakrishnan, 2017b, p. 85)
- "A rate involves two quantities. It is expressed as one quantity per unit of another quantity." (Lee et al., 2017b, p. 68)

The concept of rate in all the three textbooks is the same: rate is expressed as one quantity per unit of another quantity. But most of the examples given in the textbooks contain time as the second quantity, i.e. one quantity per unit time. This may give rise to the notion that rate must be some kind of speed. In fact, anecdotal evidence shows that there are students who subscribe to this notion that the second quantity must be time. However, all the three textbooks have also included non-examples, such as, a pack of washing powder costs \$2.30 per kilogram (Chan, 2017b), the area of surface painted is 75 m^2 per litre of paint (Fong et al., 2017b) and a car travels 15 km per litre of petrol (Lee at al., 2017b). Other counter examples include exchange rates and conversion rates, where the second quantity is not time.

Local secondary school textbooks also provide a definition of rate as shown below:

- "A rate is a comparison of two quantities by division. It is usually expressed as one quantity per unit of another quantity." (Chow, 2013, p. 23)
- "Rate is a comparison of two quantities of different kinds." (Yeo et al., 2013, p. 251)

One of the above secondary school textbooks, namely Chow (2013), follows the definition used in the three primary school textbooks, except for two differences. First, it has an additional sentence that says that a rate is a comparison of two quantities by division. Let us unpack what 'by division' means. The subsequent sentence in the textbook about 'one quantity per unit of another quantity' gives an indication that 'by division' means that one quantity is divided by another quantity. However, this idea of 'by division' is instead applied to 'ratio' (and not to 'rate') by the following two sources:

- "Two or more quantities can be compared by division and the comparison expressed as a ratio." (Fong et al., 2017a, p. 144)
- "[In mathematics, a ratio is] a comparison of two numbers calculated by dividing." (Cambridge Academic Content Dictionary, 2018b)

Perhaps what they mean 'by division' for ratio is because a ratio can be expressed as a fraction, which can be interpreted as the numerator divided by the denominator, but this applies for ratios of two quantities only, and not for ratios of three or more quantities. So we see that there are two different interpretations of 'by division': one applied to ratio and the other applied to rate. Therefore, we cannot define ratio or rate by just using the idea of 'by division' because it can be applied to either concept.

The second difference is that the secondary school textbook by Chow (2013) has added the word 'usually' in its definition of rate, which the primary school textbooks do not include. Although the textbook did not give any exceptions to the rule, let us unpack this further by looking at two examples: an item costs $2 per 100 grams, or a smartphone plan costs $10 per 2 gigabytes of data. Thus we see, first of all, that rates may not be expressed as a quantity per unit of another quantity, but as a quantity per n units of another quantity.

However, this idea of rate is still not comprehensive enough. For example, Table 1 shows the local postage rates for Singapore in 2019. This kind of rate depends on the step bracket that the weight of an envelope falls under, e.g. a standard regular envelope that weighs 10 grams or 20 grams will cost the same amount, namely $0.30, but a standard regular envelope that weighs 30 grams will cost a different amount, namely $0.37, to post locally. This kind of rate is called a step rate, which Merriam-Webster Dictionary (2018) defines as a rate that changes by regular gradations. In fact, the gradations may not even be regular, as can be seen from the different sizes of the step brackets in Table 1. Another example of step rates is the income tax rates in Singapore, which is a progressive tax based on different step brackets.

Table 1
Local postage rates (inclusive of 7% GST)

<table>
<tr><th>Weight Step Up to</th><th>Standard Regular (C5, C6 & DL size envelope)</th><th>Standard Large (Up to C4 size envelope)</th><th>Non-Standard</th></tr>
<tr><td>20 g</td><td>$0.30</td><td rowspan="3">$0.60</td><td rowspan="2">$0.60</td></tr>
<tr><td>40 g</td><td>$0.37</td></tr>
<tr><td>100 g</td><td rowspan="5"></td><td>$0.90</td></tr>
<tr><td>250 g</td><td>$0.90</td><td>$1.15</td></tr>
<tr><td>500 g</td><td>$1.15</td><td>$1.70</td></tr>
<tr><td>1 kg</td><td colspan="2">$2.55</td></tr>
<tr><td>2 kg</td><td colspan="2">$3.35</td></tr>
</table>

Another question to ask is, "Is a step rate a constant rate?" At first glance, it may appear that a step rate is not constant because the rate is different for different step brackets. However, what is different is the amount of the quantity (in this case, the postage cost) for different step brackets. If we plot the postage cost against the weight of, e.g., standard regular envelopes, we will obtain a step function. Within each step bracket, the amount of the quantity is the same, i.e. the rate (or the gradient of the step function) is actually zero (which is constant) within each step bracket. But at the boundaries of the step boundaries, the rate (or the gradient of the step function) is not defined. This may sound complicated for students as they do not learn step functions in secondary schools. Therefore, what is important is for teachers to be mindful that they do *not* give this kind of step rates as an example of a variable rate.

At higher level, students in Singapore learn about rates of change in calculus in Additional Mathematics in Secondary 4, where they use the derivative of one variable with respect to another variable to calculate the rate of change. For example, the rate of change of the area A of a circle with respect to its radius r is $\frac{\mathrm{d}A}{\mathrm{d}r} = 2\pi r$. In this case, the rate is not constant because it depends on the radius of the circle. But $\frac{\mathrm{d}A}{\mathrm{d}r}$ is still an expression of one quantity per unit of another quantity because the unit for $\frac{\mathrm{d}A}{\mathrm{d}r}$ is, e.g. cm^2 *per* cm of change in the radius of the circle.

Let us now turn our attention to the secondary school textbook by Yeo et al. (2013) where it defines rate to be a comparison of two quantities of different kinds, in contrast to ratio being a comparison of two quantities of the same kind. Although the other secondary school textbook by Chow (2013) does not explicitly state that rate compares two quantities of different kinds, all the examples of rates in the textbook also compare two quantities of different kinds. However, it is possible to use a rate to compare two quantities of the same kind.

For example, the unemployment rate measures how the number of unemployed people changes with the number of working people, and it is calculated by dividing the number of unemployed people by the number of working people, and then expressed as a percentage. For instance, the unemployment rate in Singapore in 2017 was about 2.02%. This means that there are about 2.02 unemployed people per 100 working people, or about 202 unemployed people per 10 000 working people.

We can also use rates to compare measurements of the same kind, e.g. we can find the rate of how the circumference of a circle changes with its radius as the latter increases or decreases. In this case, the rate is a constant and is equal to 2π. If a rate compares two quantities of the same kind, it usually has no units, unless one chooses, e.g., to measure the circumference of the circle in metres and its radius in centimetres.

Before we leave this section on rates, let us look at another type of measure called index, e.g. the Body Mass Index (BMI). Since the BMI of a person is obtained by dividing his or her mass, in kilograms, by the square of his or her height, in metres, the unit should be kg/m^2. Maybe because it is too troublesome to always specify its rather complicated unit, the unit is removed and we use an index (i.e. a number without any unit) instead of a rate. Another example of an index is the Straits Times Index (STI), which is a capitalisation-weighted stock market index that tracks the performance of the top 30 companies listed on the Singapore Exchange.

To summarise, I would like to propose a more general definition of rate: rate is a way of comparing how one quantity *changes* with another, unlike ratio which does not have this dynamic sense of change. This concept of rate is consistent with the idea of the rate of change of one variable with respect to another variable learnt in calculus. Also, a rate can be expressed as one quantity per n units of another quantity (which includes the case of $n = 1$ and the notation $\frac{\mathrm{d}y}{\mathrm{d}x}$) or it can be a step rate.

The next question to address is, "How is rate related to ratio?" Is ratio a special type of rate (since ratio can only compare quantities of the same kind while rate can compare quantities of the same kind or of different kinds) or is rate a special type of ratio (since rate can only compare two quantities while ratio can compare two or more quantities)?

3.3 *Ratio versus rate*

There are textbooks that define ratio to include rate. For example, Cathcart, Pothier, Vance and Bezuk (2006) wrote in an overseas textbook: "When the measuring units describing two quantities being compared are different, the ratio is called a *rate*." (p. 272). Another overseas textbook by Van de Walle, Karp and Bay-Williams (2010) also says the same thing. In fact, Wikipedia (2018b) states: "In mathematics, a rate is the ratio between two related quantities." But whenever the above mentioned textbooks and Wikipedia give examples of rates, they are never expressed using the ratio notation $a : b$; and whenever the ratio notation is used, it has no units. So why do we define ratios to include rates when rates cannot be expressed using the ratio notation? In Singapore, perhaps because the ratio notation does not have any units, the primary and secondary school textbooks have consistently used ratio to compare only quantities of the same kind. Therefore, in the local context, it is clear that ratio does not include rate because ratio cannot be used to compare quantities of different kinds but rate can. However, we need to be mindful that people in other countries may have a differing view.

The next question is, "Does rate include ratio?" Let us examine this by looking at rates that compare two quantities of the same kind and how different these rates are from ratios. For example, we have mentioned earlier in Section 3.2 that the unemployment rate in Singapore in 2017 was about 2.02%, which means that there are about 2.02 unemployed people per 100 working people, or about 202 unemployed people per 10 000 working people. This is the concept of rate as it is expressed as one quantity per n units of another quantity. However, we can convert this rate to a ratio by expressing it as the number of unemployed people to the number of working people, which is about 202 : 10 000.

On the other hand, we can also convert a ratio to a rate. In a video by Math Snacks (2012), a woman went for three dates. On the first date, she spoke only 25 words but the man spoke 175 words. So she complained to her friend on the phone that the man spoke too much: the ratio of her number of words to his number of words was 25 : 175 or 1 : 7. Then the woman made this statement, "For every word I spoke, he spoke 7 words." This second idea is the notion of rate: the man spoke 7 words per word that the woman spoke.

Sometimes, the conversion of a ratio to a rate may not make much sense. For example, the ratio of the number of boys to the number of girls in a particular class is 2 : 1. We can convert the ratio to a rate by saying that there are 2 boys for every one girl, or there are 2 boys per girl. Does this rate make sense or why do we want to compare the numbers of boys and girls in the class by using a rate and not a ratio? From the ratio 2 : 1, we can easily see whether there are more boys or girls in the class. But if we want to compare the ratios of the number of boys to the number of girls in two or more classes, it may be harder to compare if the ratios do not have the same 'base', e.g. the ratios of the number of boys to the number of girls are 3 : 2 for Class A, 5 : 2 for Class B and 4 : 3 for Class C. If we compare the first two classes, we can easily see from their ratios that the proportion (not the number) of boys in Class B is higher than the proportion of boys in Class A because the two ratios have the same 'base' 2. But if we compare Class A and Class C just by looking at the ratios, the answer is not so straightforward. However, if we express all the ratios as rates, i.e. 1.5 for Class A, 2.5 for Class B and 1.3 (rounded to one decimal place) for Class C, it is so much easier to see at a glance that that Class B has the highest proportion of boys, followed by Class A and then Class C. Therefore, whether we use ratio or rate as a way of comparing quantities depends on which form is more useful in the given context.

To summarise, the above examples illustrate that there is still a difference between ratio and rate, even when both are comparing two quantities of the same kind. Although ratio can be converted to rate and vice versa, sometimes it may be more meaningful or helpful to use one over the other. In other words, ratio and rate are two ways of comparing quantities in slightly different manners. But for quantities of different kinds, we have to use rate as a basis of comparison since ratio is used only to compare quantities of the same kind. On the other hand, a ratio which compares three or more quantities of the same kind cannot be converted to a rate. Nevertheless, rate can be used to compare three or more quantities, as in related rates of change learnt in calculus, except that at any one time, we can only write the rate between two quantities. For example, to compare three quantities x, y and t, we have three rates, namely, $\frac{dy}{dx}$, $\frac{dy}{dt}$ and $\frac{dx}{dt}$, and they are related to one another via $\frac{dy}{dt}=\frac{dy}{dx}\times\frac{dx}{dt}$.

After unpacking the meaning of ratio and rate, we are now ready to discuss what proportion, proportionality and proportional reasoning are, and how different they are from one another and from the concepts of ratio, rate and variation.

4 Comparison between Proportion, Ratio, Rate and Variation

4.1 *Concept of proportion*

The word 'proportion' has a few meanings in real life. One meaning of proportion is "the number or amount of a group or part of something when compared to the whole" (Cambridge Advanced Learner's Dictionary & Thesaurus, 2018). Wikipedia (2018c) puts it this way:

> If the two or more ratio quantities encompass all of the quantities in a particular situation, it is said that "the whole" contains the sum of the parts: for example, a fruit basket containing two apples and three oranges and no other fruit is made up of two parts apples and three parts oranges. In this case, 2/5, or 40% of the whole is apples and 3/5, or 60% of the whole is oranges. This comparison of a specific quantity to "the whole" is called a proportion.

Ratio can be used to compare one part of a whole to another part of the whole, and this is called 'part-to-part ratio'; on the other hand, ratio can also be used to compare one part of a whole to the whole itself, and this is called 'part-to-whole ratio' (Van de Walle et al., 2010). As explained above, both Cambridge Advanced Learner's Dictionary & Thesaurus (2018) and Wikipedia (2018c) call the 'part-to-whole ratio' a proportion. In the above example given in Wikipedia, it is common everyday usage to say that the proportion of apples in the basket is 40%. In fact, in Section 3.3, I have used the word 'proportion' in the same manner when talking about the proportion of boys in a class. Perhaps because of the usage of proportion in this manner, anecdotal evidence shows that many students think that ratio and proportion mean the same thing. However, there is another meaning of proportion.

4.2 *Concept of proportionality and its relation to ratio*

The Cambridge Academic Content Dictionary (2018a) also defines proportion as follows: "[In mathematics] A proportion is also an equation (= mathematical statement) that shows two ratios (= comparisons) are equal." The overseas school textbook by Cathcart et al. (2006), mentioned earlier in Section 3.3, puts it this way: "A *proportion* is a statement that two ratios are equal." (p. 273) In literature, Tourniaire and Pulos (1985) wrote, "For a mathematician, a proportion is a statement of equality of two ratios, i.e. $a/b = c/d$." But Wikipedia (2018a) uses the term 'proportionality' to describe this concept: under the topic Proportionality (mathematics), it says, "In mathematics, two variables are proportional if there is always a constant ratio between them. The constant is called the coefficient of proportionality or proportionality constant." Then it proceeds to explain what direct and inverse proportions are. In fact, as explained in Section 2, proportionality is one of the big ideas proposed for the 2020 secondary school mathematics syllabus in Singapore. Therefore, to distinguish between the two meanings of proportion, I will use the term 'proportionality' in this chapter to describe this second meaning of proportion, which consists of direct and inverse proportions.

Let us now unpack what proportionality is by looking at what is taught locally in schools. In Singapore, students first learn the topic of direct and inverse proportions in Secondary 2. In one of the textbooks, Yeo et al. (2014) start with a context of an overdue fine of 15 cents per day for each overdue book borrowed from a public library in Singapore. Then it shows a table of different amounts of fines for a book that is overdue for 1 to 10 days (see Figure 1). From the questions that are asked in the textbook after the table of values, students are guided to realise that the concept of direct proportion is as follow: if one quantity is doubled, the other quantity is also doubled; if the same quantity is tripled, the other quantity is also tripled; if the same quantity is halved, the other quantity is also halved. In other words, for direct proportion, if one quantity is multiplied by n, the other quantity is also multiplied by n. This is what we call direct *proportional reasoning*.

Direct Proportion

In Singapore, if we borrow books from a public library and are late in returning the books, we will be fined 15 cents per day for each overdue book. Table 1.1 shows the fines for an overdue book.

Number of days (x)	1	2	3	4	5	6	7	8	9	10
Fine (y cents)	15	30	45	60	75	90	105	120	135	150

Table 1.1

1. If the number of days a book is overdue increases, will the fine increase or decrease?
2. If the number of days a book is overdue is doubled, how will the fine change? *Hint:* Compare the fines when a book is overdue for 3 days and for 6 days.
3. If the number of days a book is overdue is tripled, what will happen to the fine?
4. If the number of days a book is overdue is halved, how will the fine change? *Hint:* Compare the fines when a book is overdue for 10 days and for 5 days.
5. If the number of days a book is overdue is reduced to $\frac{1}{3}$ of the original number, what will happen to the fine?

Figure 1. Concept of direct proportion (Yeo et al., 2014, p. 3; reproduced with permission from publisher)

Algebraically, if one quantity x_1 is doubled to x_2, i.e. $\frac{x_2}{x_1} = 2$, the other quantity y_1 is also doubled to y_2, i.e. $\frac{y_2}{y_1} = 2$, so $\frac{y_2}{y_1} = \frac{x_2}{x_1} = 2$. Similarly, if one quantity x_1 is tripled to x_2, i.e. $\frac{x_2}{x_1} = 3$, the other quantity

y_1 is also tripled to y_2, i.e. $\frac{y_2}{y_1} = 3$, so $\frac{y_2}{y_1} = \frac{x_2}{x_1} = 3$. Although the ratios $\frac{x_2}{x_1}$ and $\frac{y_2}{y_1}$ are *not constant* (as they can have different values such as 2 and 3), they are always equal to each other (herein lies the subtle difference between the concepts of 'being constant' and 'always equal'). In other words, if y is directly proportional to x, then $\frac{y_2}{y_1} = \frac{x_2}{x_1}$, which is a statement of *equality of two ratios*.

For inverse proportion, it is still a statement that two ratios are equal, but the two ratios are different from those for direct proportion: if y is inversely proportional to x, then $\frac{y_2}{y_1} = \frac{x_1}{x_2}$. What this means is that if one quantity x_1 is doubled to x_2, i.e. $\frac{x_2}{x_1} = 2$, then the other quantity y_1 is halved to y_2, i.e. $\frac{y_2}{y_1} = \frac{1}{2}$, so $\frac{y_2}{y_1} = \frac{x_1}{x_2}$. In other words, for inverse proportion, if one quantity is multiplied by n, the other quantity is divided by n (or multiplied by $\frac{1}{n}$). This is what we call inverse proportional reasoning.

Let us unpack this concept of proportionality further. First, we cannot just say that for direct proportion, as x increases, y increases, while for inverse proportion, as x increases, y decreases. This is because there are other types of increase or decrease, e.g. exponential change. So, to be more precise, we should say that for direct proportion, as x increases, y increases *proportionally*, and by 'increases proportionally', we mean that as x is multiplied by n, y is also multiplied by n. Similarly, for inverse proportion, as x increases, y decreases *proportionally*, and by 'decreases proportionally', we mean that as x is multiplied by n, y is divided by n. Second, the concept of the equality of two ratios in proportionality is different from the concept of equivalent ratios taught in Primary 5: equivalent ratios are more about simplifying a ratio to its simplest equivalent form, e.g. $8 : 12 = 2 : 3$. Third, as explained earlier in

Section 1, Primary 5 students are expected to use direct proportional reasoning to solve word problems. So the next question is, "How is proportional reasoning similar to or different from proportionality?"

4.3 *Proportionality versus proportional reasoning*

Let us refer to the word problem on saving money mentioned earlier in Section 1. Primary 5 students in Singapore are taught to use *direct proportional reasoning* to solve this type of word problems: if one quantity is divided by 72, then the other quantity is also divided by 72:

$$\begin{aligned} &\$72 \quad \rightarrow 18 \text{ days} \\ &\$1 \quad\ \ \rightarrow \frac{18}{72} \text{ day} \end{aligned}$$

And if one quantity is multiplied by 160, the other quantity is also multiplied by 160:

$$\begin{aligned} &\$1 \quad\ \ \rightarrow \frac{18}{72} \text{ day} \\ &\$160 \rightarrow \frac{18}{72} \times 160 \text{ days} \end{aligned}$$

The students are not expected to use the *equality of two ratios* $\frac{y_2}{y_1} = \frac{x_2}{x_1}$ to solve this kind of word problems, i.e. they are not expected to let $y_1 = 72$, $y_2 = 1$ and $x_1 = 18$, and then find x_2 using $\frac{y_2}{y_1} = \frac{x_2}{x_1}$. (In fact, if one uses the equality of two ratios, one can skip the step of finding one unit by finding the answer directly: let $y_1 = 72$, $y_2 = 160$ and $x_1 = 18$, and then find $x_2 = 40$ straightaway using $\frac{y_2}{y_1} = \frac{x_2}{x_1}$.) This method is actually taught in Secondary 2 but anecdotal evidence shows that some Secondary 2 students have difficulty deciding the values of the various variables x_1, x_2, y_1 and y_2. Therefore, the term 'proportional reasoning' refers to this kind of reasoning that is going on behind the Unitary Method, and not the formal concept of the equality of ratios $\frac{y_2}{y_1} = \frac{x_2}{x_1}$.

To summarise, the word 'proportion' has a few meanings in the real world. To avoid ambiguity, the term 'proportionality' is used in this chapter to refer to direct and inverse proportions. The concept of proportionality is the equality of two ratios: $\frac{y_2}{y_1} = \frac{x_2}{x_1}$ for direct proportion and $\frac{y_2}{y_1} = \frac{x_1}{x_2}$ for inverse proportion, and what this means is that if one quantity is multiplied by n, the other quantity is multiplied or divided by n for direct and inverse proportions respectively. This latter interpretation is what we call proportional reasoning, which is the basis of the Unitary Method. As Tourniaire and Pulos (1985) pointed out, "Although most people are probably unaware of the mathematical definition of proportions [i.e. the equality of two ratios], they do use them in familiar situations." And how they use them is actually proportional reasoning behind the Unitary Method.

However, as explained in Section 1, the above word problem on saving money is found in the chapter on rate (and not in the chapter on ratio) in all the three local primary school textbooks. So the next question is, "How is proportionality related to rate?"

4.4 *Concept of proportionality and its relation to rate*

There is another perspective of proportionality. In the previous example of library fines (see Figure 1), if we divide the amount of fine, y cents, by the number of days a book is overdue, x, $\frac{y}{x}$ is a constant, which is 15 in this case (see Figure 2). But what is $\frac{y}{x}$? The rate is actually 15 cents per day, but $\frac{y}{x} = 15$ has no units. If we want to be more precise, we can call

$\frac{y}{x}$ an index, just like the Body Mass Index (BMI) mentioned in Section 3.2. However, for the purpose of this chapter, we will just call it a rate (without units) because there is a relationship between proportion and rate as follows: the idea of $\frac{y}{x}$ being a constant means $\frac{y_2}{x_2}$ is always equal to $\frac{y_1}{x_1}$, i.e. $\frac{y_2}{x_2}=\frac{y_1}{x_1}$, which is the equality of two rates for direct proportion. In fact, we can obtain $\frac{y_2}{x_2}=\frac{y_1}{x_1}$ (equality of two rates) by manipulating $\frac{y_2}{y_1}=\frac{x_2}{x_1}$ (equality of two ratios). However, as explained in Section 4.2 (in the paragraph just before Figure 1), the ratios $\frac{x_2}{x_1}$ and $\frac{y_2}{y_1}$ in a direct proportion are actually not constant although they are always equal. What is a constant in a direct proportion is the rate. In fact, it is true that $\frac{y_1}{x_1}=\frac{y_2}{x_2}=\frac{y_3}{x_3}=\frac{y_4}{x_4}=\ldots$ (equality of two or more rates), which we cannot do the same for $\frac{y_2}{y_1}=\frac{x_2}{x_1}$ (equality of two ratios).

Complete Table 1.2.

Number of days (x)	1	2	3	4	5	6	7	8	9	10
Fine (y cents)	15	30	45	60	75	90	105	120	135	150
Rate $\left(\frac{y}{x}\right)$	$\frac{15}{1}=15$	$\frac{30}{2}=15$	$\frac{45}{3}=15$							

Table 1.2

What can we observe about the rate $\frac{y}{x}$?

What does $\frac{y}{x}$ represent? What does the constant '15' mean physically?

Figure 2. Direct proportion versus rate (Yeo et al., 2014, p. 4; reproduced with permission from publisher)

For inverse proportion, the product of the two quantities x and y, i.e. xy, is a constant. In other words, $x_2y_2 = x_1y_1$ or $\frac{y_2}{x_1} = \frac{y_1}{x_2}$ (which is still the equality of two rates). The latter can also be obtained by manipulating $\frac{y_2}{y_1} = \frac{x_1}{x_2}$ (equality of two ratios), but $\frac{y_2}{x_1} = \frac{y_1}{x_2}$ may not be so helpful as compared to $x_2y_2 = x_1y_1$ because the latter is easier to deal with as the product of the two quantities is a constant. Therefore, we can also view inverse proportion as the *equality of two products* $x_2y_2 = x_1y_1$.

To summarise, there are two perspectives of proportionality. The first perspective is the equality of two ratios: $\frac{y_2}{y_1} = \frac{x_2}{x_1}$ for direct proportion and $\frac{y_2}{y_1} = \frac{x_1}{x_2}$ for inverse proportion. The second perspective is the equality of two rates: $\frac{y_2}{x_2} = \frac{y_1}{x_1}$ for direct proportion and $\frac{y_2}{x_1} = \frac{y_1}{x_2}$ for inverse proportion, i.e. $\frac{y}{x} = k$ for direct proportion and $xy = k$ for inverse proportion, where k is a positive constant. In addition, we can view inverse proportion as the equality of two products $x_2y_2 = x_1y_1$.

Let us unpack this further. First, as explained in Section 3.1, most quantities in the real world are positive. Although these quantities can be zero at times, they are usually not negative. Therefore, k will always be positive. Although both Singapore secondary school mathematics textbooks do not explicit state that $k > 0$, in all their examples and exercises, the quantities will not have any negative values, so k ends up being positive in all cases. If k is negative, there will be a problem, which I will address in Section 4.5.

Second, if two quantities x and y are related by a constant rate such that both quantities are zero at the same time, i.e. $y = kx$, then the two quantities are in fact directly proportional to each other. (If the rate is constant but the quantities are not zero at the same time, then $y = kx + c$, where $c \neq 0$, which is not a direct proportion.) For example, the rate of the library fine is 15 cents per day, which is a constant. As we have seen earlier, this means that the amount of fine is directly proportional to the number of days. Why this understanding of the relationship between constant rate and direct proportion is so important is because when students (or even adults) encounter any constant rate, such that both quantities are zero at the same time, in school mathematics (or in real life), they can use direct proportional reasoning via the Unitary Method to solve such problems easily. As explained in Section 2, students need to learn how to identify proportional situations and this is one way to help them do so: by linking such proportional situations to constant rates and the condition that both quantities must be zero at the same time.

Second, there is a need to clarify whether an exchange rate is a constant rate. Anecdotal evidence shows that students believe that an exchange rate is not constant simply because the rate fluctuates. But what actually fluctuates is the exchange rate with time, i.e. what is not constant is the rate of change of the exchange rate with time. For example, if S$1 = US$1.30 at the time when we are at a money changer, it means that we will get US$1.30 for every S$1, i.e. the exchange rate is a constant at that point in time.

At this stage, I have shown how proportionality is related to ratio and rate. The last issue that I will examine is the similarities and differences between proportion and variation.

4.5 *Concept of variation and its relation to proportion*

As mentioned in Section 4.4, the second perspective of proportionality is the equality of two rates: $\frac{y_2}{x_2}=\frac{y_1}{x_1}$ for direct proportion and $\frac{y_2}{x_1}=\frac{y_1}{x_2}$ for inverse proportion, i.e. $\frac{y}{x}=k$ for direct proportion and $xy=k$ for inverse proportion, where k is a positive constant. The latter part of this second perspective of proportionality (i.e. the equations $\frac{y}{x}=k$ for direct proportion and $xy=k$ for inverse proportion) is what we used to call variation in the 2001 Singapore secondary school mathematics syllabus.

The definitions of direct and inverse variations from one of the local Secondary 3 textbooks (Teh & Looi, 2001) are: If y varies directly as x (we write $y \propto x$), then $y = kx$ or $\frac{y}{x} = k$, where $k \neq 0$; if y varies inversely as x (we write $y \propto \frac{1}{x}$), then $y = \frac{k}{x}$ or $xy = k$, where $k \neq 0$.

As explained in Section 4.4, the constant k in direct and inverse proportions are always positive in Singapore secondary school mathematics textbooks. But in the above definitions of direct and inverse variations, $k \neq 0$. Anecdotal evidence shows that there are many students who think that if $y = kx$ where $k < 0$, the relation is an inverse proportion or variation because as x increases, y decreases. But I have explained earlier in Section 4.2 why it is important to describe an inverse proportion as follows: as x increases, y decreases *proportionally*, and by 'decreases proportionally', we mean that as x is multiplied by n, y is divided by n. Just because y decreases as x increases for $y = kx$ where $k < 0$, we cannot straightaway jump into the conclusion that this is an inverse proportion. Instead, we need to examine whether or not the decrease is proportional. Let us consider the case where $k = -3$. For completeness sake, we will also analyse the case where $k = 3$. Figure 3 shows the graphs of $y = 3x$ and $y = -3x$ for $x \geq 0$.

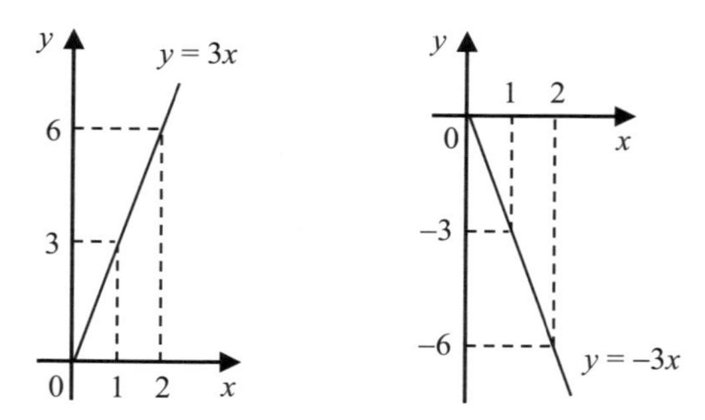

Figure 3. Direct variation (for the cases where $k = 3 > 0$ and $k = -3 < 0$)

For $y = 3x$, we observe that as x is doubled from $x = 1$ to $x = 2$, y is doubled from $y = 3$ to $y = 6$. So as x increases, y increases proportionally.

As for $y = -3x$, we observe that as x is doubled from $x = 1$ to $x = 2$, y actually doubles from $y = -3$ to $y = -6$. But since $-6 < -3$, as x increases, *y decreases but not proportionally* because y is not halved as x is doubled. So this is not an inverse proportion. But is this a direct proportion? If we stick to the original idea of direct proportion, i.e. as x increases, y increases proportionally, then this is also not a direct proportion since y actually decreases. So for the original concept of direct proportion to work, k has to be positive, which is true in most cases in the real world because most quantities are positive: for $y = kx$, if the quantities x and y are positive, then k is also positive. (If $x = y = 0$, we cannot use these values to determine whether k is positive or negative; but if $k > 0$, the equation $y = kx$ still holds for $x = y = 0$.) This is the main reason why the quantities x and y are always non-negative when dealing with proportionality in Singapore secondary school textbooks because we need to preserve the original idea of proportionality.

If $y = -3x$ is neither a direct nor an inverse proportion (where $k > 0$), then what is it? I would say this is called a direct variation (where $k \neq 0$). In mathematics, we often have to extend ideas or concepts, but during the extension, the original idea (or part of it) may break down. For example, a^n, where n is a positive integer, means that the number a is multiplied by itself n times. But when mathematicians extended positive integer indices to include zero and negative integer indices, the original idea broke down: what does it mean for the number a to be multiplied by itself 0 times or -2 times? (One may be tempted to think that if a is multiplied by itself 0 times, then the answer should be 0, but on the contrary, $a^0 = 1$ if $a \neq 0$.) Similarly, when direct proportion ($k > 0$) is extended to direct variation ($k \neq 0$), one part of the original concept of direct proportion (namely, as x is multiplied by n, y is multiplied by n) still works, but the other part (namely, as x increases, y increases) has broken down.

The same could be said for inverse proportion and variation. Figure 4 shows the graphs of $y = \frac{6}{x}$ (i.e. $k = 6$) and $y = -\frac{6}{x}$ (i.e. $k = -6$) for $x > 0$. For $y = \frac{6}{x}$, we observe that as x is doubled from $x = 1$ to $x = 2$, y is halved from $y = 6$ to $y = 3$. So as x increases, y decreases proportionally. As for

$y = -\frac{6}{x}$, we observe that as x is doubled from $x = 1$ to $x = 2$, y is still halved from $y = -6$ to $y = -3$. But since $-3 > -6$, as x increases, y *increases but not proportionally* because y is not doubled as x is doubled. So this is not a direct proportion but an inverse variation where $k < 0$.

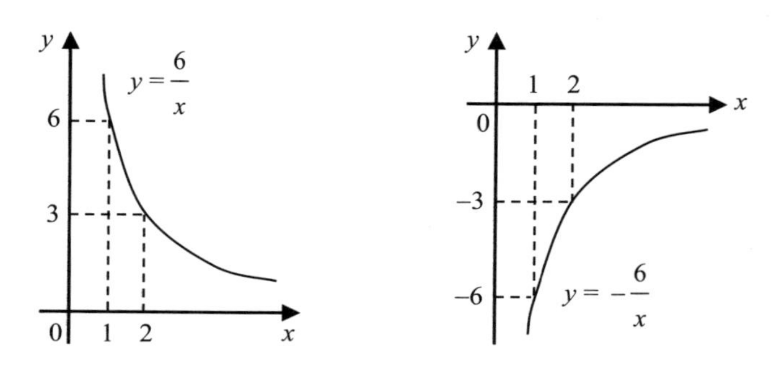

Figure 4. Inverse variation (for the cases where $k = 6 > 0$ and $k = -6 < 0$)

To summarise, variation ($k \neq 0$) is an extension of proportionality ($k > 0$). During extension, one part of the original idea of proportionality no longer remains valid while the other part is still applicable. One reason why it is necessary to extend proportionality to variation is because there are other types of variations that we need to describe. For example, y can vary directly or inversely with x^2 or $\sqrt{x}$, or z can vary directly with y^3 and inversely with $\ln x$. Finally, there are relations that are neither direct nor inverse variations, e.g. $y = mx + c$, where $c \neq 0$. If $c = 0$, the relation is a direct variation, i.e. the straight line must pass through the origin for direct proportion or variation. But if $c \neq 0$, this is a linear relation. This kind of relation is also quite common in real life, e.g. a phone plan may have a fixed monthly fee c and a variable component depending on how much talk time one has exceeded the monthly limit.

5 Implications for Teaching

After understanding the similarities, differences and connections among the concepts of ratio, rate, proportion and variation, teachers should be better equipped to teach these concepts. It does not mean that teachers have to explain to their students all the things discussed in this chapter, but they should be more prepared to answer students' queries (if any), or they should not tell students the wrong things (e.g. an exchange rate is a variable rate). In addition, I would like to point out a gap in Singapore secondary school mathematics due to the change of syllabus. In the 2001 syllabus, proportionality was taught in Secondary 1 while variation was taught in Secondary 3. But in the 2007 and 2013 syllabi, variation was combined with proportionality and taught in Secondary 2 as direct and inverse proportions. In other words, the 2007 and 2013 syllabi did away with the terminology of variation and its symbol $\propto$, but the equations $y = kx$ and $y = \frac{k}{x}$, where $k \neq 0$, still remain in the syllabus.

But one of the local secondary school textbooks only teaches the equations $y = kx$ and $y = \frac{k}{x}$ for direct and inverse proportions, without teaching the concept of equality of two ratios or proportional reasoning via the Unitary Method. Since students only learn in primary school how to use direct proportional reasoning via the Unitary Method to solve word problems involving direct proportion, these students would not have learnt how to use inverse proportional reasoning via the Unitary Method to solve word problems involving inverse proportion. For example, the following shows a word problem involving inverse proportion and its solution using the Unitary Method:

> Thirty-five workers build a house in 16 days. How many days will 28 workers working at the same rate take to build the same house?
>
> *Solution*
> 35 workers take 16 days to build a house.
> 1 worker will take $16 \times 35 = 560$ days to build the same house.
> 28 workers will take $\frac{560}{28} = 20$ days to build the same house.

In the above solution, when one quantity (in this case the number of workers) is divided by 35, the other quantity (in this case the number of days) is multiplied by 35; and when one quantity is multiplied by 28, the other quantity is divided by 28. This is the idea of inverse proportional reasoning. But if secondary school students do not learn inverse proportional reasoning, they would have to solve this word problem using the equation $y = \frac{k}{x}$ as shown below:

Let the number of workers be x and the number of days be y.

Since x and y are in inverse proportion, $y = \frac{k}{x}$, where k is a constant.

When $x = 35$, $y = 16$.

$$\therefore\ 16 = \frac{k}{35}$$

$$k = 560$$

When $x = 28$, $y = \frac{560}{28} = 20$.

The time taken by 28 workers to build the same house is 20 days.

Although there is nothing wrong in solving the word problem using this second method, students are not aware that there is another method to solve it: one that uses the simpler Unitary Method that is based upon the more fundamental concept of inverse proportional reasoning, which is commonly used in real-world situations. Therefore, teachers need to be aware of this gap and to address it if possible.

6 Conclusion

In this chapter, I have unpacked the meanings of ratio, rate, proportion (including proportionality and proportion reasoning) and variation, and showed how they are connected to one another. First, ratio and rate are two slightly different ways of comparing quantities: which one to use depends on which one is more meaningful or helpful in the given context. Second, the term ‘proportion’ has a few meanings in real life. In particular, one of its meanings (namely, a proportion is a part-whole ratio) has caused many students to think that proportion and ratio mean the same thing. However, proportion can also refer to direct and inverse proportions. This meaning of proportion is also called proportionality.

Third, the main idea of proportionality is the equality of two ratios ($\frac{y_2}{y_1} = \frac{x_2}{x_1}$ for direct proportion and $\frac{y_2}{y_1} = \frac{x_1}{x_2}$ for inverse proportion), and its interpretation (i.e. if one quantity x is multiplied by n, the other quantity y is multiplied or divided by n) is called proportional reasoning.

Fourth, a second perspective of proportionality is the equality of two rates ($\frac{y_2}{x_2} = \frac{y_1}{x_1}$ for direct proportion and $\frac{y_2}{x_1} = \frac{y_1}{x_2}$ for inverse proportion), which is also the same as viewing the relations in terms of the equations $\frac{y}{x} = k$ and $xy = k$, $k > 0$, for direct and inverse proportions respectively. Fifth, the latter equations have been extended in the concept of variation to $k \neq 0$. But during extension to include the case where $k < 0$, one part of the original idea of proportionality (namely, as one quantity x increases, the other quantity y increases or decreases proportionally) is no longer valid while the other part (namely, if one quantity x is multiplied by n, the other quantity y is multiplied or divided by n) is still applicable.

Lastly, I have pointed out that both direct and inverse proportional reasoning via the Unitary Method (rather than the equality of two ratios or two rates, or formulating the relation in terms of an equation) is used frequently in the real world. In particular, if a situation involves a constant rate between two quantities such that both quantities are zero at the same time, the relation between the two quantities is a direct proportion, and so we can use direct proportional reasoning via the Unitary Method to find an unknown amount of one quantity given the amount of the other quantity. Therefore, there is a need for students to understand the connections among the concepts of ratio, rate, proportion and variation so that they can apply them correctly not only in school mathematics but in the real world.

References

Ball, D. L., Thames, M. H., & Phelps, G. (2008). Content knowledge for teaching: What makes it special? *Journal of Teacher Education, 59*, 389-407.

Behr, M., Harel, G., Post, T., & Lesh, R. (1992). Rational number, ratio, and proportion. In D. Grouws (Ed.), *Handbook of research on mathematics teaching and learning* (pp. 296-333). New York: McMillan.

Cambridge Academic Content Dictionary (2018a). *Proportion.* Retrieved from https://dictionary.cambridge.org/dictionary/english/proportion

Cambridge Academic Content Dictionary (2018b). *Ratio.* Retrieved from https://dictionary.cambridge.org/dictionary/english/ratio

Cambridge Advanced Learner's Dictionary & Thesaurus (2018). *Proportion.* Retrieved from https://dictionary.cambridge.org/dictionary/english/proportion

Cambridge Business English Dictionary (2018). *Ratio.* Retrieved from https://dictionary.cambridge.org/dictionary/english/ratio

Cathcart, W. G., Pothier, Y. M., Vance, J. H., & Bezuk, N. S. (2006). *Learning mathematics in elementary and middle schools* (4th ed.). Upper Saddle River, NJ: Pearson.

Chan, C. M. E. (2017a). *Targeting mathematics 5A*. Singapore: Star Publishing.

Chan, C. M. E. (2017b). *Targeting mathematics 5B*. Singapore: Star Publishing.

Charles, R. I. (2005). Big ideas and understandings as the foundations for elementary and middle school mathematics. *Journal of Mathematics Education Leadership, 7*(3), 9-24.

Chow, W. K. (2013). *Discovering mathematics 1B* (2nd ed.). Singapore: Star Publishing.

Fielding-Wells, J., Dole, S., & Makar, K. (2014). Inquiry pedagogy to promote emerging proportional reasoning in primary students. *Mathematics Education Research Journal, 26*, 44-47.

Fong, H. K., Gan, K. S., & Ramakrishnan, C. (2017a). *My pals are here! Maths 5A* (3rd ed.). Singapore: Marshall Cavendish.

Fong, H. K., Gan, K. S., & Ramakrishnan, C. (2017b). *My pals are here! Maths 5B* (3rd ed.). Singapore: Marshall Cavendish.

Hart, K. (1981). *Children's understanding of mathematics 11-16*. London: John Murray.

Heinz, K., & Sterba-Boatwright, B. (2008). The when and why of using proportions. *Mathematics Teacher, 101*, 528-533.

Hoffer, A., & Hoffer S. (1988) Ratios and proportional thinking. In T. Post (Ed.), *Teaching Mathematics in Grades K-8* (pp 285-312). Boston, MA: Allyn & Bacon.

Lamon, S. J. (1999). *Teaching fractions and ratios for understanding: Essential content knowledge and instructional strategies for teachers.* Mahwah. NJ: Erlbaum.

Lanius, C. S., & Williams, S. E. (2003). Proportionality: A unifying theme for the middle grades. *Mathematics Teaching in the Middle School, 8*, 392–396.

Lee, N. H., Koay, P. L., Collars, C., Ong, B. L., & Tan, C. S. (2017a). *Shaping Maths Coursebook 5A* (3rd ed.). Singapore: Marshall Cavendish.

Lee, N. H., Koay, P. L., Collars, C., Ong, B. L., & Tan, C. S. (2017b). *Shaping Maths Coursebook 5B* (3rd ed.). Singapore: Marshall Cavendish.

Lesh, R., Post, T., & Behr, M. (1988). Proportional reasoning. In J. Hiebert & M. Behr (Eds.), *Number concepts and operations in the middle grades* (pp. 93-118). Hillsdale, NJ: Erlbaum.

Lo, J-J., & Watanabe, T. (1997). Developing ratio and proportion schemes: A story of a fifth grader. *Journal for Research in Mathematics Education, 28*, 216-236.

Math Snacks (2012). *Bad date.* Retrieved from https://www.youtube.com/watch?v=BZ1M01YBKhk

Merriam-Webster Dictionary (2018). *Step rate.* Retrieved from https://www.merriam-webster.com/dictionary/step%20rate

Ministry of Education. (2018). *Secondary mathematics syllabuses (draft).* Singapore: Curriculum Planning and Development Division.

Ojose, B. (2015). Proportional reasoning and related concepts: Analysis of gaps and understandings of middle grade students. *Universal Journal of Educational Research, 3*, 104-112.

Shulman, L. S. (1986). Those who understand: Knowledge growth in teaching. *Educational Researcher, 15*(2), 4-14.

Staples, M. E., & Truxaw, M. P. (2012). An initial framework for the language of higher-order thinking mathematics practices. *Mathematics Education Research Journal, 24*, 257-281.

Teh, K. S., & Looi, C. K. (2001). *New syllabus mathematics 3* (5th ed.). Singapore: Shinglee.

Tourniaire, F., & Pulos, S. (1985). Proportional reasoning: A review of the literature. *Educational Studies in Mathematics, 16*, 181-204.

van de Walle, J. A., Karp, K. S., & Bay-Williams, J. M. (2010). *Elementary & middle school mathematics* (7th ed.). Boston, MA: Pearson.

van Dooren, W., De Bock, D., Hessels, A., Janssens, D., & Verschaffel, L. (2005). Not everything is proportional: effects of age and problem type on propensities for overgeneralisation. *Cognition and Instruction, 23*, 57-86.

Wikipedia (2018a). *Proportionality (mathematics).* Retrieved from https://en.wikipedia.org/wiki/Proportionality_(mathematics)

Wikipedia (2018b). *Rate (mathematics).* Retrieved from https://en.wikipedia.org/wiki/Rate_(mathematics)

Wikipedia (2018c). *Ratio.* Retrieved from https://en.wikipedia.org/wiki/Ratio

Yeo, J. B. W., Teh, K. S., Loh, C. Y., Chow, I., Neo, C. M., & Liew, J. (2013). *New syllabus mathematics 1* (7th ed.). Singapore: Shinglee.

Yeo, J. B. W., Teh, K. S., Loh, C. Y., Chow, I., Neo, C. M., Liew, J., & Ong, C. H. (2014). *New syllabus mathematics 2* (7th ed.). Singapore: Shinglee.

Chapter 13

Teaching towards Big Ideas: Deepening Students' Understanding of Mathematics

CHUA Boon Liang

Teaching towards big ideas is one of the key shifts in the 2020 Singapore Secondary Mathematics Syllabus (Draft version). As the notion of Big Ideas in Mathematics is relatively fresh to most mathematics teachers in Singapore, this chapter serves to start a discussion about this topic of Big Ideas in Mathematics. It presents various definitions of big ideas found in the literature, then discusses why teaching towards big ideas is essential to students' learning of mathematics, and offers a glimpse into the big ideas identified in the 2020 Singapore Secondary Mathematics Syllabus (Draft version). The chapter ends with instructional practice and strategies for mathematics teachers to move forward to teach towards big ideas.

1 Introduction

Many students should be familiar with the formula for finding the circumference of a circle: *circumference* = $\pi \times$ *diameter*. But how well do they understand the constant π? Ask any primary or secondary school student to explain what π is. Very likely the student would respond by citing the value of π, which is approximately $\frac{22}{7}$ or 3.142, rather than pointing out that it is the ratio of the circumference of any circle to the diameter. Giving such a response reveals a lack of understanding about π by the student. However, seeing π as a value rather than a ratio is not uncommon because mathematics teachers tend to introduce the formula for the circumference of a circle just as a tool for computation, stopping short of explaining that π is the ratio of the circumference to the diameter. So students who learn the circumference formula in this way will treat it simply as a means to find either the circumference or the diameter of a circle when given the other, and will miss out on the key mathematical idea underpinning the formula: that is, the *direct proportionality* between the circumference of a circle and its diameter.

Other than underpinning the formula for calculating the circumference of a circle, the concept of *proportionality* also occurs in many different mathematical topics including, scale drawing, gradient of a slope, similarity of figures and trigonometry. This concept is so important in mathematics that many mathematics educators (e.g., Charles, 2005; Watson, Jones, & Pratt, 2013) have regarded it as a big idea in mathematics.

What makes a concept qualify as a big idea? Besides proportionality, what other big ideas in mathematics are there? How big are these big ideas? Why are these big ideas important? How can mathematics teachers move forward to teach towards the big ideas? These questions are worth pursuing and discussing.

The purpose of this chapter is to offer a personal perspective on the topic of Big Ideas in Mathematics. The next section, Section 2, reviews various definitions of big ideas found in the mathematics education literature, argues why they are crucial in helping students develop and deepen the understanding of mathematics, and outlines the big ideas in the 2020 Singapore Secondary Mathematics Syllabus (Draft version). In Section 3, the importance of a well-recognised big idea in mathematics, *Proportionality*, is described through manifestations of its prevalent presence in topics across the different mathematics content strands covered in the different year levels in the current Singapore Secondary Mathematics Syllabus. This section purports to raise teachers' awareness of the significance of proportionality in mathematics in order to appreciate its status as a big idea. Finally, Section 4 suggests some instructional practice and strategies for mathematics teachers to teach towards the big ideas.

2 Big Ideas in Mathematics

2.1 *What are Big Ideas in Mathematics?*

The notion of big ideas is not a new concept and has been advocated in curriculum design frameworks such as Understanding by Design (see, e.g., Wiggins & McTighe, 2012) and Teaching for Understanding (see, e.g., Blythe & Associates, 1998). Specifically to the learning of mathematics, *Big Ideas in Mathematics* (henceforth referred to as BIM for the singular form and BIMs for the plural form in this chapter) is a recurring notion that has not only been talked about (see, for example, Watson et al., 2013; Woodbury, 2000), but been promoted by mathematics educators in textbooks for quite a while (see, for example, Van de Walle, 2004; Boaler, 2017). One of the earliest mentions of BIMs appears in the Principles and Standards for School Mathematics (NCTM, 2000), in which it calls for mathematics teachers to "*understand the big ideas of mathematics and be able to represent mathematics as a coherent and connected enterprise*" (p. 17). Over the subsequent years, different interpretations of BIMs have emerged.

Charles (2005) defines a BIM as "*a statement of an idea that is central to the learning of mathematics, one that links numerous mathematical understandings into a coherent whole*" (p. 10). As the definition shows, BIMs aim to help students garner mathematical knowledge through connection and meaning-making in new situations. Take, for example, the notion of *equivalence* – considered a BIM by many educators (e.g., Charles, 2005; Watson et al., 2013) because of its manifestation in many mathematical ideas such as decimals, fractions and percentages, and in mathematical processes such as solving linear equations and simplifying expressions. Consider the following three numbers: 0.5, $\frac{1}{2}$ and 50%. All three assume different forms but still represent the same value. They are thus said to be equivalent to one another. In this illustration, not only should students take note of the three different ways of expressing a number, more crucially they should realise the commonality amongst those three cases – that is, the same value is expressed differently. In a similar vein, the three algebraic linear equations, $5x + 1 = 7 + 3x$, $2x + 1 = 7$ and $2x = 6$, are said to be equivalent because they have the same solution $x = 3$.

Unlike the earlier approaches of BIMs above, some educators looked at BIMs beyond the coherence and connection of mathematical ideas across different content strands. According to Schweiger (2006), BIMs are fundamental ideas that are "anchored in everyday activities" (p. 68), in addition to being recurring ideas across both the content strands and the year levels. Kuntze et al. (2011) characterised BIMs as ideas that should have a high relevance for building knowledge about mathematics as a science and support meaningful communication and mathematics-related arguments, further to having a high mathematics-related potential of encouraging learning of conceptual knowledge with understanding.

In Singapore, the Ministry of Education (MOE) places an explicit emphasis on teaching towards BIMs in the 2020 Secondary Mathematics Syllabus (Draft version) so as to develop in students a deeper and more robust understanding in mathematics and better appreciation of mathematics. BIMs are described as ideas that are central to the mathematics discipline, bring coherence and connect ideas from different strands and year levels (MOE, 2018). Eight BIMs are identified in the syllabus and are related to the following four themes that permeate and connect concepts from the different content strands: Properties and relationships (PR), Operations and algorithms (OA), Representations and communications (RC), and Abstractions and applications (AA). The eight BIMs are Functions, Invariance, Notations, Diagrams, Measures, Equivalence, Proportionality and Models, forming the acronym FIND ME PM. Table 1 shows how the eight BIMs are linked to the four themes. As the table shows, Measures and Models are placed under AA, Invariance and Proportionality under PR, and Notations and Diagrams under RC. Functions is related to both AA and PR while Equivalence to both PR and OA.

To sum up, BIMs permeate different content strands, with some integrated into more than topics. They also connect different mathematical topics at different year levels into a coherent whole, and contribute to building knowledge and developing a deep understanding of mathematics and its applications.

Table 1
Themes and BIMs in the 2020 Singapore Secondary Mathematics Syllabus (Draft version)

<table>
<tr><th>Themes</th><th colspan="2">Big Ideas in Mathematics</th></tr>
<tr><td rowspan="2">Abstractions and applications</td><td>Measures</td><td rowspan="3">Functions</td></tr>
<tr><td>Models</td></tr>
<tr><td rowspan="2">Properties and relationships</td><td>Invariance</td></tr>
<tr><td>Proportionality</td><td rowspan="2">Equivalence</td></tr>
<tr><td>Operations and algorithms</td><td></td></tr>
<tr><td>Representations and communications</td><td colspan="2">Diagrams
Notations</td></tr>
</table>

2.2 *Why are Big Ideas in Mathematics important?*

Mathematics should be meaningful, relevant and useful to the learners. Building the learners' mathematics content knowledge on BIM allows them to see mathematics as a coherent and connected whole rather than a set of disconnected concepts, skills and facts. The connections of mathematical ideas within and between topics then deepen their understanding of the mathematical ideas. When learners develop a deep understanding of mathematics, they not only grow in confidence and develop motivation in learning mathematics, but also appreciate the value and beauty of mathematics.

Apart from learners, BIMs have benefits for teachers too. A conscious and intentional effort on the part of the teachers in teaching towards BIMs can ground them firmly in mathematics content knowledge. Teachers who have a robust understanding of the mathematics that they are teaching will know how to connect the concepts and skills developed at each level to BIMs. Subsequently, the teachers can then facilitate learners' establishment of linkages between different mathematical ideas within and across content strands and levels. Furthermore, the teachers can pique the learners' interest, and motivate them to enjoy learning mathematics and to appreciate the relevance and usefulness of mathematics as well.

In the next section, one of the BIMs identified in the 2020 Singapore Secondary Mathematics Syllabus (Draft version) is presented alongside a discussion of its importance. Then how this BIM is featured across different content strands in the syllabus is described. The section closes with some suggestions for teaching towards this BIM.

3 A Big Idea in Mathematics: Proportionality

3.1 *What is proportionality and its applications?*

Proportionality is a theme that underpins many topics in the secondary mathematics curriculum, including scale drawing, similarity of figures, gradient of a linear function graph, trigonometry and probability. Its ubiquitous application is easily felt in our everyday lives, ranging from reading a map or the floor plan of a flat to adapting a cake recipe for twice the intended number of persons, and finding the price of fish, meat and vegetable that is charged by weight. In some parts of the world, the beauty of an object is defined by using proportionality. For instance, the Chinese use *san-ting-wu-yan* (三庭五眼) (San-ting-wu-yan, 2015) and the Westerners use the golden ratio (Bourne, 2018) to judge whether a person's face is beautiful or not. In the case of *san-ting-wu-yan*, a human face is considered beautiful if the length from the forehead hairline to the tip of the chin is divided evenly into three portions (三庭*san-ting*) with the middle portion from the brow to the tip of the nose. Additionally, the width of the face is divided evenly into five portions (五眼*wu-yan*) with the eyes and the nose covering the middle three portions. Other than these commonly seen real-life examples, some professionals such as visual artists and interior designers employ proportionality in their work as well. Figure 1(a) shows the meticulous scale drawing of an abstract painting with geometric patterns, and Figure 1(b) shows an enlargement of the mathematical calculations involving proportionality scribbled on the top right corner of the scale drawing. The precise calculations enabled the late Singapore Cultural Medallion artist Anthony Poon to create the huge painting with aesthetic beauty on canvas, as shown in Figure 1(c).

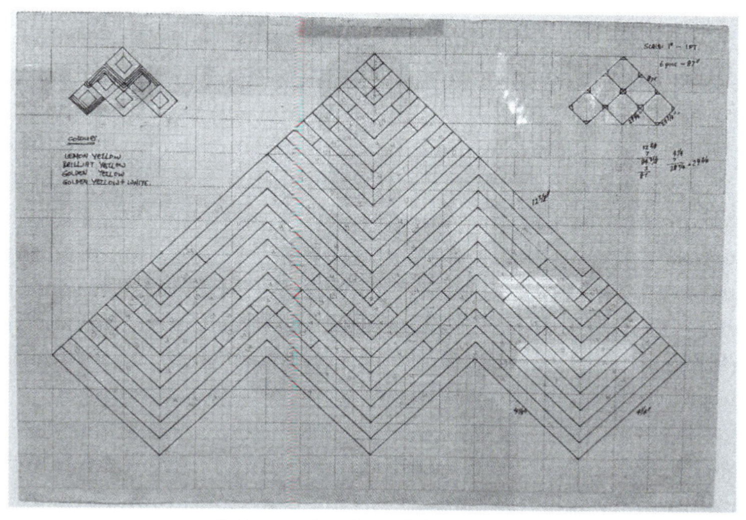

(a) Scale drawing of the artwork

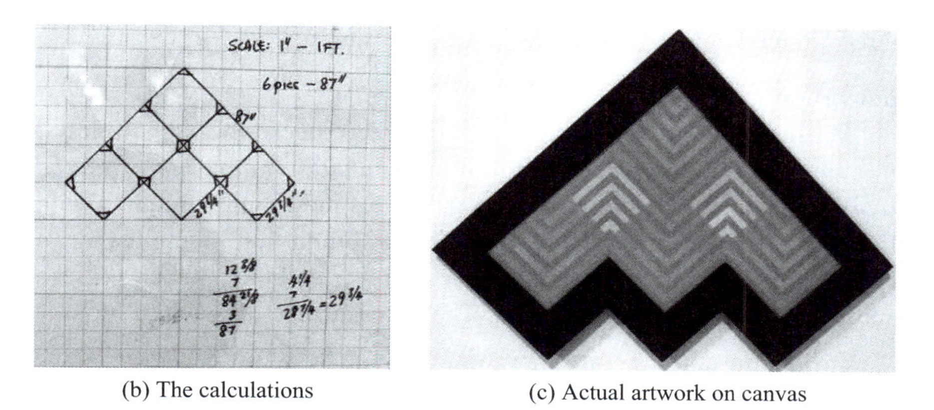

(b) The calculations

(c) Actual artwork on canvas

Figure 1. Artwork by the late Singaporean artist Anthony Poon

The prevalent use of proportionality in real life thus requires everyone to demonstrate a good knowledge of this notion and be able to apply it. But has one ever asked what proportionality really means? How is it different from proportion, which is probably a more familiar term to many people? If asked about the difference between the two terms by somebody, how might one respond? Though both terms appear to be similar, there is a subtle difference between both concepts.

Interestingly, the definition of proportionality as used in mathematics is hardly offered in dictionaries. Many give the meaning of proportion instead. According to the Cambridge dictionary, proportion is defined as the number or amount of a group or a part of something when compared to the whole, or when compared to another. Merriam-Webster dictionary describes proportion as the relation of one part to another or to the whole with respect to magnitude, quantity, or degree. By treating proportion as a comparison of part to part or part to whole, both dictionaries seem to be describing a ratio. However, Langrall and Swafford (2000) describe proportion as "*the statement that two ratios are equal in the sense that both convey the same relationship*" (p. 255). For example, the ratio of two red apples to three green apples is the same as that of ten red apples to fifteen green apples. In other words, the ratio 2 : 3 is said to be in the same proportion as the ratio 10 : 15. This is because both ratios hold the same multiplicative relationship that for every two red apples, there are three green apples. Hence, the statement that $\frac{2}{3} = \frac{10}{15}$ expresses a proportion.

What then is proportionality? Charles (2005) describes it as follows: *If two quantities vary proportionally, that relationship can be represented as a linear function* (p. 18). This definition is aligned with Langrall and Swafford's (2000) definition of proportion. Using the apple example above, denote the number of red apples and the number of green apples as p and q respectively. The two quantities, p and q, vary proportionally and are related by the equation: $\frac{p}{q} = \frac{2}{3}$ or $p = \frac{2}{3}q$. Expressed in the form $y = mx$ where m is a constant value, this equation represents a linear function. Hence, the equation expresses a proportionality. A noteworthy point emerging from Charles' definition is that the comparison of the two quantities when expressed as a fraction is always a constant. In a similar vein to Charles' definition, the Singapore MOE describes proportionality as a relationship between two quantities that allows one quantity to be computed from the other based on multiplicative reasoning (MOE, 2018). For example, find p when q is known and vice versa when the relationship that $p = \frac{2}{3}q$ is established. In this latter definition, the focus rests on the utility of the relationship, instead of the nature of the relationship. This way of explaining proportionality appears to be more student-friendly and easier for school students to understand. To sum up, proportion conveys the relationship of equality between two ratios whereas proportionality expresses the relationship connecting the two quantities that are in proportion.

3.2 *Proportionality in the 2020 Singapore secondary mathematics syllabus*

Students in Singapore learn about ratio in Primary 5 and encounter the idea of proportionality when they are asked to find one quantity given the other quantity and their ratio. In Primary 6, they learn that the circumference of a circle is proportional to the diameter of the circle, and the constant of proportionality is represented by π.

At the secondary level, proportionality is applied in topics across all three content strands. The students make use of proportion when they read scales on a Cartesian plane and find the gradient of a linear function under Number and Algebra. Then under Measurement and Geometry, they encounter proportionality in similarity of figures, scale drawing, arc length and area of a sector of a circle, and basic trigonometric ratios. Proportionality is also manifested in Statistics and Probability when the students learn statistical graphs such as pie chart, bar chart and histogram with equal class width, as well as in computing the probability of an event happening.

As the previous two paragraphs make clear, the notion of proportionality is a major idea in the learning of mathematics, with wide applications in several mathematics topics across different content strands. It is therefore not surprising to see proportionality being regarded as a big idea in mathematics education.

4 Teaching towards Proportionality

4.1 *Determining proportionality*

Secondary school students in Singapore are generally competent in solving word problems on proportion. Consider the 2013 GCE O Level question which stated that the frequency of a note produced by a string is proportional to the square root of the string's tension. The question also provided the information that the string produces a note with a frequency of 360 Hertz when the tension is 64 Newtons. Students were asked to find an equation connecting the frequency and tension first, then use it to compute the tension in the string when a note of 540 Hertz is produced. The examiners' report indicated that many students were able to correctly write the relationship between the variables and answer the computation question well (Cambridge International Examinations, 2014). A research study on proportional reasoning involving 14-year-old students in Singapore has reported similar findings as well (Ramakrishnan, 2016). Despite the students' success in procedural fluency, do they really have a good grasp of proportionality? In Ramakrishnan's study, the students were found to correctly describe two variables that were in an inverse proportion relationship in one question, but in a different question, not many were able to identify the graphical representation of two variables in an inverse proportion relationship. It can thus be suggested that many students, although adept in computing a quantity in a proportion question, still somewhat cannot reason and figure out the graphical representation of inverse proportion relationship. In a separate study on high school students' proportional reasoning by McLaughlin (2003), the students were found to be competent in solving most of the proportion problems but they did not understand what they were doing. It is therefore reasonable to require students to gain a good grounding in proportionality when it is introduced at the lower secondary level. Otherwise, they may face obstacles in understanding mathematics taught at a higher level (Langrall & Swafford, 2000).

If teachers were to teach towards a BIM, they must have a robust understanding of that BIM that they are dealing with in their teaching. This is so that they will be able to make, and also help students make, connections amongst various mathematical ideas across the different content strands. Hence, acquiring a good understanding of the BIM is crucial. To begin, teachers should examine critically and then highlight the key characteristics of the BIM to students. In the remaining section, a suggestion for developing a deep understanding of proportionality is offered for consideration.

To draw students' attention to the key characteristics of proportionality, teachers can present students with real-world situations and then get them to identify those that depict proportionality. Consider the following four examples in Figure 2 below, each depicting a situation that students have likely encountered in real life. Ask the students to determine whether or not the two quantities in each situation describe direct proportionality.

Example 1

The table below shows the number of units of electricity used (n) and the total cost of an electricity bill (C).

n in kWh	50	100	150	200	250	300
C in cents	16.975	25.95	34.925	43.90	52.875	61.85

Explain whether the total cost of an electricity bill is proportional to the number of units of electricity used.

(a) Electricity bill

Example 2

The cost of photocopying a page from a book is 6 cents. Ai Mei pays $1.80 for photocopying 30 pages. Bing Hong pays $3 for 50 pages. Cai Lin pays $4.50 for 75 pages.

Explain whether the cost of photocopying is proportional to the number of pages printed.

(b) Photocopying

Example 3

A supermarket sells three sizes of a particular brand of coffee powder.

The three diagrams below show the prices of the coffee powder.

\$2.90 per jar (50g) $5.50 per jar (100g) $9.90 per jar (200g)

Explain whether the price of coffee powder is proportional to the mass of coffee powder.

(c) Coffee Powder

Example 4

A shop sells both cupcakes and cookies. The price list for cupcakes is shown in the diagram below.

A cupcake costs \$3.20, and a box of three cupcakes costs \$9, a box of six costs \$18, and a box of 12 costs \$36.

Explain whether the cost of cupcakes is proportional to the number of cupcakes sold.

(d) Cupcakes

Figure 2. Four situations in real life

In the classroom discussion, what should teachers get students to take notice about direct proportionality from these four examples? The students should be able to detect from all the four examples that one quantity increases as the other quantity increases. However, does this characteristic sufficiently describe direct proportionality? This is where the teachers should engage the students in examining the examples to spot the differences between them. For instance, the total cost of the electricity bill in Example 1 increases uniformly by 8.975 cents with every increase of 50 kWh of electricity used. This indicates that the two quantities follow a linear rule. Similarly, the cost of photocopying in Example 2 can be shown to increase uniformly with the number of pages printed as well. Therefore, the two quantities are also connected by a linear rule. With the quantities assuming a linear relationship in the two examples, one may then think that these two examples agree with Charles' (2005) definition of proportionality, and display proportionality. However, only Example 2, and *not* Example 1, exhibit proportionality. Thus what distinguishes Example 2 from Example 1 as a case of proportionality? Crucially, the teachers must prompt students to consider the relationship *between* the two variables as well as *within* the variables.

In Example 2, the between-quantity relationship shows that the corresponding terms of the two quantities vary in such a way that they can be expressed as the same fraction. Paying \$1.80 to photocopy 30 pages as well as paying \$3 for 50 pages and \$4.50 for 75 pages all convey the same multiplicative relationship: 6 cents for every page. That is, 1.80/30 = 3/50 = 4.50/75 = 0.06. In other words, \$1.80 for 30 pages is in the same proportion as both \$3 for 50 pages and \$4.50 for 75 pages. By looking at the within-quantity relationship, the six terms in the three specific cases mentioned above can also be compared. If the number of pages printed increases 1.5 times from 50 to 75, so does the cost of photocopying: that is, 50 : 75 = 3 : 4.50. Both ratios can be simplified to 2/3 as a fraction, indicating they are in the same proportion as well.

× 1.5

Number of pages printed	50	75
Cost of photocopying in \$	3	4.50

× 1.5

As the above elaboration shows clearly, the between-quantity and within-quantity relationships in Example 2 each lead to the same value, a characteristic that is not observed in Example 1, and also Example 3 and Example 4. Hence, the three key characteristics of proportionality that the teachers must get the students to recognise are as follows:

(1) the constant obtained by expressing the ratios as a fraction in both the between-quantity and within-quantity relationship is the critical defining characteristic of direct proportionality,

(2) proportionality implies linearity, as shown in Charles' (2005) definition of proportionality, but the converse is not true, and

(3) the graph of the linear function representing the proportionality must pass through the origin.

Crucially, what the last two characteristics mean is that when two quantities vary proportionally, the between-quantity relationship can be represented as a linear function of the form $y = kx$, where x and y represent the two quantities and k is a constant. The graph of this function will pass through the origin. However, when the quantities are connected by a linear equation, the quantities may not necessarily vary proportionally. This is because the linear equation may assume the form of $y = kx + c$, where c is a non-zero real number. Such a graph does not pass through the origin. In a nutshell, proportionality requires the linear graph to pass through the origin whereas linearity does not.

4.2 *Connecting proportionality to gradient and trigonometry*

As Woodbury (2000) points out, teaching towards a big idea connotes "a sense of movement towards a larger conceptual understanding" (p. 230). So teachers must connect the big idea with other mathematical ideas to help students not only develop a deep understanding of the big idea and related mathematical concepts, but also make sense of what they learn in class better. This section presents two mathematical ideas that connect with the larger concept of proportionality.

In the current Singapore Secondary Mathematics syllabus for lower secondary levels, proportionality is manifested in the concept of gradient of a line. To help students better appreciate the gradient concept, linking gradient to proportionality is important. Teachers should introduce the definition of the gradient measure as a ratio of the vertical distance to the horizontal distance, and then get students to work out the gradient using this definition. The gradient of a line can be measured as follows: pick two points on the line, then determine the vertical distance and horizontal distance between the two points, and, finally, compute the ratio of the vertical distance to the horizontal distance. In this calculation process, a common question that the teachers get asked by students is: "how far apart must the two points be on the line?" To answer this question, this is where the teachers can engage the students in investigating and discussing whether the gradient of a line depends on the choice of the two points. Through careful selection of appropriate examples, the students should realise that the gradient of a line is independent of the choice of the two points. In other words, the distance apart between the two points does not matter when measuring the gradient of a line because, for the same line, the gradient remains constant, and so the ratio of the vertical distance to the horizontal distance is constant. This means that the vertical distance is always proportional to the horizontal distance regardless of the two points that are chosen. Unfortunately, some teachers tend to gloss over the definition of gradient, skip the method of calculating gradient using the definition, and eagerly introduce the following method of using the coordinates of two points on the line, (x_1, y_1) and (x_2, y_2), to compute the gradient:

$$\text{gradient of a line} = \frac{y_2 - y_1}{x_2 - x_1}$$

While the latter method allows students to compute the gradient easily, it focuses too much on substituting the coordinates into the formula to obtain a value for the gradient, and then in the process, draws the students' attention away from the proportionality characteristic that underpins the formula. As a result, the essence of the big idea of proportionality is lost when the students do not notice the link between gradient and proportionality. When this link is not established, students may not be able to recall and derive the gradient formula if they fail to memorise it.

Another manifestation of proportionality occurs in the topic of trigonometry. All secondary school mathematics teachers should be familiar with the definitions of the three basic trigonometric ratios: sine, cosine and tangent of an acute angle in a right-angled triangle. They should also be competent in demonstrating the direct application of these definitions to find the length of a side of the triangle and an acute angle. However, not many have paused to wonder why those three are called ratios and have realised that the three trigonometric ratios are actually independent of the size of the right-angled triangle. Thus they may not have highlighted this characteristic to their students. Crucially, through teaching towards BIMs, the teachers get to become more aware of the connections amongst mathematical ideas, which in this case, the link connecting proportionality and the three basic trigonometric ratios. Discussing such connections in class helps the students make sense of the various mathematical ideas and appreciate the importance of the BIMs.

4.2 *Connecting proportionality to other mathematical ideas*

As mentioned in Section 3.1, proportionality is one significant BIM with wide ranging applications across the various content strands. Due to space constraint in this chapter, it is not possible to describe the other mathematical ideas that are linked to proportionality.

To give an idea how "big" and important proportionality is, Table 2 lists the mathematical ideas in the current Singapore Secondary Mathematics syllabus where proportionality is applied. The identification of these mathematical ideas across the different content strands demonstrates how the big idea of proportionality is deconstructed into several single ideas to encourage mathematics teachers to draw attention to the proportionality characteristic in each of the instances during the discussion in class.

To further deepen the students' understanding and enhance their appreciation of proportionality, a good teaching activity is to generate a concept map together with the students to help them see and remember better the connections among the mathematical ideas. The concept map may look bare initially but it gets filled up progressively as the various mathematical ideas are introduced. Figure 3 shows an example of how a complete concept map on proportionality might look like.

Table 2

Linking proportionality and other mathematical ideas

Numbers and Algebra	Measurement and Geometry	Statistics and Probability
• Reading of scales on the Cartesian Plane • Gradient of a linear function	• Similarity of figures • Scale drawing • Enlargement • Arc length, area of sector • Trigonometric ratios	• Pie chart, bar chart, histogram with equal class width • Probability

Ratio → Proportionality ← Rate

Proportionality → Direct Proportionality

Proportionality → Inverse Proportionality

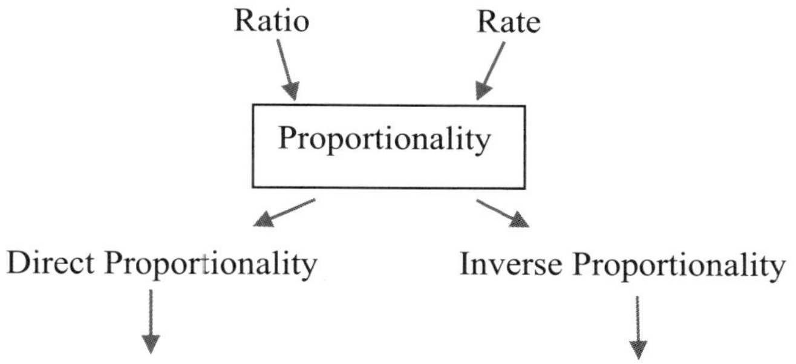

Direct Proportionality ↓

Numbers and Algebra	Measurement and Geometry	Statistics and Probability
• Reading of scales on the Cartesian Plane • Gradient of a linear function	• Similarity of figures • Scale drawing • Enlargement • Arc length, area of sector • Trigonometric ratios	• Pie chart, bar chart, histogram with equal class width • Probability

Inverse Proportionality ↓

Examples
• The length and breadth of a rectangle with fixed area • The speed and time taken when distance travelled is constant • The number of people completing a task and the time needed

Figure 3. Concept map on proportionality

As the concept map shows, proportionality is introduced after ratios and rates are learned because proportion is concerned about the equality of two ratios involving quantities of the same or different units of measure. The concept map also reveals the two types of proportion situations covered in the syllabus: direct proportionality or inverse proportionality. Direct proportionality is more widely applied in mathematics than inverse proportionality. For this reason, this chapter focuses only on direct proportionality.

5 Conclusion

The 2020 Singapore Secondary Mathematics Syllabus (Draft version) is getting mathematics teachers to teach towards BIMs. The purpose of this chapter is to introduce the notion of BIMs and then start mathematics teachers thinking about the BIMs to prepare them to face the new syllabus with some confidence in 2020. The teachers should find time to mull over and make sense of the definitions of a BIM as given in Section 2.1. Familiarise themselves with the eight BIMs identified in the 2020 Singapore Secondary Mathematics Syllabus (Draft version). To help recall all the eight of them, use the acronym, *FIND ME PM*, or refer to the mathematics syllabus if necessary. It would also be worthwhile to find out how the BIMs compare with those that have already been identified by schools that embrace the *Understanding by Design* framework for curriculum design. The discussion on proportionality in Section 3 serves to describe the importance of proportionality in the mathematics syllabus, and explain why it is big enough to be regarded as one of the BIMs by many mathematics educators. Finally, the purpose of Section 4 is to start a discussion about the way forward in teaching towards BIMs through the example of proportionality. This chapter suggests building a deep understanding of the BIM as well as the smaller mathematical ideas connected to the BIM. The teachers should be mindful about the essence of the BIM when developing the smaller ideas so that they can consciously find opportune junctures during the lesson to bring out the big idea that connects the smaller idea. When this happens, the students will be able to see and appreciate mathematics as a coherent and connected whole.

Acknowledgement

The author gratefully acknowledge Mr. Tan Kuo Cheang for his constructive comments given in the early stages of writing this chapter.

References

Blythe, T., & Associates. (1998). *The teaching for understanding guide.* San Francisco, CA: Jossey-Bass.

Boaler, J., Munson, J., & Williams, C. (2017). *Mindset mathematics: Visualising and investigating big ideas, Grade 4.* San Francisco: Jossey-Bass.

Bourne, M. (2018). *The math behind the beauty.* Retrieved from https://www.intmath.com/numbers/math-of-beauty.php

Cambridge International Examinations. (2014). *Report on the November 2013 Examination.* Cambridge: UK.

Charles, R. I. (2005). Big ideas and understandings as the foundation for elementary and middle school mathematics. *Journal of Mathematics Education Leadership*, *7*(3), 9-24.

Kuntze, S., Lerman, S., Murphy, B., Kurz-Milcke, E., Siller, H.-S., & Winbourne, P. (2011). Professional knowledge related to big ideas in mathematics. In M. Pytlak, T. Rowland, & E. Swoboda (Eds.), *Proceedings of CERME7* (pp. 2717-2726). Rzeszów, Poland: ERME.

Langrall, C. W., & Swafford, J. (2000). Three balloons for two dollars: Developing proportional reasoning. *Mathematics Teaching in the Middle School*, *6*(4), 254-261.

McLaughlin, S. (2003). *Effect of modeling instruction on development of proportional reasoning 1: An empirical study of high school freshman.* Retrieved from http://modeling.asu.edu/modelling-HS.html.

Ministry of Education, Singapore. (2018). *2020 secondary mathematics syllabuses (draft).* Singapore: Curriculum Planning and Development Division.

National Council of Teachers of Mathematics (NCTM). (2000). *Principles and standards for school mathematics*. Reston, VA: NCTM.

Ramajrishnan, R. (2016). *Mathematical reasoning of Secondary Two students in proportion.* Unpublished Masters dissertation. Singapore: National Institute of Education, Nanyang Technological University.

San-ting-wu-yan. (2015). Retrieved from http://www.baike.com/wiki/%E4%B8%89%E5%BA%AD%E4%BA%94%E7%9C%BC.

Schweiger, F. (2006). Fundamental ideas: A bridge between mathematics and mathematics education. In J. Maaβ & W. Schlöglmann (Eds.), N*ew mathematics education research and practice* (pp. 63-73). Rotterdam: Sense.

van de Walle, J. A. (2004). *Elementary and middle school mathematics: Teaching developmentally* (5th ed.). Boston: Pearson.

Watson, A., Jones, K., & Pratt, D. (2013). *Key ideas in teaching mathematics: Research-based guidance for ages 9 – 19*. Oxford: Oxford University Press.

Wiggins, G., & McTighe, J. (2012). *The understanding by design guide to advanced concepts in creating and reviewing units.* Alexandria, VA: ASCD.

Woodbury, S. (2000). Teaching toward the big ideas of algebra. *Mathematics Teaching in the Middle School*, *6*(4), 226-231.

Chapter 14

Teaching Pre-University Calculus with Big Ideas in Mind

TOH Tin Lam

Both informal calculus (taught at the secondary and pre-university level) and formal analysis (taught at the undergraduate level) serve their function in providing a student with a complete calculus education. This paper illustrates, from the perspective of Big ideas of mathematics and "Big idea of calculus education", how selected calculus topics can be taught at the pre-university level with a view to build up students' repertoire of concept images in preparing them for university calculus education. Some common mistakes in the teaching and learning of calculus are also discussed.

1 Introduction

Singapore is one of the several countries that teach calculus in the secondary and pre-university mathematics courses, probably due to its British roots. Among the many important reasons of including calculus in the school mathematics curriculum, calculus is a gateway to many disciplines in studies at the higher level. Not only that, it is a basis for modelling and problem solving in real-world applications (e.g. Tall, Smith & Piez, 2008). This paper discusses (1) the current state of student and teacher knowledge on calculus, (2) how calculus can be taught according to the true intent of introducing this in our curriculum, and most importantly, (3) a re-look at the big ideas in calculus and how the gap between (1) and (2) can be bridged.

2 Big Ideas in Calculus

As early as the beginning of this century, the National Council of Teachers of Mathematics (NCTM) began stressing the importance of *big ideas of mathematics*. According to the *Principles and Standards for School Mathematics* (NCTM, 2000), "[t]eachers need to understand the big ideas of mathematics and be able to represent mathematics as a coherent and connected enterprise (p. 17). Charles (2005) gave a formal definition of a Big idea:

> A Big idea is a statement of an idea that is central to the learning of mathematics, one that links numerous mathematical understandings into a coherent whole (p. 10).

In brief, one can easily appreciate that Big ideas is about connections of various topics and concepts across mathematics. Charles (2005) further stressed the importance of Big Ideas for teaching; a teacher's mathematical content knowledge and teaching practices should be grounded in Big Ideas.

What then is/are the Big idea(s) of calculus? Sawyer's (1961) seminal introductory calculus textbook describes through examples that calculus deals with the study of changes, and ways of representing and describing the changes. The changes are usually represented abstractly as variables (for example, time is a variable in describing the motion of a particle). In the process of doing thus, it is inevitable that one needs to understand the variables and their relation, and the big idea of a function – a rule that assigns to each member of a variable a unique member of another set (in our case, the set of all real numbers). Thus, two ideas that connect across many topics of mathematics easily surface – that of a variable and functions.

In teaching calculus with Big idea in mind, teachers will unlikely present the various calculus concepts "only as computational procedures" but more as a "potential source of conceptual meaning" (Jones, Lim & Chandler, 2016). However, anecdotal evidence from the Singapore classroom shows that calculus is usually taught at the secondary and pre-university level with a focus on schemata and formulae. Consequently, students might not have a sound understanding of the various calculus concepts that they are required to know at the level. In fact, this trend is not unique to Singapore; this phenomenon was also reported in the German mathematics classroom (Blum, 2000). From the perspective of Big idea, it is likely that teachers will approach the teaching of calculus entirely differently.

2 Calculus in the School Mathematics Curriculum

Calculus in the mathematics curriculum can be classified under two broad categories: (1) informal calculus; and (2) formal calculus (*real analysis*) (Tall, 1992). Informal calculus involves acquiring concepts related to functions and graphs, their differentiability and integrability, and applications in contextual problems. Formal calculus builds on informal calculus and develops a rigorous treatment of concepts involving limits and extended to continuity, differentiability and integrability. The main difference between informal and formal calculus lies with the rigors of treating the concepts – the rigors of the formal involves mathematical proofs, which are usually dispensed with in informal calculus. Secondary and pre-university calculus falls under the former category.

Here we are not judging the merit of either forms of calculus (and should not be!); in fact, the *"Big idea" of calculus teaching* should be built on the idea that both forms of calculus serve to complement one another. *Informal calculus* at the secondary and pre-university level is intended to facilitate students to appreciate the application of calculus in the real-world without the rigorous sophistry of calculus. At the *informal calculus* level, students should be exposed to a wide range of calculus concepts appropriate at the various phases of cognitive maturity appropriate to their level of cognitive development. In this way, various types and levels of "concept images" of calculus formed by the students are enriched at each stage in preparation for the next higher stage of calculus education, and eventually they are prepared for the rigorous *formal analysis* with a deeply connected set of concept images in calculus.

At the superficial level, there is an apparent inconsistency in the approach of both types of calculus. This obviously has been identified as one cause of confusion for calculus students as they move from pre-university to tertiary calculus (Toh, 2009a; 2009b).

2.1 *Concept image and calculus*

According to Tall and Vinner (1981), every person has their own mental representation, termed concept image, of concepts and processes they have experienced. For example, the concept image associated with the concept of differentiation involves finding the gradient of tangent at a particular point, the limit process of secant lines converging to tangent, the process of finding the derivative of a given function etc. As an individual goes through the various stages of education (e.g. from secondary to pre-university to undergraduate education), one builds up more concept images pertaining to a particular concept. All these concept images associated with a particular concept (which may be contradictory to one another) are stored in separate compartments. They generally do not mutually reinforce each other in understanding the concept. Confusion in learning occurs when these contradictory concept images are evoked simultaneously.

The term *concept definition*, first coined by Tall and Vinner (1981), refers to a form of words used by students in learning a particular mathematical concept. It could be the formal definition of the mathematical concept (i.e., that which is generally accepted by mathematicians and those in the mathematics community), or could be words reinvented by the learners to define the concept images in their mind, which is referred by Tall and Vinner as informal concept definition.

The usual assumption generally held among teachers is that (formal) concept definition, which controls the content of the concept image, is sufficient for students' acquisition of concepts (Vinner, 1994). However, Vinner and Dreyfus (1989) pointed out that concept image is usually not built on concept definitions, but is determined by numerous examples that students encounter and their related learning experience. Thus, the disparity between the teachers' expectation of the learning processes and students' actual learning could also be another underlying cause of students' learning difficulties in mathematics concepts in general (and calculus in particular).

3 Technology in Calculus Classroom

Information and Communication Technology (ICT) has been introduced into the Singapore classroom more than two decades ago. Anecdotal evidence from the Singapore classroom shows that little change has been made to the teaching of calculus since then. In the Topical Study Group (TSG) 16 of ICME11, Hohenwarter, Hohenwarter, Kreis and Lavicza (2008), quoting the study by Cuban, Kirpatrick and Peck (2001), lamented that despite the availability of free software, the process of incorporating technology in classrooms was "slow and complex". Anecdotal evidence in the Singapore classrooms also show that teachers rarely tap on ICT to enhance students' learning of calculus. Building on the purpose of informal calculus teaching, and the Big idea of mathematics and calculus that we know, we believe that more could be done in the mathematics classroom regarding the use of ICT. We shall discuss with two illustrations on the idea of differentiability and integrability.

3.1 *Differentiation and the concept of differentiability*

There are many conceptualizations of "slope" (Nagle, Moore-Russo, Vighetti, & Martin, 2013), including the calculus conception as a "limit, derivative, measure of instantaneous rate of change for any (even nonlinear) functions; tangent line to a curve at a point" (Nagle et al, 2013). Studies have also shown that both school students and teachers have very limited conceptual understanding of slope (Stump, 2001). This could at least be partially attributed to the lack of opportunity in the lower secondary level, perhaps due to time pressure, to enrich students with "concrete" ideas of slope, such as "rise and run" of stairs and steps (e.g. Smith, Seshaiyer, Peixoto, Suh, Bagshaw & Collins, 2013). The notion of slope as derivative already poses much difficulty to students at the secondary level. It is thus not surprising that most teachers at the pre-university level will think that the concept of differentiability, which is a much more abstract concept, has no place in the secondary and pre-university mathematics (which deals with informal calculus). The only sensible thing, to the teachers, is emphasize the differentiation formula or at most the derivation of the differentiation formulae by the first principle algebraically.

If we understand the objective of *informal calculus* within the entire calculus education for a student from secondary to university, the routine formulae and computational procedures should not be the main emphasis. Rather, informal calculus should build up students' concept images of as many of the related calculus concepts as possible, so as to prepare them with repertoire of concept images in preparation for formal analysis in the university education.

Tall (1985) introduced his approach of doing calculus in understanding differentiability as "local linearity" of the graph of the function. The traditional method of teaching differentiation involves the visualization of a series of secant lines moving "nearer and nearer" to the tangent, but this limiting concept poses much difficulty to student learning. Presenting the same knowledge using a visual way exploring the local linearity with the help of ICT (a graph plotter) to magnify into a particular point has its advantage.

Consider an example of a function which is everywhere differentiable, say $y = x^2$. With the use of graph plotter, it is easy to obtain the graph of the function over any domain. The magnifying function of the graph plotter allows one to zoom in at any point as much as one desires. For example, consider what happens when one zooms in at the point (1, 1) on the graph $y = x^2$.

It is interesting to observe that a magnification of the point (1, 1) shows that linearity of the graph occurs in a sufficiently small neighbourhood of the point. The "straight line" that appears on the right in Figure 1 gives the value of the gradient of the graph at that point (or the gradient of the "tangent" at that point).

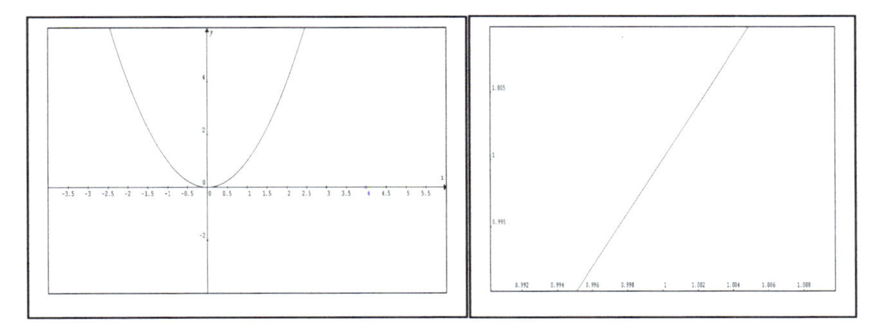

Figure 1. Plot of the graph of $y = x^2$ and a magnification at the point (1, 1)

In fact, this local linearity property will show at any point on this graph. One could easily link this local linearity property to the rigorous definition of differentiability of a function – the gradient of the "straight line" gives the precise meaning of the gradient of the graph at a particular point.

To further link to the Big idea of function – for which the chapter on calculus is about, and to capitalize on the use of Graphing Calculator (GC), we can use a tabular form to represent the values of the function at selected points. For example, in studying the behavior of the function around the point (1, 1), one obtains the following table (Figure 2) if one chooses a unit width increase of 0.1 unit around the point $x = 1$.

X	Y1	
.7	.49	
.8	.64	
.9	.81	
1	1	
1.1	1.21	
1.2	1.44	
1.3	1.69	

X=.7

Figure 2. Table of values of $y = x^2$ from 0.7 to 1.3 with a unit width of 0.1

As one takes a smaller neighbourhood around $x = 1$ and considers table of values, one would see that for every 0.0001 unit increase in the value of x, the value of y increases proportionally by 0.0002 unit, which is twice the increase in x (Figure 3). This is easily linked to the Big Idea on proportionality (Charles, 2005) – an attribute of a linear function that is represented by a table of values. This proportional increase in y values over proportional increase in x values is characteristic of a linear function.

X	Y1	
.9997	.9994	
.9998	.9996	
.9999	.9998	
1	1	
1.0001	1.0002	
1.0002	1.0004	
1.0003	1.0006	

X=.9997

Figure 3. Table of values of $y = x^2$ from .9997 to 1.0003 with a unit width of .0001

If one zooms in at any point on the graph of the function $y = x^2$, one will observe this local linearity property, both by the graphical or the tabular representation. Local linearity is a property of the function being differentiable at the point.

What happens if one were to zoom in on a point of the graph of a function which is non-differentiable there? Consider the graph of $y = |x|$ at the point (0, 0) and its magnification at that point (Figure 4).

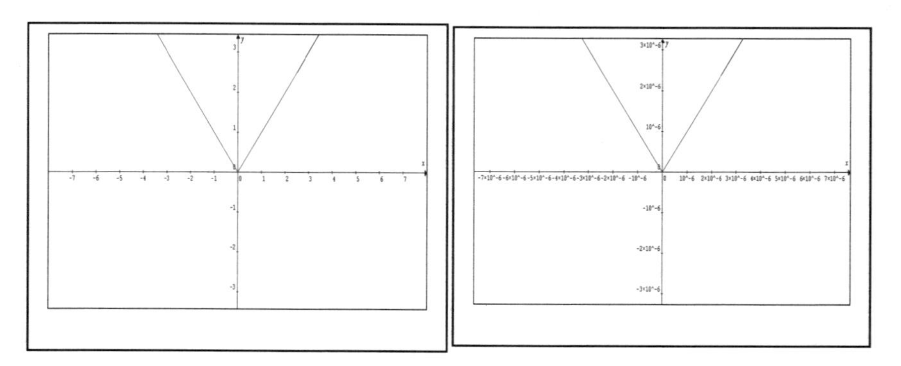

Figure 4. Plot of the graph of $y = |x|$ and a magnification at the point (0, 0)

Unlike the previous case, here any magnification at the point (0, 0) will not cause the linearity to occur. The "sharp point" at (0, 0) does not vanish even upon much magnification. This is a pictorial demonstration of a non-differentiable point (0, 0). An obvious advantage of the local linearity approach using ICT over the traditional dynamic approach of a secant line is that the former allows students to have an intuitive understanding of the notions of (1) gradient *at a point* of a given graph of a function, and (2) differentiability of the function at a particular point.

3.2 *Integration as antiderivative and area under the graph*

Integration has been a difficult concept for many students. Since the 1980s, there have been many studies on students' understanding of integration, and it was generally recognized that students have much difficulty in understanding integrals. This could have been attributed to, among many other reasons, that teachers could have taught integration and Riemann integral (as area under the graph) with an emphasis on the computational procedure, rather than a source of conceptual meaning in integral calculus (Jones, 2015).

In the Singapore context, most students are able to compute a definite integral using the Fundamental Theorem of Calculus without distinguishing between the notion of a Riemann integral and an antiderivative. The notion of Riemann sums is only introduced to prepare students for typical high-stake national examinations.

Integration is dealt with completely differently at the informal calculus and formal analysis in mathematics education. Integration in informal calculus begins with understanding integration as the inverse operator of differentiation. Definite integrals are then evaluated with the help of the Fundamental Theorem of Calculus (Kouropatov, 2008). Evaluating area under the graph of a function is treated as an "application of integration". On the other hand, *formal analysis* begins with Riemann integration, which is defined as the limit of a sequence of Riemann sums. Indefinite integral is then treated as antiderivative, and the two integrals are consorted through the Fundamental Theorem of Calculus.

Here we propose the use of technology (in our case, graph plotter) to enhance student learning of integral. Sealey (2014) proposed a framework for developing students' understanding of definite integral. She identified four categories (or layers) of understanding the definite integral concept: (1) product (intuitively, area of rectangle as the product of length and breadth); (2) summation (of rectangles gives an approximation to the area under graph); (3) limit (of the sums of rectangles, or Riemann sum); and (4) function (in that the definite integral can be seen as a function of its upper limit, which can be treated as a variable).

We argue here that a simple graph plotter is equipped with many sophisticated features which are instrumental in leading the students into a deeper appreciation of integration identified as the four layers identified by Sealey (2014). Consider Figure 5, which shows the graph of $y = x^2$ and which has an option for integration in the menu bar.

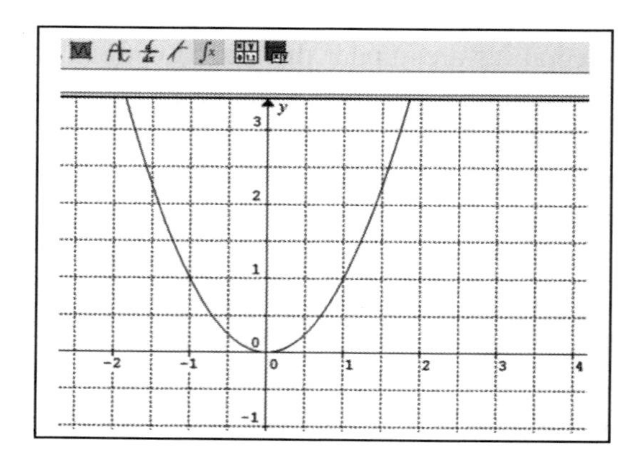

Figure 5. Screen of a Graph Plotter with integration option

It is not difficult to obtain from the graph the meaning of the definite integral by selecting the option (Figure 6). The pedagogical value, however, of this tool lies in the options that it provides (Figure 7).

As shown in Figure 7, the tool allows the user to manipulate the different ways of estimating the area under the graph, with different number of segments used to perform the estimation. By using small number of segments (e.g. $n = 1$), one is able to appreciate the product

$f(x)\Delta x$ in the Riemann sum. Choosing larger number of segments, one will appreciate graphically that $\sum f(x)\Delta x$ as the sum of rectangles which approximate the area under the graph. Choosing extremely large number of segments illustrates the limiting process of the sum of rectangles. The user is being allowed to alter the lower and upper limits of the integral, demonstrating that the integral is a function of the limits. This reinforces the Big idea of integral as a function – a value which varies with varying input. In other words, a simple graph plotter is able to address the four layers of understanding integration and link the user to the Big idea of function.

Here we highlight that knowing the Big idea of calculus in mind, a teacher would not teach integration by focusing almost exclusively on the purely computational approach of applying the Fundamental Theorem of Calculus. The teacher will attempt to achieve a more balanced view of integration as (1) anti-derivative; and (2) area under the graph; being fully aware that the informal calculus at the pre-university level helps to build a rich concept image among the students in preparation for higher level calculus.

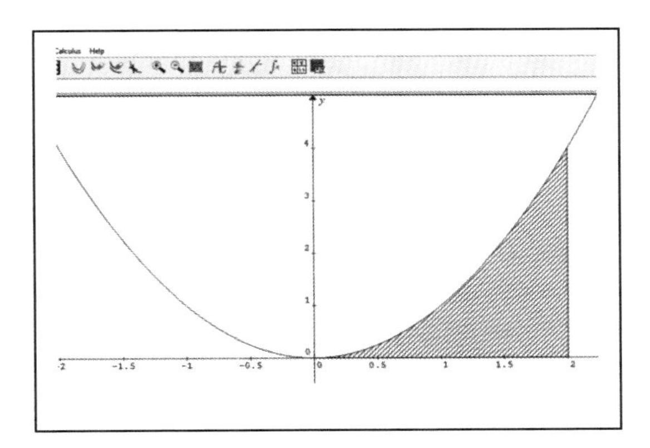

Figure 6. Screen of a Graph Plotter showing the area bounded between the graph, the line $x = 2$ and the x-axis.

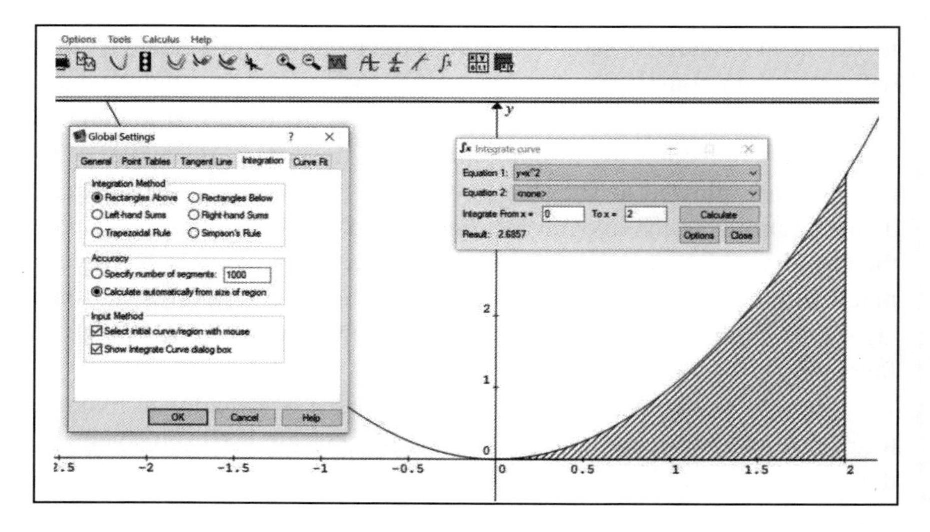

Figure 7. Screen of a Graph Plotter showing the options for integration

4 Concavity of a Graph in Relation to Stationary Points

There are relatively fewer studies on students' understanding of concavity (and points of inflection) compared to students' understanding of slopes (or first derivative). However, the importance of concavity cannot be denied since without it, students will have difficulty to "interpret whether a quantity would be increasing at an increasing rate, increasing at a decreasing rate..." (Carlson, Jacobs, Coe, Larsen & Hsu, 2002, cited in Jones, 2018). Baker, Cooley and Trigueros (2000) found that some students associated concave up and concave down with increasing and decreasing functions respectively. Other studies on concavity found students' misconceptions about points of inflection (Tsamir & Ovodenko, 2013). This section presents two teaching points related to the second derivative and concavity.

4.1 *Nature of stationary points*

In formal analysis in studying graphs of (usually differentiable) functions: A point (a, f(a)) on the graph of y = f(x), (where a is an interior point in the domain of f), is said to be a maximum (resp. minimum) point if $f(a) \geq f(x)$ (resp. $f(a) \leq f(x)$) for all x within a neighbourhood of a.

In the Singapore Additional mathematics syllabus (Ministry of Education, 2012), it was explicitly stated that students must be able to use the second derivative test to discriminate between maximum and minimum point, but the emphasis on the interpretation of the second derivative as the concavity was absent. Thus, teachers should note that at the secondary level, the second derivative test in the Additional mathematics curriculum turns to be a computational procedure without much relational understanding with the related calculus concept. This will not be able to enhance students in building up a richer repertoire of concept image in preparation for higher mathematics. Students entering pre-university might still not have sufficiently rich concept image for A-level mathematics curriculum, or could even have much misconceptions!

Here I urge for a return to an emphasis on the interpretation of the second derivative (as concavity) in connecting to the nature of stationary points. With ample examples and illustrations (especially with easy access to graph plotters and other ICTs), students would already have a picture of a maximum and minimum point on graphs of functions (see Figure 8). It is thus not difficult to pictorially associate the occurrence of a maximum point with a graph which concaves downwards; while a minimum point with one that concaves upwards, hence the sign of the second derivative.

This will enable students to make sense of the test of the nature of the stationary points with the second derivative test.

This turns to our next question: for a stationary point, if the second derivative vanishes, i.e., $f''(a) = 0$, the test is inconclusive. Must we abandon the second derivative test and look for alternative means (e.g. the first derivative test)?

For one with a rich repertoire of concept images related to stationary points, it is not difficult to explain why a minimum point, for example, can have its second derivative vanished (in fact, they can even extend the second derivative test!). Consider the graph of the function $y = x^8$ (Figure 9). It has a minimum point at (0, 0), where its second derivative vanishes.

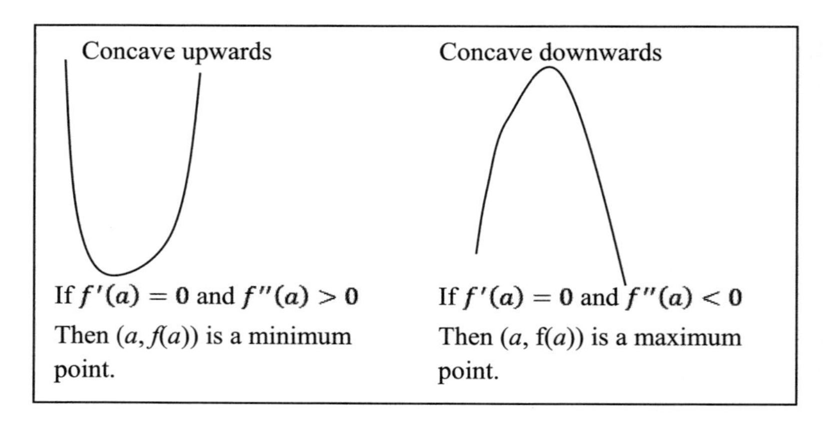

Figure 8. Second derivative tests on the nature of the stationary points

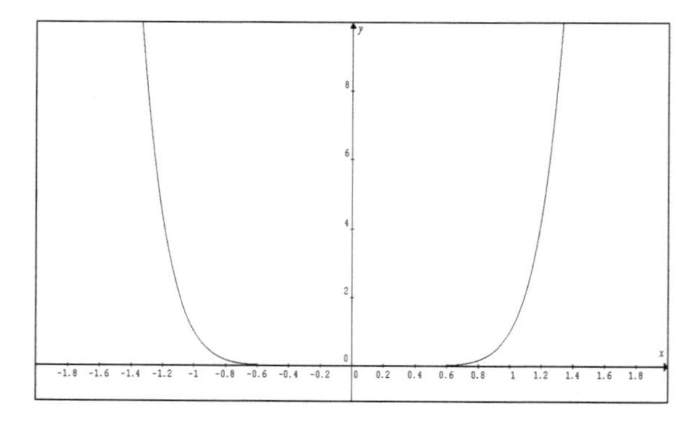

Figure 9. Graph of $y = x^8$ with a minimum point at (0,0)

It is obvious in this case that any point in the neighbourhood of $x = 0$, the graph shows the property of upward concavity. Even though at $x = 0$ its second derivative vanishes, it is still within a neighbourhood where the graph shows upward concavity. At $x = 0$, the rate of change of the gradient is "momentarily zero" (in the sense that a particle being momentarily at rest does not necessarily mean that it is at rest over an interval of time) but at all other points in the neighbourhood the rate of change of gradient is positive.

The same argument goes for maximum point argument. In fact, we can have a generalized second derivative test:

Generalized second derivative test:
If $f'(a) = 0$ and $f''(a) = 0$ but there is an interval $(a - k, a + k)$ such that $f''(x) > 0$ for all x in the interval $(a - k, a + k)$ other than a, then $(a, f(a))$ is a minimum point.

If $f'(a) = 0$ and $f''(a) = 0$ but there is an interval $(a - k, a + k)$ such that $f''(x) < 0$ for all x in the interval $(a - k, a + k)$ other than a, then $(a, f(a))$ is a maximum point.

4.2 *Inflection points*

Inflection points are most difficult to deal with in school mathematics. In the Singapore Additional Mathematics, inflection points are classified under the same category as the "stationary points" in the school mathematics (Ministry of Education (MOE), 2012, p. 46). The implication is that only stationary points of inflection are treated in the O-Level syllabus. Even in the A-Level H2 Mathematics curriculum, it was explicitly stated that "finding non-stationary points of inflection" (Singapore Examination and Assessment Branch, 2017, p. 7) is not required. Consequently, the concept image of an inflection point as one about which there is a change of concavity is not dealt with in both the secondary and A-Level syllabuses.

For a teacher who does not see the "big idea" of calculus teaching, there is a possibility that an emphasis on the concept image or the notion of a point of inflection could be replaced by efficiency procedural computation to find the point of inflection. For example, to find a point of inflection, equate $f'(x) = 0$ and $f''(x) = 0$ (note that there are other points which satisfy the two equations but are not stationary points!). This is reflected in the research by Tsamir and Ovodenko (2013), who found that many of his students have erroneous understanding about inflection points, e.g. the first derivative must be zero in order to be an inflection point.

When one emphasizes the geometrical interpretation of a point of inflection as one about which there is a change of concavity, to distinguish between stationary and non-stationary points of inflection becomes redundant. One can even have a point of inflection that is not differentiable. For details, see the chapter on applications of calculus in Toh (2008).

5 Incorrect Concept Image

This section presents two examples of incorrect concept images among pre-university students, which I witnessed during my encounter with them. These are instances in which teachers introduce calculus concepts with incorrect concept images. Somehow, the computational aspects in solving calculus problems by the students are not affected.

5.1 *Asymptotes*

Consider the process of finding the asymptote(s) of the hyperbolic functions represented by the equation $\frac{y^2}{b^2} - \frac{x^2}{a^2} = 1$, where a and b are positive. A quick way to obtain the asymptote is to ignore the number 1, that is, obtaining the equation $\frac{y^2}{b^2} - \frac{x^2}{a^2} = 0$, for which we simplify to $y = \pm\frac{b}{a}x$, which yields the correct asymptotes of the graphs. The rationale that was explained to me was that "when x and y are extremely large, a difference by 1 does not matter, hence we may ignore this difference." However, one should note that this gives rise to an incorrect concept image of asymptotes.

When this argument is applied to the problem in Toh and Toh (2016), it does not yield the correct answer:

Consider the equation $y = \frac{x^2+2x+2}{x+1}$. *Suppose we divide both the numerator and denominator by x, we obtain* $y = \frac{x+2+2/x}{1+1/x}$. *By letting* $x \to 0$, *we obtain* $y \to x+2$, *so that* $y = x + 2$ *is an asymptote.*

One can refer to Toh and Toh (2016, p.12) to read a full explanation of the mistake in the above argument. In explicating the notion of asymptotes, it is crucial that teachers highlight that the "closeness" of the functions and their asymptotes can be made *arbitrarily close* (that is, as close as you want it to be). Without demonstrating the two functions can be made arbitrarily close, the concept image of an asymptote is not being brought out in full.

Consider the graphs of the functions $y = x^2$ and $y = x^2 + 1$. If we examine the two graphs for small values of x, it is unlikely that one will conclude that the two functions are close to each other. On the other hand, however, if one were to examine the two graphs for a large range of values of x and y, the two graphs will appear very close to each other (Figure 10). However, there is a fixed distance of 1 unit between the two graphs $y = x^2$ and $y = x^2 + 1$ for all values of x. They cannot be made arbitrarily close, hence it is incorrect to say that one of them is an asymptote of the other function.

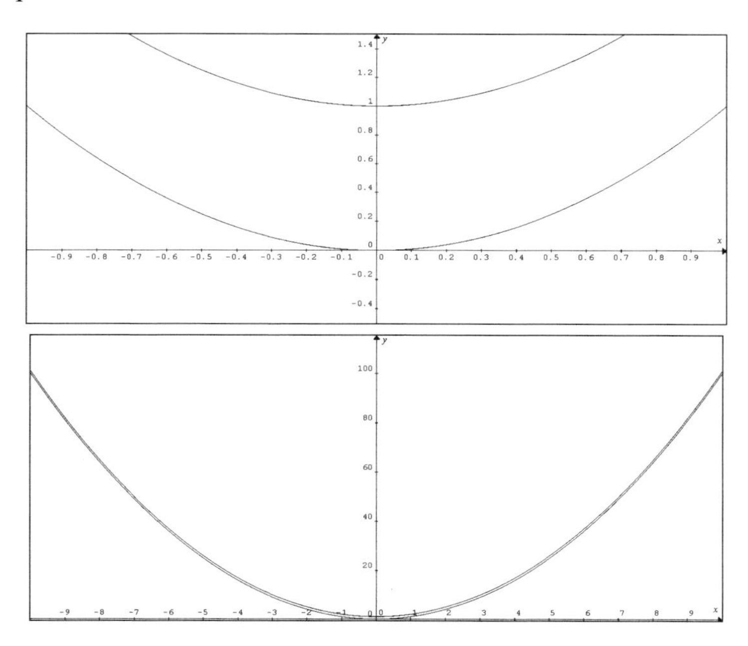

Figure 10. Graphs of the functions $y = x^2$ and $y = x^2 + 1$ in a close-up view and a larger view.

As most of the graphs that pre-university students encounter are simple rational functions, where both numerator and denominators are polynomials of degree at most two, it becomes easy to generalize incorrectly that the graphs of *any* function do not cut their asymptotes. Although the property that an asymptote does not cut the function for many simple graphs, it should not be taken as the requirement (either a property or the definition) of an asymptote. One can easily create an example where a function cuts its asymptote indefinitely many times (for example, Figure 11).

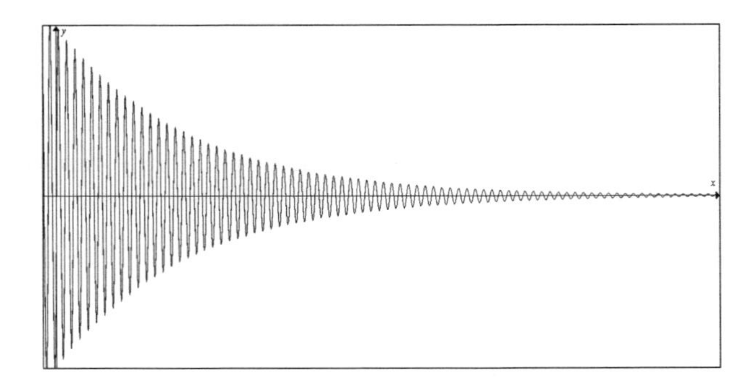

Figure 11. The graph of the functions $y = e^{-x}\sin x$ and its asymptote $y = 0$.

6 Conclusion

In this chapter, we have presented several concepts related to pre-university mathematics, which we believe are usually not well addressed in the school mathematics curriculum. This chapter offers, based on mathematics education literature and Singapore classroom anecdotal evidence, of how these calculus concepts can be understood from Big ideas of mathematics and also within the larger framework of a complete calculus education. I would like to reiterate the importance of the role of informal calculus at the pre-university level in a student's education in developing in students a rich repertoire of "concept images" associated with various calculus concepts so as to equip them with the prior knowledge and experience in pursuing formal analysis in the university.

References

Baker, B., Cooley, L., & Trigueros, M. (2000). A calculus graphing schema. *Journal for Research in Mathematics Education, 31*(5), 557-578.

Blum, W. (2000). Perspektiven fur den Analysisunterricht. *Der Mathematikunterricht, 46*(4/5), 5-17.

Carlson, M. P., Jacobs, S., Coe, E., Larsen, S., & Hsu, E. (2002). Applying covariational reasoning while modeling dynamic events: A framework and a study. *Journal for Research in Mathematics Education, 33*(5), 352-378.

Charles, R. I. (2005). Big ideas and understandings as the foundation for elementary and middle school mathematics. *Journal of Mathematics Education Leadership*, *7*(3), 9-24.

Cuban, L., Kirkpatrick, H., & Peck, C. (2001). High access and low use of technologies in high school classrooms: Explaining the apparent paradox. *American Educational Research Journal, 38*(4), 813-834.

Hohenwarter, M., Hohenwarter, J., Kresis, Y., & Lavicza, Z. (2008). Teaching and learning calculus with free dynamic mathematics software Geogebra. *TSG16: Research and development in the teaching and learning of calculus. ICME11, Monterrey, Mexico 2008.*

Jones, S. R. (2015). Areas, anti-derivatives, and adding up pieces: Definite integrals in pure mathematics and applied science contexts. *The Journal of Mathematical Behavior, 38*, 9-28.

Jones, S. R. (2018). Students' application of concavity and inflection points to real-world contexts. *International Journal of Science and Mathematics Education.* doi:10.1007/s10763-017-9876-5

Jones, S. R., Lim, Y. R., & Chandler, K. R. (2016). Teaching integration: How certain instructional moves may undermine the potential conceptual value of the Riemann Sum and the Riemann Integral. *International Journal of Science and Mathematics Education, 15*, 1075-1095. doi:10.1007/s10763-016-9731-0

Kouropatov, A. (2008). *Approaches to the integral concept: The case of high school calculus.* Paper for YESS-4, Tel Aviv University. Retrieved from http://www.yess4.ktu.edu.tr/YermePappers/Anatoli Kouropatov.pdf

Ministry of Education (2012). *Additional mathematics (O and N(A) levels) teaching and learning syllabus.* Singapore: Author.

Nagle, C., Moore-Russo, D., Viglietti, J., & Martin, K. (2013). Calculus students' and instructors' conceptualizations of slope: A comparison across academic levels. *International Journal of Science and Mathematics Education, 11*, 1491-1515.

National Council of Teachers of Mathematics (2000). *Curriculum and evaluation standards for school mathematics*. Reston (VA): NCTM.

Sawyer, W. W. (1964). *What is calculus about?* Washington: Mathematical Association of America.

Sealey, V. (2014). A framework for characterizing students' understanding of Riemann sums and definite integrals. *The Journal of Mathematical Behavior, 33*, 230-245.

Singapore Examinations and Assessment Branch (2017). Mathematics H2 (2017) (Syllabus 9740). Singapore: Author.

Smith, T. M., Seshaiyer, P., Peixoto, N., Suh, J. M., Bagshaw, G., & Collins, L. K. (2013). Exploring slope with stairs and steps. *Mathematics Teaching in the Middle School, 18*(6), 370-377.

Stump, S. (2001a). Developing preservice teachers' pedagogical content knowledge of slope. *The Journal of Mathematical Behavior, 20*, 207-227.

Tall, D. (1985). The gradient of a graph, *Mathematics Teaching 111*, 48-52.

Tall, D. (1992). Students' difficulties in calculus. In K-D. Graf, N. Mahara, N. Zehavi, & J. Ziegenbalg (Eds.), *Proceedings of Working Group 3 at ICME-7, Quebec 1992* (pp. 13 – 28). Berlin: Freie Universitat Berlin.

Tall D., Smith, D., & Piez, C. (2008). Technology and Calculus. In M. K. Heid & G. M Blume (Eds), *Research on Technology and the Teaching and Learning of Mathematics, Volume I* (pp. 207-258). Charlotte: Information Age.

Tall, D., & Vinner, S. (1981). Concept image and concept definition in mathematics with particular reference to limits and continuity. *Educational Studies in Mathematics, 12*, 151-169.

Toh, T. L. (2008). *Calculus for Secondary School Teachers* (2nd Ed.). Singapore: McGraw-Hill (Asia).

Toh, T. L. (2009a). On in-service mathematics teachers' content knowledge of calculus and related concepts. *The Mathematics Educator, 12*(1), 69-86.

Toh, T. L. (2009b). Teaching of Calculus in Singapore Secondary Schools. *Mathematical Medley, 35*(1), 16-23.

Toh, T. L., & Toh, P. C. (Eds.). (2016). Ask Dr Math. *MathsBuzz, 16*(1).

Tsamir, P., & Ovodenko, R. (2013). University students' grasp of inflection points. *Educational Studies in Mathematics, 88*(3), 409-427.

Vinner, S. (1994). Research in teaching and learning mathematics at an advanced level. In D. Tall (Ed.), *Advanced mathematical thinking* (2nd ed,). Dordrecht: Kluwer.

Vinner, S., & Dreyfus, T. (1989). Images and definitions for the concept of function. *Journal for Research in Mathematics Education, 20*(4), 356-366.

Chapter 15

Making Big Ideas Explicit Using the Teaching through Problem Solving Approach

Pearlyn LIM Li Gek

The term "Big ideas" in mathematics education refers to ideas that help to link related concepts from different strands and levels. This paper presents lesson ideas focusing on two big ideas on measure and variance, using the teaching through problem solving approach. In this approach, students are given a problem to work productively on, before they present their solutions to the problem. Teachers act as facilitators and they are reminded to focus on the big ideas of mathematics and to make these ideas explicit during discussion. Lesson ideas shared are from the strand of measurement and are suitable for the primary classroom.

1 Introduction

Mathematics involves concepts and systems that are interrelated and it is expedient for a mathematics curriculum to ensure that the links between mathematical ideas as well as the links between mathematics and other disciplines are made clear to learners (Australian Curriculum Assessment and Reporting Authority, 2013). The term "big ideas" has been used to refer to central mathematical ideas that link ideas from different strands and levels (MOE, 2018). Charles (2005) defines a big idea as

> a statement of an idea that is central to the learning of mathematics, one that links numerous mathematical understandings into a coherent whole (p. 10).

Big ideas can be used to organize broad areas of knowledge in a domain and determine strategies for solving a wide range of problems (Niemi, Vallone & Vendlinski, 2006).

The benefits of understanding big ideas are many, including improved problem solving skills, greater ease in mastering new facts and procedures, more flexible knowledge use and better transfer (Niemi et al., 2006). Niemi et al. (2006) reported a study by Chi, Feltovich and Glaser (1981), which found that experts sorted problems printed on index cards based on central concepts and principles whereas novices performed the sorting based on physical features of the problems. The organization of knowledge around big ideas demonstrated by experts enabled them to solve problems with linked concepts more effectively than novices, whose focus on surface features might hinder them from coming up with successful solution strategies. Another benefit of teaching big ideas is that it lessens the cognitive load by reducing the amount of information that needs to be remembered (Lambdin, 2003).

Different big ideas in mathematics have surfaced in literature. For instance, Charles (2005) identified 21 big ideas, including big ideas about numbers, data collection and representation and patterns. Charles (2005) opined that it was not necessary for mathematicians and mathematics educators to agree on a set of big ideas, but the identified big ideas should be broad enough to encapsulate a number of related ideas, yet be useful to practitioners like teachers and test developers. Askew (2013) examined Charles' (2005) 21 big ideas and suggested five that could possibly improve learning if they were paid attention to. He particularly stressed equivalence as well as meanings and symbols for their mathematical or pedagogical importance. Siemon, Bleckly and Neal (2012) found that Number was the most difficult aspect in school mathematics to teach and learn and discussed how the Australian mathematics curriculum emphasized big ideas in Number at various school stages. In Singapore, the Ministry of Education has come up with a list of eight big ideas for its secondary school curriculum, namely equivalence, proportionality, measures, functions, notations, invariance, diagrams and functions (MOE, 2018).

For this paper, the focus is on the big idea about measures. In MOE (2018), the big idea about measures refers to using numbers as measures to quantify a property of real-world or mathematical objects, so that these properties can be analysed, compared and ordered. In Charles (2005), the term "big idea on measurement" was given a similar description. Additionally, Charles (2005) provided examples of concepts of measurement that could be linked into a coherent whole. For instance, measurement involves a selected attribute of an object (e.g. length, area, mass, volume). The following concepts of measurement are also included in Charles (2005): (1) The object being measured can be compared against a unit of the same attribute. (2) The longer the unit of measure, the fewer units it takes to measure the object and vice versa. (3) The appropriate measurement unit can be determined by the magnitude of the attribute to be measured and the accuracy needed.

The Ontario Ministry of Education (2008) used the term "big ideas of measurement". The document listed attributes, units, measurement sense and measurement relationship as big ideas of measurement. It also suggested how teachers can help students gain understanding of measurement. The suggestions include investigating measurement problems in real-life settings, extending their knowledge of measurement units and their relationships and investigating the relationships between and develop formulas for area and perimeter.

This paper will also allude to the big idea about invariance. The big idea about invariance refers to the property of a mathematical object remaining unchanged when the object undergoes some form of transformation (MOE, 2018).

In particular, two lesson ideas that explicate the big ideas of measures and invariance will be discussed. The two lessons are based on the teaching through problem solving (TTPS) approach. TTPS involves teaching mathematics concepts through problem solving contexts and it allows students to develop higher-level thinking during mathematical problem solving (Bostic, Pape & Jacobbe, 2016). In TTPS, students are assigned a problem to solve at the beginning of the lesson. The problem is usually open-ended, allowing for more than one solution method and could possibly engage students in reasoning. Students then work on the problem for an allocated amount of time while the teacher monitors the problem solving process and notes the students who have come up with different solutions. A whole-class discussion is conducted after students have sufficient time to work on the problem. Students present their varied solutions and compare the ideas, noting the similarities and differences in each solution. From the discussion, students learn new mathematical concepts or procedures (McDougal & Takahashi, 2014). Takahashi (2006) illustrated the approach using Figure 1.

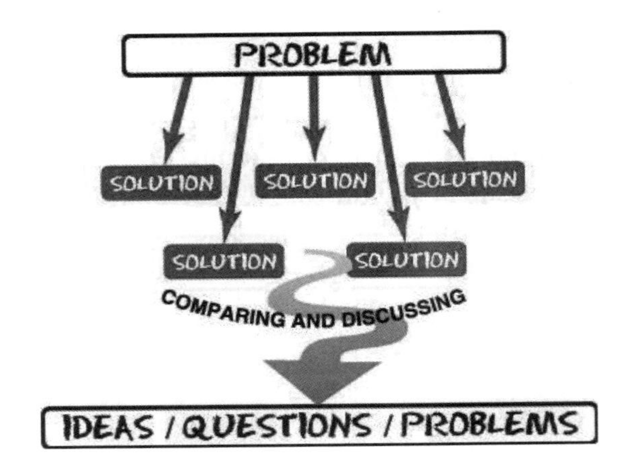

Figure 1. Teaching through problem solving (Takahashi, 2006)

As can be seen from the description of TTPS, there is greater involvement of students in TTPS than in didactic teaching approaches. Students are involved in solving a problem before the concept or skill identified for that lesson is taught. They are also responsible for presenting their solutions and discussing the multiple solutions offered. The teacher facilitates the acquisition of knowledge through directing the discussion, but is not the disseminator of information. This implies that teachers need to ask effective questions and to do this, they need to have a strong understanding of how mathematical contents are connected through big ideas.

Additionally, Takahashi (2017) categorised teaching into three levels. According to Takahashi (2017), level 1 teaching is when teachers teach by telling students basic mathematical ideas like concepts and procedures while level 2 is when teachers teach by explaining the why behind the mathematical ideas. Level 1 and level 2 seem to resemble Skemp's (1976) ideas of instrumental and relational understanding respectively. Level 1 and level 2 teaching can be easily executed by robots. In level 3 teaching however, teachers provide students with the opportunity to understand the mathematical ideas and aim to nurture independent learners instead of recipients of knowledge (Takahashi, 2017). They do this by effective questioning and selecting examples for rich discussion. TTPS is a teaching approach that meets the goals of level 3 teaching.

TTPS may be effective in addressing big ideas as students make sense of new ideas by linking them coherently to existing knowledge in TTPS (Lambdin, 2003). Also, the use of open-ended problems in TTPS allows students to produce multiple solutions or solution methods. Through the discussion and comparison of different solution methods, students may be able to connect different mathematical ideas used by various students. The comparison of methods also helps students to identify correct solutions as well as those which are more efficient with greater effectiveness (McDougal & Takahashi, 2014).

In the next two sections, lesson ideas based on TTPS focussing on the big ideas of measurement and invariance will be discussed. The lesson idea in Section 2 relates to the concept of area while the lesson idea in Section 3 relates to the area of a triangle.

2 Concept of Area

In Singapore, the concept of area is introduced to students at Primary 3. Current Primary 3 textbooks provide opportunities for students to have hands-on experiences with the concept of area, like covering a desk with books and finding out how many books can fill the desk (Fong, Ramakrishnan & Choo, 2015). Collars, Koay, Lee and Tan (2015) also include an activity where students can discover the formula for area of a rectangle or square. Nonetheless, the ideas cited are still very much directed. As opposed to directed tasks found in current textbooks, TTPS could be used. The lesson could begin with students performing the following problem solving task that has been adapted from Iitaka, Fujii and Takahashi (2012). Figure 2 shows the game that precedes the problem solving task.

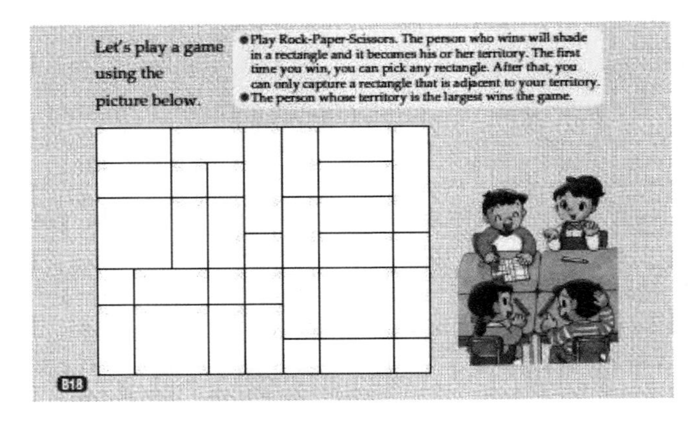

Figure 2. Game used before problem solving task (Iitaka et al., 2012)

In Figure 2, students are given the instruction to play rock-paper-scissors. The winner shades in a rectangle and it becomes his or her territory. On the first win, the player can choose any rectangle. Subsequently, the player can only capture a rectangle that is adjacent to his territory. The game continues until all the rectangles are filled. At the end of the game, students are asked to compare the sizes of their territories. Alternatively, students can be given a hypothetical outcome (Figure 3) and asked to compare the size of the different territories.

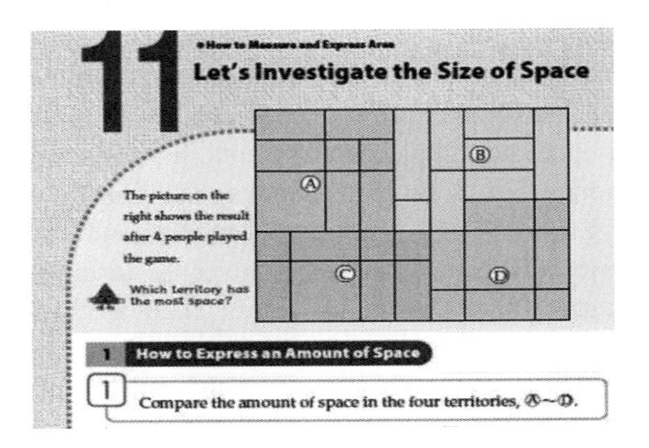

Figure 3. Problem solving task from Iitaka et al. (2012)

It is recommended that this problem solving task be assigned before students are taught the concept of area in school. However, students should have learnt comparison and be able to use words such as "greater than" and "smaller than". Teachers can ensure that students understand the meaning of the word "territory" before students start on the task.

The problem in the task can be solved in a number of ways. First, students can divide the territories into unit squares. They can then count the number of unit squares to determine which territory is larger. Alternatively, students can match the pieces which are the same, then compare the sizes of the remaining pieces. A similar method would be to cut out each territory and stack the territories to be compared. The overlapping parts can be ignored, and only the sizes of the parts which are not overlapping are compared. Students may use erroneous methods to compare, like using a ruler to measure the length of the territories. When this happens, teachers can ask students questions about the appropriateness of the method.

From this task, it is hoped that students will develop understanding of the big idea about measures with regard to the attribute of area. For example, students should be able to recognise that appropriate units, like unit squares, need to be used to measure area and length measures like the ruler are not appropriate for area measurement (Nitaback & Lehrer, 1996). Furthermore, units used to measure area should be similar (Nitaback & Lehrer, 1996), thus the most efficient method of comparing the sizes of the territories is to divide the territories into unit squares. It is also hoped that students can recognise the principle of unit iteration (Nitaback & Lehrer, 1996), where the units of measurement need to be placed side by side without gaps or overlaps. Finally, measurement is characterised by additivity (Nitaback & Lehrer, 1996). Thus, after dividing the territories into unit squares, the area is found by adding the number of unit squares.

Later in the unit, students can be led to discover that the area of a square or rectangle can be found by multiplying its length and its breadth, thus connecting area to multiplication. This can be accomplished by having students find the area of a square or rectangle by counting the number of unit squares that make up the shape, then exploring the relationship between the breadth and length of the shape and its area.

As explained above, the task is capable of eliciting the big idea about measures, particularly the measure of area, from students without the teacher explicitly imparting the content to them. The big idea includes the use of appropriate units for area and unit iteration. Teachers are important as they need to facilitate the discussion with effective questioning and steer students towards a correct understanding of the concepts to be learnt.

3 Area of a Triangle

The next example is on area of a triangle. From examining textbooks used in Singapore (Fong, Gan & Ramakrishnan, 2017; Lee, Koay, Collars, Ong & Tan, 2017), the topic "Area of a Triangle" in Primary 5 usually begins with an introduction to the terms base and height in relation to a triangle. There is an emphasis that the base of a triangle has to be perpendicular to its height, after which students will do exercises whereby they need to identify the height when given the base of a triangle and vice versa. In addition, Lee et al. (2017) requires students to draw the height of a triangle when the base is given. Following that, students derive the formula to find the area of a triangle. Diagrams showing how the parts of a triangle can be repositioned to fit half of its related rectangle are presented in the textbooks (Figure 4a and 4b).

The textbooks support level 2 teaching and Figures 4a and 4b show students clearly that the area of a triangle is half that of its related rectangle and consequently that the area of a triangle can be calculated by the formula ½ x base x height. However, in order to allow students to explore the relationship between the area of a triangle and the area of a rectangle on their own, teachers can teach through problem solving. A problem solving task requiring students to draw triangles with areas that are half that of a given rectangle could be given even before teachers begin teaching the topic of area of triangle. An example of such a task is shown in Figure 5.

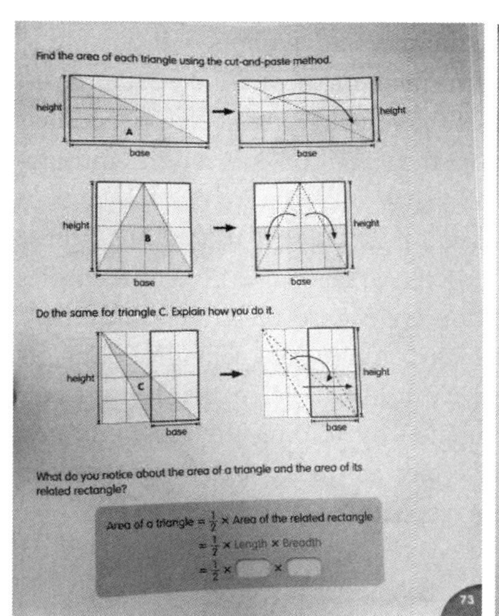

Find the area of each triangle using the cut-and-paste method.

Do the same for triangle C. Explain how you do it.

What do you notice about the area of a triangle and the area of its related rectangle?

Area of a triangle $= \frac{1}{2} \times$ Area of the related rectangle

$= \frac{1}{2} \times$ Length $\times$ Breadth

$= \frac{1}{2} \times \square \times \square$

73

Figure 4a. Deriving area of triangle (Lee et al., 2017)

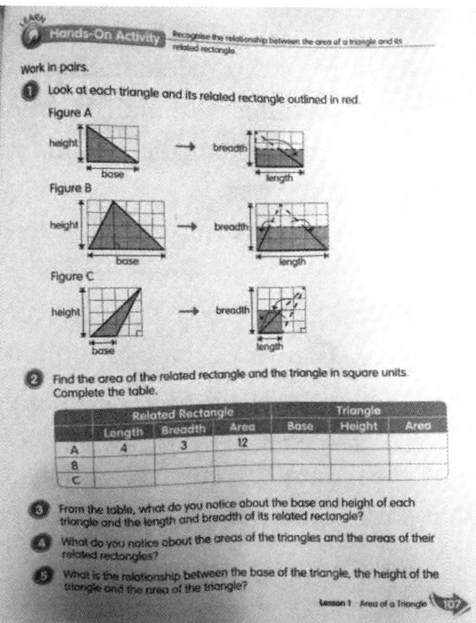

Hands-On Activity — Recognise the relationship between the area of a triangle and its related rectangle.

Work in pairs.

1. Look at each triangle and its related rectangle outlined in red.

Figure A

Figure B

Figure C

2. Find the area of the related rectangle and the triangle in square units. Complete the table.

	Related Rectangle			Triangle		
	Length	Breadth	Area	Base	Height	Area
A	4	3	12			
B						
C						

3. From the table, what do you notice about the base and height of each triangle and the length and breadth of its related rectangle?
4. What do you notice about the areas of the triangles and the areas of their related rectangles?
5. What is the relationship between the base of the triangle, the height of the triangle and the area of the triangle?

Lesson 1 Area of a Triangle 107

Figure 4b. Deriving area of triangle (Fong et al., 2017)

Figure 5. Example of a problem solving task

Through performing this task, it is hoped that students will be able to use their prior knowledge on fractions and on area of a rectangle to construct triangles with half the area of the rectangle. It is also hoped that students can derive the formula for the area of a triangle through performing the task.

Most students may be able to draw a right-angled triangle and three acute triangles with areas that are half the area of the given rectangles since they have learnt fractions previously. Figure 6a to 6d show these triangles. Figure 6a can be obtained by folding the rectangle along its diagonal. Figure 6b can be obtained by folding the rectangle into half vertically, while figure 6d can be obtained by folding the rectangle into half horizontally. To draw figure 6c is not as direct, but it can be seen that part A and part B are half of their respective related rectangles.

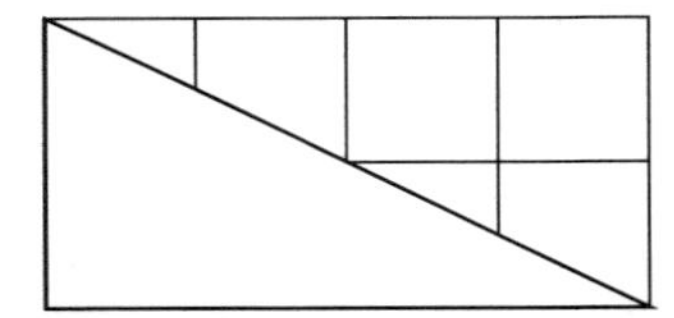

Figure 6a. Right angled triangle with an area of 4 units^2

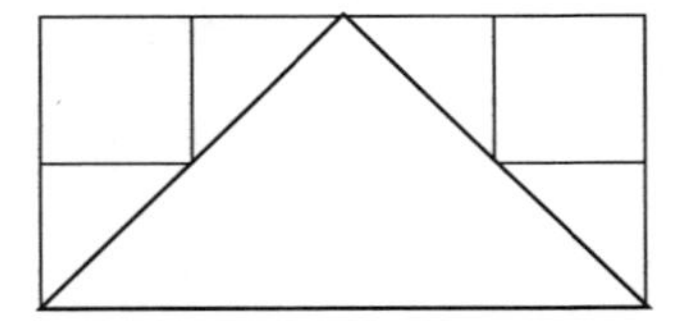

Figure 6b. First acute triangle with an area of 4 units^2

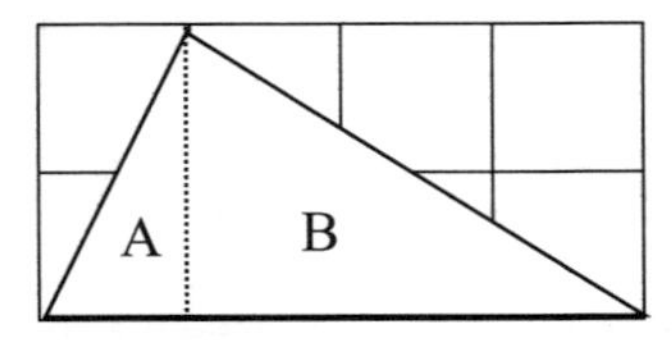

Figure 6c. Second acute triangle with an area of 4 units^2

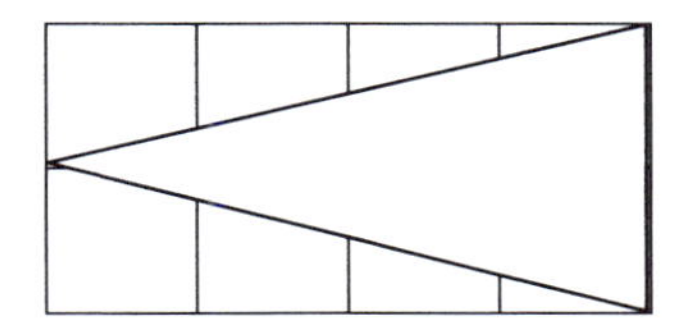

Figure 6d. Third acute triangle with an area of 4 units^2

It may be more difficult for students to draw obtuse triangles since in order to do so, the triangle has to be extended out of the given rectangle. An example of an obtuse triangle that can be drawn is shown in Figure 7. The teacher could give cues and encourage students to "think out of the box". Sufficient time should be given to students to construct as many triangles as possible.

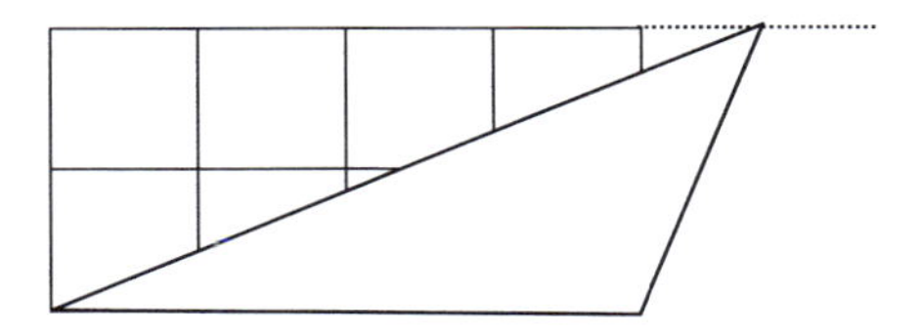

Figure 7. Obtuse triangle with an area of 4 units^2

As written in the task instructions, students are not only required to draw the triangles, they also have to explain why the area of each triangle is half that of the given rectangle. Students can do the explanation in a few ways. For example, they can count the number of squares occupied by the triangle and show that the area of triangle is 4 unit^2, as shown in Figure 8. They can also match the area of the triangle with the remaining area of the rectangle, as shown in Figure 9.

It is also possible to cut the triangle and rearrange the parts to show that the triangle can fit into 4 unit squares. The rearrangements of a right-angled triangle, an acute triangle and an obtuse triangle are similar to what was shown in Figures 4a and 4b, except that the parts are cut out and rearranged, instead of being shown statically in diagrams.

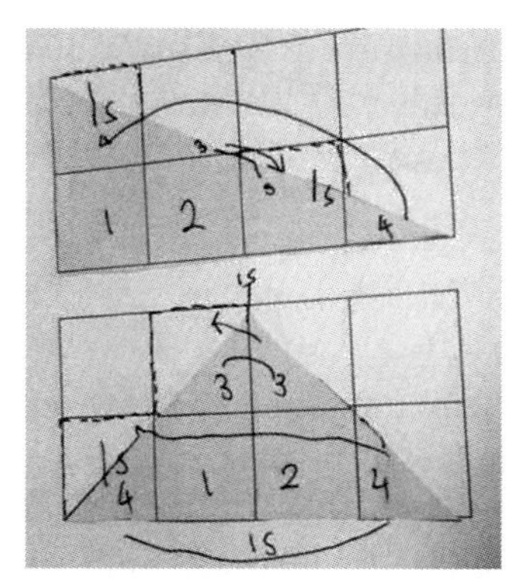

Figure 8. Counting squares to show that the area of the triangle is 4 units^2.

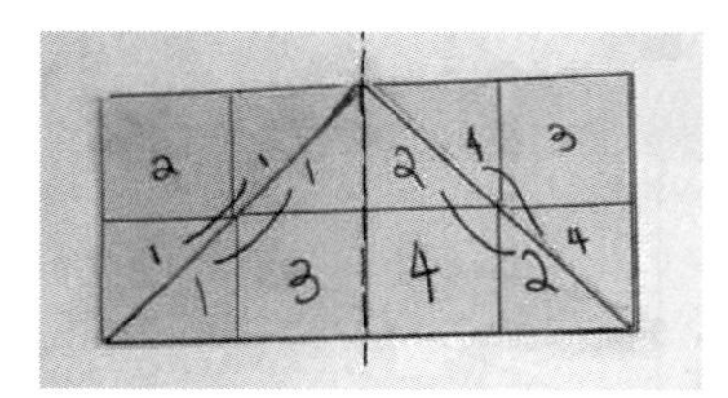

Figure 9. Showing that the area of the triangle is the same as the area of the remaining part of the rectangle.

During the discussion, students can be asked to observe each triangle and answer questions such as "What does this triangle have in common with the given rectangle?" Students may respond that the base of the triangle is equal to the length of the rectangle and the height of the triangle is equal to the breadth of the rectangle. As students have not learnt the words "base" and "height" in relation to a triangle, some students may refer to the base and height of the triangle as its sides or as the length and breadth of the triangle. The teacher may use the picture of a mountain to elicit the words "base" and "height". Subsequently, the teacher should lead students to explore the big idea about invariance that as long as the base and the height of the triangle remain the same, the area of the triangle remains the same. This means that if the base is the side that is 4 units long, the vertex of the triangle that is opposite from the triangle's base can be shifted anywhere along the line that is 2 units away from the base. Figure 10a illustrates this. Similarly, if the base is the side that is 2 units long, the vertex of the triangle that is opposite from the triangle's base can be shifted anywhere along the lines that is 4 units away from the base. Figure 10b illustrates this. Hence, an infinite number of triangles can be drawn.

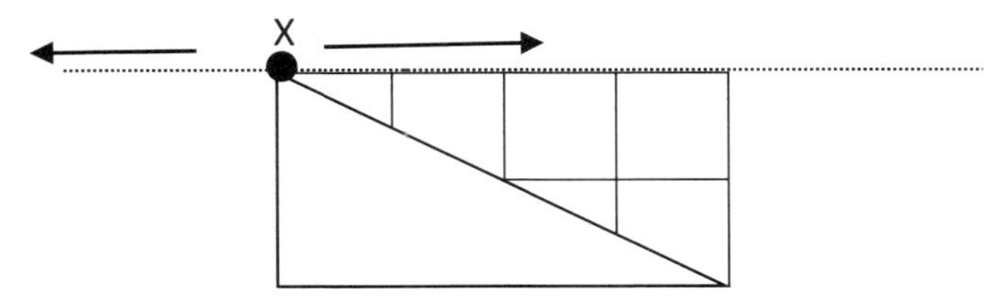

Figure 10a. The vertex marked X could be moved anywhere along the dotted line.

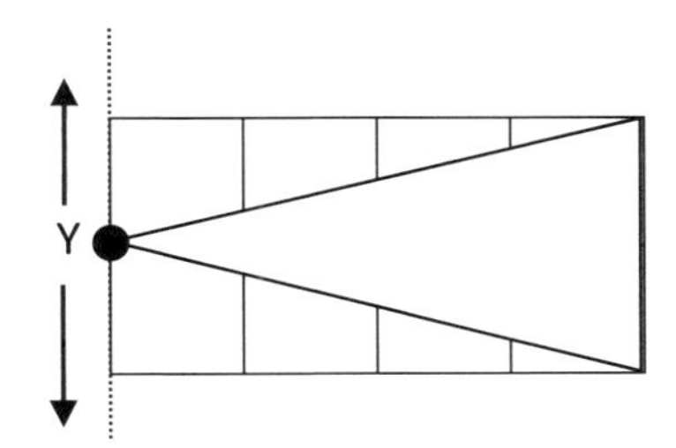

Figure 10b. The vertex marked Y could be moved anywhere along the dotted line.

To extend their learning, students could be asked to draw triangles with other dimensions and be led to discover that the relationship between the area of a triangle and the area of its related rectangle can be generalised. Students should subsequently be able to deduce that the area of a triangle can be calculated by the formula ½ x base x height. To accomplish this, questions such as "Compare the area of the triangles you have drawn to the areas of their related rectangles. What do you observe about the area of the triangles in comparison with the area of the rectangle? Is this always true? Draw more triangles with different bases and heights. Can you find an example where this is not true? Since the area of a rectangle can be calculated by using the formula length x breadth, and the area of the triangle is half of that, what is the formula to find the area of a triangle?"

This big idea of measurement relationship could also be exuded with level 2 teaching. However, Japanese mathematics educators believe that level 2 teaching does not equip students to develop mathematics proficiency with understanding. They believe that

> students should be given a reasonable amount of independent work, such as problem solving, in order to develop the knowledge, the understanding, and the skills of mathematics (Takahashi, 2017, p. 48).

A number of studies support this opinion, as their findings show that students who have been taught through problem solving tend to have higher levels of conceptual understanding and problem solving skills compared to students who have been taught didactically (Fuson, Caroll & Drueck, 2000; Wood & Sellers, 1997).

Furthermore, a carefully chosen problem may be used to draw out more than one big idea. Using the same problem solving task, the big idea of invariance can also be elicited. Students can be guided to see that the area of the triangle would remain the same as long as its base and height stays unchanged. The triangle can also be rotated, reflected or translated, but its area does not change.

4 Conclusion

In this chapter, two lesson ideas based on TTPS have been shared. The examples show that TTPS is capable of drawing out big ideas, in particular the big ideas about measures and invariance. Figure 11 shows the ideas that have been discussed in this chapter and how they are connected.

Besides the big ideas of measures and invariance, the diagram shows that area can be linked to multiplication and 2D shapes. Area of triangles can also be linked to fractions. Moreover, area is only one example of measures. The big idea of measures applies to other attributes of measurement like length, area and volume too.

The advantages of using TTPS have also been highlighted in the chapter and it is hoped that teachers would adopt different approaches in teaching.

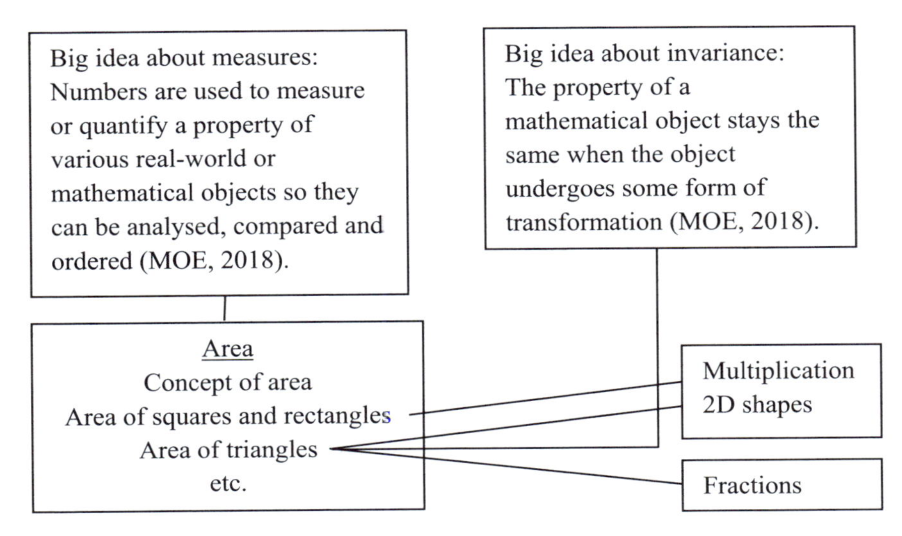

Figure 11. Ideas mentioned in this chapter and how they have been connected

References

Askew, M. (2013). Big ideas in primary mathematics: Issues and directions. *Perspectives in Education, 31*(3), 5-18.

Australian Curriculum Assessment and Reporting Authority (2013). *Australian curriculum: Mathematics.* Retrieved from http://www.australiancurriculum.edu.au

Bostic, J., Pape, S., & Jacobbe, T. (2016). Encouraging sixth-grade students' problem solving performance by teaching through problem solving. *Investigations in Mathematics Learning, 8*, 30-58.

Collars, C., Koay, P. L., Lee, N. H., & Tan, C. S. (2015). *Shaping maths coursebook 3B* (3rd ed.). Singapore: Marshall Cavendish Education.

Charles, R. I. (2005). Big ideas and understandings as the foundations for elementary and middle school mathematics. *Journal of Mathematics Education Leadership, 7*(3), 9-24.

Chi, M. T. H., Feltovich, P., & Glaser, R. (1981). Categorization and representation of physics problems by experts and novices. *Cognitive Science, 5*, 121-152.

Fong, H. K., Gan K. S., & Ramakrishnan, C. (2017). *My pals are here! Maths 5A* (3rd ed.). Singapore: Marshall Cavendish Education.

Fong, H. K., Ramakrishnan, C., & Choo, M. (2015). *My pals are here! Maths 3B* (3rd ed.). Singapore: Marshall Cavendish Education.

Fuson, K. C., Caroll, W. M., & Drueck, J. V. (2000). Achievement results for second and third graders using the standards-based curriculum everyday mathematics. *Journal for Research in Mathematics Education, 31,* 277-295.

Iitaka, S., Fujii, T., & Takahashi, A. (2012). *Tokyo Shoseki's mathematics international grade 4*. Tokyo: Tokyo Shoseki.

Lambdin, D. V. (2003). Benefits of teaching through problem solving. In F. K. Lester, Jr. (Ed.), *Teaching mathematics through problem solving: Prekindergarten-grade 6* (pp. 3-13). Reston, VA: NCTM.

Lee, N. H., Koay, P. L., Collars, C., Ong, B. L., & Tan, C. S. (2017). *Shaping maths coursebook 5A* (3rd ed.). Singapore: Marshall Cavendish Education.

McDougal, T., & Takahashi, A. (2014). Teaching mathematics through problem solving. *Independent Teacher*. Retrieved from https://www.nais.org/magazine/independent-teacher/fall-2014/teaching-mathematics-through-problem solving

Ministry of Education. (2018). *2020 secondary mathematics syllabuses (draft)*. Singapore: Author.

Niemi, D., Vallone, J., & Vendlinski, T. (2006). *The power of big ideas in mathematics education: Development and pilot testing of POWERSOURCE assessments* (CSE Report 697). Retrieved from https://files.eric.ed.gov/fulltext/ED494280.pdf

Nitaback, E., & Lehrer, R. (1996). Developing spatial sense through area measurement. *Teaching Children Mathematics, 2*(8), 473.

Ontario Ministry of Education (2008). *Measurement grades 4 to 6: A guide to effective instruction in mathematics: Kindergarten to grade 6*. Ontario: Queen's Printer for Ontario.

Siemon, D., Bleckly, J., & Neal, D. (2012). Working with the big ideas in number and the Australian Curriculum: Mathematics. In B. Atweh, M. Goos, R. Jorgensen, & D. Siemon, (Eds.), *Engaging the Australian National Curriculum: Mathematics – Perspectives from the field* (pp. 19-45). Online Publication: Mathematics Education Research Group of Australasia.

Skemp, R. (1976). Relational understanding and instrumental understanding. *Mathematics Teaching, 77*, 20-26.

Takahashi, A. (2006). Characteristics of Japanese mathematics lessons. *Tsukuba Journal of Educational Study in Mathematics, 25,* 37-44.

Takahashi, A. (2017). Lesson study: The fundamental driver for mathematics teacher development in Japan. In B. Kaur, O. N. Kwon, & Y. H. Leong (Eds.), *Professional development of mathematics teachers* (pp. 47-61). Singapore: Springer.

Wood, T., & Sellers, P. (1997). Deepening the analysis: Longitudinal assessment of a problem-centered mathematics program. *Journal for Research in Mathematics Education, 28*, 163-186.

Chapter 16

Mathematical Vocabulary: A Bridge that Connects Ideas

Berinderjeet KAUR WONG Lai Fong
TOH Wei Yeng Karen TONG Cherng Luen

Research has shown that competent and effective teachers of mathematics place emphasis on their students connecting mathematical ideas. A sound grasp of mathematical vocabulary is a pre-requisite for students to do so. In this chapter, we draw on data from the Enactment Project that is presently underway in Singapore secondary schools. We first examine data, specifically related to mathematical vocabulary, from the survey that attempted to document how widespread the pedagogies of competent experienced mathematics teachers of secondary schools were in schools across Singapore. Next, we elaborate Marzano's six steps for effective vocabulary instruction and through examples, drawn from lessons of competent experienced mathematics teachers in the Project, illustrate how teachers may facilitate the development of mathematical vocabulary during instruction.

1 Introduction

Mathematical vocabulary affords access to mathematical ideas. Therefore, the importance of developing and clarifying students' understanding of mathematical words is critical for mathematics teachers. Thompson and Rubenstein (2000) noted that although mathematical vocabulary shares word meanings with English, it is controlled and uniquely embedded in the field of mathematics. They also stated that knowledge of mathematical vocabulary is necessary for mathematics achievement but insufficient for attaining computational competence. Mathematical vocabulary is also inextricably bound to students' conceptual understanding of mathematics (Capraro, Capraro, & Rupley, 2010). One would assume that mathematical vocabulary is taught at some level during mathematics classes; however, language development is often overlooked by Math teachers (Riccomini & Witzel, 2010).

A sound grasp of mathematical vocabulary is necessary for students to connect mathematical ideas that congeal big ideas, such as equivalence, proportionality, etc. "A big idea is a statement of an idea that is central to the learning of mathematics, one that links numerous mathematical understandings into a coherent whole" (Charles, 2005, p.10). For example, "solving equations' is part of the big idea of equivalence" (Charles, 2005, p. 14). Students will encounter solving linear, quadratic, simultaneous and trigonometric equations as part of their secondary school mathematics curriculum at appropriate grade levels. Students must first comprehend 'what is an equation', before they learn that solving an equation involves isolating the variable and expressing it as a range of numbers or as an expression in terms of another variable. A target expression approach would then involve suitable algebraic manipulations (see Chapter 8 of this book).

Research has shown that competent and effective teachers of mathematics place emphasis on their students connecting mathematical ideas (Ma, 1999). Hiebert and his colleagues (1997) noted that "We understand something if we see how it is related or connected to other things we know" (p. 4). Though teachers may not draw on explicit direct instruction to develop mathematical vocabulary during mathematics lessons, they often draw on the meanings of words in English to exemplify mathematical terms. At times, such actions do not necessarily help students comprehend the mathematical essence of the words (Dunston & Tyminski, 2013). This is so as there is a need to link the vocabulary to the symbolic language of mathematics.

Research has also shown that language is a pivotal component of mathematics success (Seethaler, Fuchs, Star, & Bryant, 2011) and a student's general knowledge of mathematical vocabulary can predict mathematical performance (van der Walt, 2009). In particular for students who are weak in language, explicit instruction of vocabulary terms and their meanings in context helps to remove guesswork on the part of the students (Riccomini, Smith, Hughes, & Fries, 2015).

Marzano (2004) recommends six steps for educators to maximize student learning of essential vocabulary based on components of evidence-based instructional strategies that aid in achieving positive academic outcomes across content areas. These strategies are a) explicit instruction, b) stimulating prior knowledge, c) repetition, d) differentiating instruction and e) cooperative learning.

In this chapter, we draw on the data from the Enactment Project (Kaur, Tay, Toh, Leong, & Lee, 2018), that is presently underway in Singapore. We first examine data, specifically related to mathematical vocabulary, from the survey that attempted to document how widespread the pedagogies of competent experienced mathematics teachers of secondary schools were in schools across Singapore. Next, we elaborate Marzano's (2004) six steps for effective vocabulary instruction and through examples, drawn from lessons of competent experienced mathematics teachers in the Enactment Project, illustrate how teachers may facilitate the development of mathematical vocabulary during instruction.

2 The Enactment Project

The project is about the interactions between secondary school mathematics teachers and their students, as it is these interactions that fundamentally determine the *nature* of the actual mathematics learning and teaching that take place in the classroom. It also examines the content through the instructional materials used – their preparation, use in classroom and as homework. Such studies are crucial for the Ministry of Education (MOE) in Singapore and schools to gain a better understanding of what works in the instructional core in their classrooms and schools. This is critical for the development of their education system.

The project has two phases, the first and second. The first phase is the video segment and the second one is the survey segment. The survey segment is dependent on the findings of the video segment. The video segment documents the pedagogy of competent experienced secondary mathematics teachers while the survey segment aids in establishing how uniform the pedagogy of these competent experienced teachers is in the mathematics classrooms of Singapore schools. The video segment of the study is adopting the complementary accounts methodology developed by Clarke (1998 & 2001), a methodology which is widely used in the study of classrooms across many countries in the world as part of the Learner's Perspective Study (Clarke, Keitel, & Shimizu, 2006). This methodology recognizes that only by seeing classroom situations from the perspectives of all participants (teachers and students) can we come to an understanding of the motivations and meanings that underlie their participation. It also facilitates practice-oriented analysis of learning. For the survey, the project is adopting a self-report questionnaire to collect data on teachers' enactment of their "teacher-intended" curriculum.

Thirty competent experienced teachers (10 Express course of study, 4 Integrated Programme, 8 Normal (Academic) course of study and 8 Normal (Technical) course of study) and 447 (in each class from 8-20 students, who volunteered to be the focus students were interviewed) students in their classrooms participated in the video segment of the project. In the context of the project, a competent experienced teacher is one who has taught the same course of study for a minimum of 5 years, and is recognized by the school / cluster as a competent experienced teacher who has developed an effective approach of teaching mathematics. These teachers were nominated by their respective school leaders and the research team followed up on the nominations and interviewed the teachers. A strict requirement for participation in the study was that the teacher had to teach the way she / he did all the time, i.e., no special preparation was expected. For the survey segment of the project, 689 secondary school mathematics teachers from 109 schools in Singapore participated in the project.

3 The Survey

The survey is based on the findings in phase 1. It is an online survey comprising three sections: pedagogical structure and student-teacher interaction (60 items); enactment of the different facets of the "pentagon framework" (MOE, 2012) undergirding the secondary mathematics curriculum (78 items); and instructional materials (226 items, of which a participant needs only respond according to one of the subjects: Additional Mathematics, Elementary Mathematics or Normal Academic Mathematics). In this chapter, we focus on the first section of the survey.

3.1 *Items and participants*

The 60 items in the first section were divided into two parts of 36 and 24 items respectively. Every item was linked to a form and purpose of *classroom talk* (rote, recite, instruct, expose, discuss, scaffold dialogue, narrate, explain, analyse, speculate, explore, evaluate, argue, justify), a *phase of instruction* (development, seatwork or review) and a *model of instruction* (traditional instruction, direct instruction, teaching for understanding or co-regulated learning strategies).

The first part had items which elicited responses about what the teacher did in class. An example of such an item is as follows:

> I ask students to recall past knowledge.
> [*rote, review, traditional instruction*]

The second part had items which elicited responses about what the teacher wanted the student to do in class. An example of such an item is as follows:

> I get my students to explain how their solutions or how their answers are obtained.
> [*explain, development, teaching for understanding*]

Participants were required to respond on a Likert Scale of 1 (Never/Rarely) to 4 (Always) for all the 60 items.

Of interest to us in this chapter are two items from the first part which are as follows:

1. I focus on mathematical vocabulary (such as equations, expressions) to help my students build mathematical concepts.
2. I focus on mathematical vocabulary (such as factorise, solve) to help my students adopt the correct skills needed to work on mathematical tasks.

Both items belong to *exposition talk* in the *development phase* of a lesson and *teaching for understanding* model of instruction.

The survey respondents were Singapore secondary school teachers from 4 different academic courses: Integrated Programme (60), Express (388), Normal Academic (151), and Normal Technical (90). These four academic courses are broadly based on academic achievement of students in the Primary School Leaving Examination. The most academically inclined students are in the Integrated Programme and the least inclined ones are in the Normal Technical course of study. The on-line survey was administered to teachers following their written consent to participate in the project.

3.2 *Data and findings*

Table 1 shows the responses of the survey participants to the two items that are of interest to this chapter. From Table 1, it is apparent that more than 90% of teachers in the Integrated Programme, Express and Normal (Academic) courses of study frequently or always focused on mathematical vocabulary, both for development of concepts and also skills. For the Normal (Technical) course of study, more than 80% of teachers did the same as their counterparts in the other three courses of study. Though the findings generally are noteworthy and suggest good instructional practice for mathematics in our secondary schools, of concern is the lower percentage of teachers focusing on mathematical vocabulary in the Normal (Technical) course of study. Generally, students in this course of study demonstrate weaker command of the language of mathematics instruction, i.e. English. This calls for attention to the development of mathematical vocabulary during mathematics instruction for these students.

Table 1

Frequency of responses to the two items on mathematical vocabulary in the survey

Item 1 I focus on mathematical vocabulary (such as equations, expressions) to help my students build mathematical concepts.					
Course of Study	Frequency N (%)				
	Never / Rarely	Sometimes	Frequently	Always	Total
IP	0(0)	3(5)	17(28)	40(67)	60(100)
Express	3(1)	26(7)	160(41)	199(51)	388(100)
N(A)	3(2)	8(5)	76(50)	64(43)	151(100)
N(T)	0(0)	16(18)	37(41)	37(41)	90(100)
Item 2 I focus on mathematical vocabulary (such as factorise, solve) to help my students adopt the correct skills needed to work on mathematical tasks.					
Course of Study	Frequency N (%)				
	Never / Rarely	Sometimes	Frequently	Always	Total
IP	0(0)	0(0)	20(33)	40(67)	60(100)
Express	1(0)	15(4)	165(43)	207(53)	388(100)
N(A)	1(0)	6(4)	76(51)	68(45)	151(100)
N(T)	0(0)	11(13)	39(43)	40(44)	90(100)

In the next section, we explore some insights and examples of vocabulary instruction demonstrated by competent experienced teachers who participated in the first phase of the enactment project.

4 Some Insights and Examples of Vocabulary Instruction

In this section, we first elaborate Marzano's (2004) six steps for effective vocabulary instruction. Next, through examples, drawn from lessons of competent experienced mathematics teachers in the Enactment Project, we illustrate how teachers facilitated the development of mathematical vocabulary during their instruction.

4.1 *Marzano's six steps for effective vocabulary instruction*

Figure 1 shows the concept map based on the six recommendations by Marzano (2004) for effective vocabulary instruction.

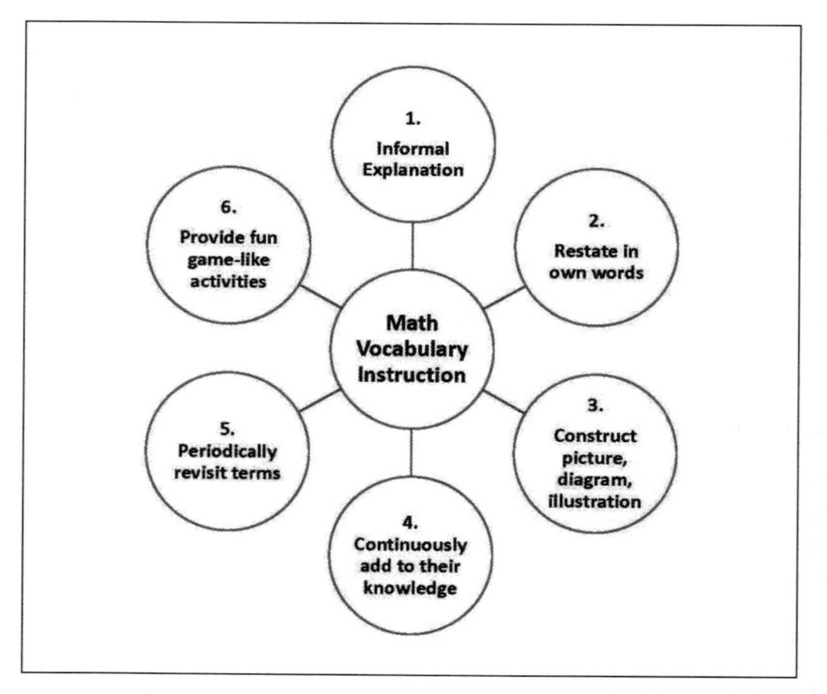

Figure 1. Concept map based on six recommendations by Marzano (2004) for effective vocabulary instruction (Riccomini et al., 2015, p. 240)

The six steps for effective vocabulary instruction, advocated by Marzano (2004), are as follows:

- First, through direct or indirect means, introduce students to new vocabulary by explaining what it means and with examples of its usage in contexts relevant to students learning of mathematics. For students, this starts the process of connecting the new meaning to their prior knowledge.
- Second, provide students with opportunities to re-state the teacher-provided descriptions, explanations or examples in their own words. This helps to reinforce the connections to their prior knowledge.
- Third, strengthen their linkage to prior knowledge by asking them to construct a picture, symbol, or graphic representation of the new vocabulary.
- Fourth, provide students with opportunities to re-engage with the vocabulary frequently so as to develop deeper understanding necessary for mathematical reasoning and communication.
- Fifth, through small group or peer-to-peer discussions on specific terms, further develop a deeper understanding and also eradicate any past misconceptions that may be present.
- Sixth, to facilitate long term retention, provide opportunities for students to revisit essential and already learnt terms through fun activities such as games and riddles.

4.2 *Examples of mathematical vocabulary instruction*

In this section we draw on the lessons of three teachers, T1, T2 and T3, in the first phase of the Enactment Project to illustrate how teachers may facilitate meaningful vocabulary instruction in their mathematics lessons.

4.2.1 Pythagoras' theorem and trigonometrical ratios

T1 taught a class of 36 grade 10 Normal Technical students Pythagoras' theorem and trigonometrical ratios (sine, cosine and tangent) of a right-angled triangle over a span of 6.5 hours of instructional time. The mathematical abilities of students in the class of T1 were from the 20th percentile of their cohort. At the start of the first lesson, which was of 60 minutes in duration, T1 told the students that they would learn new mathematical terms during the lesson and also revisit some terms they had already learned.

Teacher wrote on the board:

- Pythagoras' theorem
- hypotenuse
- square
- square root

The new terms were: Pythagoras' theorem and hypotenuse, and the terms they would revisit were: square and square root. Following an investigative activity, during which students verified that the area of a square on the longest side of a right angled triangle was the sum of the areas of the squares on the remaining two sides, T1 drew the attention of his students to the word "square" and asked them to use their calculators and find the square of numbers. With inputs from students, T1 drew on the board a right-angled triangle and labelled its sides a, b and c. Next, T1 wrote the relationship $c^2 = a^2 + b^2$ and pointed out that this relationship between a, b and c is Pythagoras' formula (theorem). T1 also drew the attention of the class to the longest side c and introduced the term hypotenuse.

For the rest of the lesson, T1 demonstrated on the board how the hypotenuse of a right-angled triangle could be found given the other two sides and students were engaged in seatwork to do similar tasks. During this period of time, the words Pythagoras' theorem, hypotenuse, square and square root were reinforced by T1 whilst engaging students in explaining how they worked through the solutions of the tasks which were assigned as seatwork. T1 articulated clearly that when $x^2 = 89$, to find the value of x, you used your calculator and found the square root of 89.

During the third lesson, which was a prelude to trigonometric ratios of right-angled triangles, T1 taught students how to name the sides of a right-angled triangle specific to an angle in the triangle. Two new words opposite and adjacent were introduced. To reinforce their learning, students were given a short exercise (see Figure 2) to do in class and to clarify their errors if any during the discussion period that ensued.

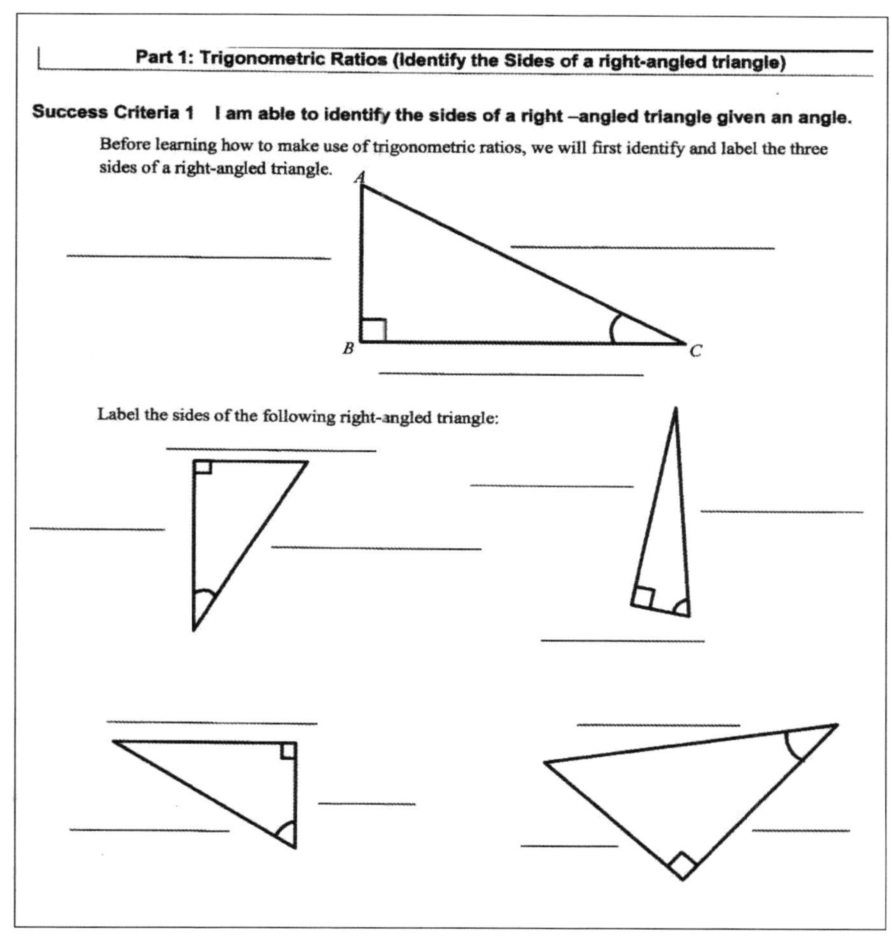

Figure 2. An exercise to name the sides of a right-angled triangle given an angle

It is apparent from the approach, adopted by T1 in developing mathematical vocabulary of his students, that a direct means was used to introduce new words and also engage them in the recall of past vocabulary. T1 ensured that students were constantly engaged in using the words during their classroom discourse. This helped to develop correct use of appropriate algorithms, such as Pythagoras' theorem, and also skills, such as to find squares, square roots and ratios (later on when students did tasks involving trigonometry). Furthermore, through the multiple exercises, students were engaged in using Pythagoras' theorem and also stating it at the start of their solution process, identifying the hypotenuse and finding squares and square roots helped to strengthen their new vocabulary with their past one.

4.2.2 Quadratic equations and graphs

T2 taught a class of 28 grade 8 Integrated Programme students Quadratic Equations and Graphs over a span of 8 hours of instructional time. The mathematical abilities of students in the class of T2 were from the 80th percentile of their cohort. During the first lesson, which was of 60 minutes in duration, T2 reviewed her students' understanding of a quadratic expression and a quadratic equation. In grade 7, students had worked with such expressions and factorised them. The following 3 episodes illuminate the discourse in T2's lesson that facilitated students' understanding of mathematical terms necessary for the topic.

Episode 1 - Clarification of "power of 2" and "square"
T2 began her lesson by asking her students, "Who can remember what a quadratic expression is?"

Responses given by the students such as:

P1: I think it's power of 2.
T2: What is power of 2?
P1: Square.
T2: What thing is square? "Power of 2", "square".
Can you come and write an example of a quadratic expression?
[P1 wrote 5^2 on the whiteboard]
T2: 5 squared is a quadratic expression? Is this correct?
[P1 wrote $x^2 + x + 7$ on the whiteboard]
T2: P1 says this is a quadratic expression.
[T2 points to $x^2 + x + 7$ on the whiteboard]

How many of you agree? Just raise your hands … Ok, you can put down your hands.

T2: Can somebody else give me another expression? Ok, come, P2.

[P2 wrote $y^2 - 9y$ on the whiteboard]

The ensuing classroom discourse between teacher and students led to the clarification and articulation that in a quadratic expression the variable which was x in one case and y in another has "a power of 2".

<u>Episode 2</u> – The general form of a quadratic expression

With reference to the two examples of quadratic expressions:

$$x^2 + x + 7$$
$$y^2 - 9y$$

T2 posed to the class the question: "What makes a quadratic expression?"

T2: P5, what are quadratic expressions?

P5: I don't know how to explain.

T2: [T2 hears P6's response] Ok, come. P6 wants to explain.

P6: I think the indices are in descending order.

T2: The indices are in

P6: descending order

T2: [Referring to $x^2 + x + 7$] So there's a square, and then this is power 1, and then this is a constant.

P6: Yah

The ensuing class discussion led to the articulation by students that a quadratic expression can be represented as:

$$ax^2 + bx + c, \text{ where } a, b \text{ and } c \text{ are constants}$$

and that a cannot be zero but b and c can be equal to zero as shown in:

$$y^2 - 9y \text{ where } c = 0;$$
$$x^2 - 9 \quad \text{where } b = 0.$$

Episode 3 – What is the difference between an equation and an expression?

T2: Alright now, what is the difference between an equation and an expression?

Responses from students such as

- an equation has an equal sign
- $x^2 + x + 7 = 0$
- $3x + x = 4x$
- $3x(x + 1) = 0$
- $(x + 7)(3x - 4) = 0$

which were discussed led to the clarification of the difference between an expression and an equation. The abundant opportunities to use correct mathematical vocabulary such as coefficients, constants, factors, and solutions were also capitalised by T2 throughout the classroom discourse.

T2 concluded this episode by explaining that an equation is formed by two equivalent sets of expressions.

T2: OK, let's look at this [T2 refers to $x^2 - x + 3 = 0$]. So over here [T2 draws a box around $x^2 - x$ on the left hand side of the equation], what do you call this without the equal sign? What do you call x square minus x plus 3?

Chorus: Expression.

T2: What can you say about these two expressions when you see this sign [T2 draws an arrow to point at the equal sign between the expressions $x^2 + 6$ and $x + 3$]?

Chorus: They are equal.

T2: They are equal. So when you have two equal expressions, alright, then it forms an equation. This expression [$x^2 + 6$] is equal to this expression [$x + 3$]. So now, this expression [$x^2 - x + 3$] is equal to 0. Another expression but the expression is a constant zero. So when you have two

> equivalent expressions—this expression [$x^2 + 6$] is equal to another expression [$x + 3$]—so that's why, because they are equivalent, you have this equal sign. It becomes an equation. It can be expression [$x^2 + 6$] equal to another expression [$x + 3$], it can be expression [$x^2 - x + 3$] equal to a constant which is also another expression. So this becomes an equation. It has an equal sign.

It is apparent from the approach, adopted by T2 in reviewing and deepening understanding of mathematical vocabulary of her students, that a dialogic means was used. T2 harnessed her students' prior knowledge as spring boards for clarifying understanding of key mathematical terms that were essential for the study of the topic at hand. Through the examples and non-examples that were provided by the students during the classroom discourse, T2 guided her students in deepening their understanding of mathematical concepts as well as developing sound mathematical vocabulary for communication.

4.2.3 Solving algebraic linear equations

T3 taught a class of 33 grade 7 Normal Academic students Solving Algebraic Linear Equations over a span of 6 hours of instructional time. The mathematical ability of students in the class of T3 were at the 30th percentile of their cohort. At the start of the first lesson, which was of 60 minutes in duration, T3 told the students that they would be learning how to solve linear algebraic equations.

The lesson began with a group activity on evaluating algebraic expressions with given values for the variables.

T3: You will be given sixteen cards. On the cards, there are algebraic expressions. For example, you may get things like these [T3 wrote some expressions on the board].

T3: What I want you to do is to evaluate. [T3 wrote the word "Evaluate" on the board]

T3: Do you still remember what is the meaning of evaluate?

P1: Find the value.

T3: Ok, P1 says find the value. So find the value of the expression.

The objective of the activity was to get students to find two expressions, in y, that have the same value given a certain value of y, and therefore, students learnt the mathematical vocabulary "equation" and "solution of an equation". Throughout the lesson, T3 highlighted the difference between "algebraic expressions" and "algebraic equations".

It is apparent that T3, through direct or indirect means, introduced students to new mathematical vocabulary by explaining what it means and with examples of its usage. T3 had also planned her lessons such that new mathematical vocabulary was introduced to the students in a meaningful sequence, building on students' prior knowledge and reinforcing the connections. For examples, students learnt how two expressions are equated to form an equation; they learnt what "solution of an equation" was before learning "solving an equation".

T3 also used Spelling Lists to help students remember mathematical terms (and their spelling). T3 identified a list of mathematical vocabulary essential for the learning of the topic, and students were given this list to learn the spelling. The following illustrates how a spelling test was carried out in the second lesson.

[T3 wrote "$3x^2 + x - 3y + 4$" on the board]

T3: This is an example of an algebraic expression.
Spell "algebraic expression".

T3: The letter x [T3 pointed at "x"] and letter y [T3 pointed at "y"] here are known as variables.
Spell "variables" .

T3: Here are one [T3 pointed at "$3x^2$"], two [T3 pointed at "x"], three [T3 pointed at "$- 3y$"], four [T3 pointed at "4"] terms in this expression.
Spell "terms".

T3: [T3 pointed at "3"]
This term in front of x^2 is known as the coefficient of x^2.
Spell "coefficient".
[T3 pointed at "$- 3$"]

This negative three in front of y is known as the coefficient of y.
Spell "coefficient".
And the coefficient of x is one.

T3: [T3 pointed at "4"]
This number here in the algebraic expression is called a constant.
Spell "constant".

T3: [T3 wrote "$2x + x + y = 3x + y$]

T3: We can simplify this to three x plus y because "$2x + x = 3x$".
[T3 underlined "$2x + x = 3x$" and pointed at it]

T3: These are like terms.
Spell "like terms".

It is apparent that T3, understanding the students' low proficiency in English language, provided opportunities for students to revisit essential and already learnt terms through this approach, that is, the spelling test, to facilitate students' long-term retention.

5 Discussion and Concluding Remarks

The findings of the two items from the survey signal to us that there is a gap in the pedagogy of a small but possibly significant number of mathematics teachers in our secondary schools. As shown by research that language is a pivotal component of mathematics success, this gap must be addressed so that students achieve their potential in the learning of mathematics. This implies that mathematics educators at the National Institute of Education (NIE) as well as teacher leaders in schools and other Professional Development providers must recognise that mathematical vocabulary instruction is necessary for students to comprehend concepts, execute skills and most importantly connect their mathematical ideas.

From the three examples, which are by no means exhaustive and perfect, it is apparent how diverse vocabulary instruction may be. The diversity may be due to teachers' own pedagogy, language and mathematical ability and prior knowledge of their students. It is also apparent from the approaches of T1, T2 and T3 that Marzano's (2004) suggested six steps are not linear in form. From the three examples, some gaps in vocabulary instruction are also apparent.

T1 worked around the first three steps, but he scaffolded his students' learning by providing them with images of right-angled triangles to label (see Figure 2) in contrast to students themselves drawing and labelling the images. He also did not directly address the duality of the word "square". When students were verifying Pythagoras' theorem they were working with the construction of a square (shape) on the sides of a right angled triangle but when they transferred their findings to a relationship, the areas of the respective squares led to $a^2 = b^2 + c^2$ and subsequently the focus shifted to the square (index) of numbers. A development that followed was finding the square root, which was merely using the calculator as a computational tool. A missed opportunity to connect mathematical ideas was given the area of a square, how you would find the length of a side. Such rich connections allow for deepening of students' learning of mathematical inter-related concepts through vocabulary.

It was apparent from T2's lesson that students had vague understandings of mathematical terms they needed for their lessons on quadratic equations and graphs. The dialogic approach adopted to re-engage with the mathematical terms allowed for deeper understanding and also opportunities to surface misconceptions that were harnessed for deeper learning. This shows that on-going development of mathematical vocabulary is necessary for students. In the class of T3, vocabulary development involved spelling the words and recognition of mathematical symbols. This is yet another key aspect of mathematical vocabulary instruction, particularly with students weak in language and mathematics.

Needless, to say, as mathematics educators, we must use language as a tool to connect mathematical ideas. This language comprises words and symbols, which are specific to contexts. For example, an "equation" can also be an "expression" as in the following context: if "$x = 3$, then $y \geq x^2$." Therefore, as part of mathematics instruction, we must place adequate focus on vocabulary development in varying contexts, through appropriate ways to engage our students in making sense of the world around them mathematically.

Acknowledgement

This chapter is based on the Programmatic Research Project: A Study of the Enacted School Mathematics Curriculum funded by Singapore Ministry of Education (MOE) under the Education Research Funding Programme (OER 31/15 BK) and administered by National Institute of Education (NIE), Nanyang Technological University, Singapore. Any opinions, findings, and conclusions or recommendations expressed in this material are those of the author(s) and do not necessarily reflect the views of the Singapore MOE and NIE.

References

Capraro, R. M., Capraro, M. M., & Rupley, W. H. (2010). Semantics and syntax: A theoretical model for how students may build mathematical misunderstandings. *Journal of Mathematics Education, 3*(2), 58-66.

Charles, R. I. (2005). Big ideas and understandings as the foundations for elementary and middle school mathematics. *Journal of Mathematics Education Leadership, 7*(3), 9-24.

Clarke, D. J. (1998). Studying the classroom negotiation of meaning: Complementary accounts methodology. In A. Teppo (Ed.), *Qualitative research methods in mathematics education: Journal for Research in Mathematics Education Monograph Number 9*. Reston, VA: NCTM, 98-111.

Clarke, D. J. (Ed.). (2001). *Perspectives on practice and meaning in mathematics and science classrooms*. Dordrecht, Netherlands: Kluwer Academic Press.

Clarke, D. J., Keitel, C., & Shimizu, Y. (Eds.). (2006). *Mathematics classrooms in twelve countries: The insider's perspective*. Rotterdam: Sense Publishers.

Dunston, P. J., & Tyminski, A. M. (2013). What's the big deal about vocabulary? *Mathematics Teaching in the Middle School, 19*(1), 38-45.

Hiebert, J., Carpenter, T. P., Fennema, E., Fuson, K., Wearne, D., Murray, H., … Human, P. (1997). *Making sense: Teaching and learning mathematics with understanding.* Portsmouth, N.H: Heinemann.

Kaur, B., Tay, E. G., Toh, T. L., Leong, Y. H., & Lee, N. H. (2018). A study of school mathematics curriculum enacted by competent teachers in Singapore secondary schools. *Mathematics Education Research Journal, 30*, 103-116.

Ma, L. (1999). *Knowing and teaching elementary mathematics: Teachers' understanding of fundamental mathematics in China and the United States*. Mahwah: N.J.: Erlbaum.

Marzano, R. J. (2004). *Building background knowledge for academic achievement.* Alexandria, VA: Association for Supervision and Curriculum Development.

Ministry of Education. (2012). *The teaching and learning of 'O' Level, N(A) Level & N(T) Level mathematics*. Singapore: Author.

Riccomini, P. J., & Witzel, B. S. (2010). *Response to intervention in mathematics.* Thousand Oaks, CA: Corwin Press.

Riccomini, P. J., Smith, G. W., Hughes, E. M., & Fries, K. M. (2015). The language of mathematics: The importance of teaching and learning mathematical vocabulary. *Reading and Writing Quarterly, 31*(3), 235-252.

Seethaler, P. M., Fuchs, L. S., Star, J. R., & Bryant, J. (2011). The cognitive predictors of computational skill with whole versus rational numbers: An exploratory study. *Learning and Individual Differences, 21*, 536-542.

Thompson, D. R., & Rubenstein, R. N. (2000). Learning mathematics vocabulary: Potential pitfalls and instructional strategies. *Mathematics Teacher, 93*, 568-574.

van der Walt, M. (2009). Study orientation and basic vocabulary in mathematics in primary school. *South African Journal of Science and Technology, 28*, 378-392.

Chapter 17

Engaging Students in Big Ideas and Mathematical Processes in the Primary Mathematics Classrooms

CHENG Lu Pien Vincent KOH Hoon Hwee

Big ideas in mathematics have been variously described. In this chapter, we describe some Big ideas, in particular, multiplicative thinking that emerged as we worked with some teachers in Singapore to design journal writing tasks in their lower primary mathematics classrooms. We conclude the chapter by presenting the model used by the teachers to design and implement journal writing tasks, guided by Big ideas. Our reflection on the work with teachers provide possible guidelines in the development of meaningful tasks and activities to support students in connecting the big ideas.

1 Introduction

Big ideas "can be identified through a careful content analysis" (Charles, 2005, p. 10). However, many must be identified by "listening to students' recognizing common areas of confusion, and analysing issues that underlie that confusion" (Schifter, Russell, & Bastable, 1999, p. 25). This chapter reports our observation of some Big ideas as we worked in partnership with a primary school for the professional development of their lower primary mathematics teachers.

Indeed, Big ideas have become a topic of interest in mathematics education (Siemon, Blecky, & Neal, 2012, p. 19). However, Big ideas in mathematics have been variously described (Hurst, 2015). For example, Schifter and Fosnot (1993) defined them as "the central, organizing ideas of mathematics – principles that define mathematical order" (p. 35) from the discipline of mathematics. Fosnot (with Dolk) built from her work with Schifter suggested that "Big Ideas also mark a shift in learner's reasoning" (Askew, 2013, p. 7). Charles (2005) defines a big idea as "a statement of an idea that is central to the learning of mathematics, one that links numerous mathematical understandings into a coherent whole" (p. 10). He offered 21 Big mathematical ideas for primary and middle-school mathematics. His intention was that a focus on Big ideas will enable teachers to "connect topics across grades; they know the concepts and skills developed at each grade and how those connect to previous and subsequent grades" (Charles, 2005, p. 11).

Siemon et al. (2012) discussed the Big Ideas in Number (trusting the count, place-value, multiplicative thinking, partitioning and proportional reasoning and generalising) and the Australian Curriculum for mathematics. Their suggestion was that a focus on Big Ideas could help to address the need to 'thin out' the crowded curriculum. The Big "ideas presented by Siemon et al. are embedded in Charles' 'big ideas' in various ways" (Hurst & Hurrell, 2014, p. 4). Lord and Kiddle (2016) led a regional Mathematics Specialist Teacher programme structured around Big ideas of mathematical thinking, representation, generalisation, pattern and proportionality. Ashlock (2006) emphasized on helping students understand the Big ideas of "meanings of numerals, equals and equivalence, properties of operations and compensation principles" (p. iv). Hurst (2015) differentiated between Big content ideas (for teachers to deconstruct their own conceptual structures) and Big process ideas (e.g. problem solving, reasoning, inferring and constructing arguments etc.).

Despite the different definitions, "a Big idea should be both culturally, that is mathematically, significant as well as individually and conceptually significant" (Askew, 2013, p. 7). From our reflection of some Big ideas that emerged from our work with teachers in a school in Singapore, we hope to provide some insight to designing meaningful classroom tasks.

2 Background on the Professional Development in this Study

Developing mathematical reasoning is one of the important aims of mathematics education. In Singapore, mathematical reasoning, communication and connection are some of the process skills to be developed during the teaching and learning of mathematics (MOE, 2012). The *Principles to Actions: Ensuring Mathematical Success for All* listed the implementation of tasks that promote reasoning and problem solving (NCTM, 2014) as one of the eight Mathematics Teaching Principles to promote deep learning of mathematics. Indeed, mathematical tasks are "important vehicles for building student capacity for mathematical thinking and reasoning" (Stein, Grover & Henningsen, 1996). We worked with one typical primary school in Singapore to design meaningful and worthwhile tasks to "engage students at a deeper level by demanding interpretation, flexibility, the shepherding of resources, and the construction of meaning" (Stein, Grover & Henningsen, 1996, p. 459). Specifically, the teachers in the school were interested in using journal writing prompts as a vehicle to develop mathematical reasoning, that is, the journal writing prompts demand students to communicate and sometimes justify their procedures and understandings in written and/or oral form. Through journal writing prompts, the teachers also seek to "provide a valuable means to facilitate a personalized and making-of-meaning approach to learning mathematics" (Borasi & Rose, 1989, p. 347) so that students can reach deep and personal meaning of the material they are learning.

The first year of our work with the school focused on the design of mathematics journal writing prompts to empower Primary 1 students the ability to reason, justify and represent their mathematical ideas. The first author taught the Primary 1 mathematics teachers how to design content, process and affective prompts (Dougherty, 1996, p. 557), as shown in Table 1, during their common planning slots.

Table 1.
Journal writing prompts adapted from Dougherty (1996)

Goal of content prompts	**Possible content prompts:**
• Focus on mathematical topics and their relationships	• Define what a concept means
Goals of process prompts	**Possible process prompts:**
• Reflect on why solution strategies were chosen or preferred • Reflect on ways in which one learns. • Become stronger problem solvers as learners begin to understand their own problem-solving approaches by engaging in the process of explaining why they choose a method.	▪ What is your favourite method? Why?"
Goal of affective prompts	**Possible affective prompts:**
Find out how students view themselves as e.g. problem solvers, mathematics students.	I feel ...... when I learn mathematics because .....

The teachers then worked collaboratively to create journal writing prompts for their Primary 1 students. Next, they implemented those prompts to their students and shared their experiences implementing those prompts with the team. During the sharing sessions, the team examined the students' responses to the journal writing prompts to delve more deeply into students' mathematical reasoning and thinking. The teachers were encouraged to pen their reflections on the whole process of designing and implementing the journal writing prompts. In the second year, the teachers continued to design and implement mathematics journal writing prompts for their Primary 2 students. The journal writing prompts were completed by the students at the end of each topic. All the topics for the journal writing prompts which the team worked on for the study were identified by the teachers according to what they thought were challenging for their students, either from test scores or from their understanding of their students' progress and learning trajectories. The school continued with the design and implementation of journal writing prompts on their own in the third year when their students moved on to Primary 3 and the teachers invited the first author back to the school occasionally for feedback on the design and implementation of the journal writing prompts.

3 Big Ideas that Emerged from the Design of Journal Writing Prompts

Figures 1, 2 and 3 are examples of some of the journal writing prompts designed by the team.

<table>
<tr>
<td>Task 1
This is Bala's working.
Tens Ones
3 2
- 1 8
2 6
His working is wrong. What is wrong with the working?
(a) Explain why the working is wrong?
(b) Show the correct working.</td>
<td>Task 2
Use the "count on" and "make ten" strategies to show how you add the three numbers below. You may use numbers, words or pictures.
Method 1: Count On
8 + 6 + 5 = ____

Method 2: Make Ten
8 + 6 + 5 = ____
Which is your favourite method? Why?</td>
</tr>
<tr>
<td colspan="2">Task 3
Siti wants to write a word problem using the cubes shown.

Write a division story. Use the cubes to help you. Then solve the problem.
Modified from My Pals are Here! Maths Workbook 1B (Fong, Ramakrishnan & Lau, 2013, p. 84).</td>
</tr>
</table>

Figure 1. Samples of content and process prompts for Primary 1 students

Task 4

(1) Write or draw what multiplication means to you.
(2) Write or draw what division means to you.

Task 5

Alice measured the arrow. She said that the length of the arrow is 9 cm.

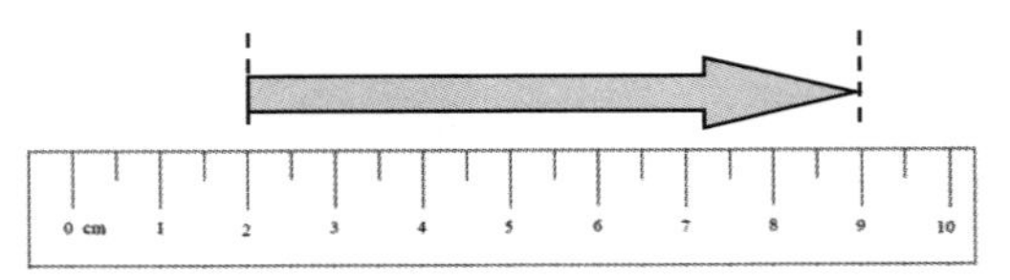

(a) Is she correct?
(b) Explain your answer.
Modified from *Shaping Maths Activity Book 2A* (Collars, Koay, Lee, & Tan, 2014a, p. 53).

Task 6

If you have $10, how would you spend your money?
You may draw or write to illustrate.
Modified from *Shaping Maths Activity Book 2B* (Collars, Koay, Lee, & Tan, 2014b, p. 64).

Task 7

Who will get more of the chocolate? Use diagram to show your answer.

Modified from *My Pals are Here! Maths Workbook 2B* (Fong, Ramakrishnan & Choo, p. 125).

Figure 2. Samples of content and process prompts for Primary 2 students

Task 8

Tommy counted the wheels of the vehicles parked at a carpark. There were 8 cars and 6 trucks at the carpark. How many wheels did Tommy count altogether?
He counted ______ wheels altogether.
Explain your working.
(A picture of a car with 4 wheels and a truck with 6 wheels is provided in the worksheet.)

Task 9

The total mass of the water bottle and the apple is shown on the scale below. The mass of the apple is 120g. What is the mass of the water bottle?
Answer: ___
Explain your working.
(A picture of a scale that reads 380g with an apple and water bottle is provided in the worksheet. Students are required to read the scale).

Task 10

(i) 20 trees were planted along one side of Nanyang Street. The trees were planted 10 m apart.

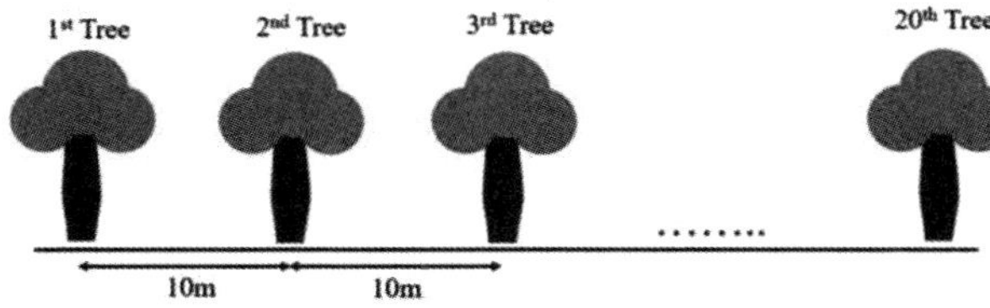

1st Tree 2nd Tree 3rd Tree .... 20th Tree
10m 10m

What is the total distance from the first tree to the last tree?
Answer: _________
Explain your working.

(ii) The distance between the first and last tree was 380m. How many trees were there? Answer: _____
Explain your working.
Modified from *Shaping Maths Activity Book 3B* (Collars, Koay, Lee, & Tan, 2015, pp. 23-24).

Figure 3. Samples of content and process prompts for Primary 3 students

As we examined the content and process journal prompts designed for the Primary 1, 2 and 3 students, we noticed some common mathematical ideas that the team needed to address every year. The ideas are similar to some of those mentioned in Siemon et al. (2012), in particular, multiplicative thinking and partitioning. The development of Big ideas of numbers by Siemon et al. was also adapted by Hurst and Hurrell (2014). Other ideas we noticed were part-whole relationship, representation such as symbolic expressions, drawings, written words or diagrams. In this chapter, we will describe briefly part-whole relationships and partitioning ideas but elaborate more on multiplicative thinking.

Big process ideas articulated by Hurst (2015) – problem solve, reason, infer, construct arguments and identify patterns – came through quite strongly in the journal writing prompts as the prompts were designed to engage the students in mathematical thinking and reasoning.

The teaching of part-whole relationships and the importance of developing the logic of considering the whole and its parts simultaneously were elaborated in Cathcart, Pothier, Vance and Bezuk, (2006, p. 90). In particular, the relationship of numbers to 5 and 10 is an important part-part-whole relationship where Ten-frame can be used to help children understand this relationship (Cathcart et al., 2006). Building on children's understanding of part-whole relationships, Task 2 for Primary 1 students required the students to separate/decompose a group (6) into parts (2 and 4) before combining/compose 8 (part) and 2 (part) to make 10. Lastly students add up the 3 parts 10, 4 and 5 to make 19 (whole) as shown in Figure 4a.

Task 9 for Primary 3 students requires the students to reason out the mass of the water bottle by applying part-whole relationship after reading the total mass of the water bottle and the apple in the scale provided in the diagram. The mass of the apple is 120g is given in the journal writing prompt. When Primary 2 students decide how they would spend their $10 for Task 6, they could also used the part-whole relationships to reason out their decisions (See Figure 4b).

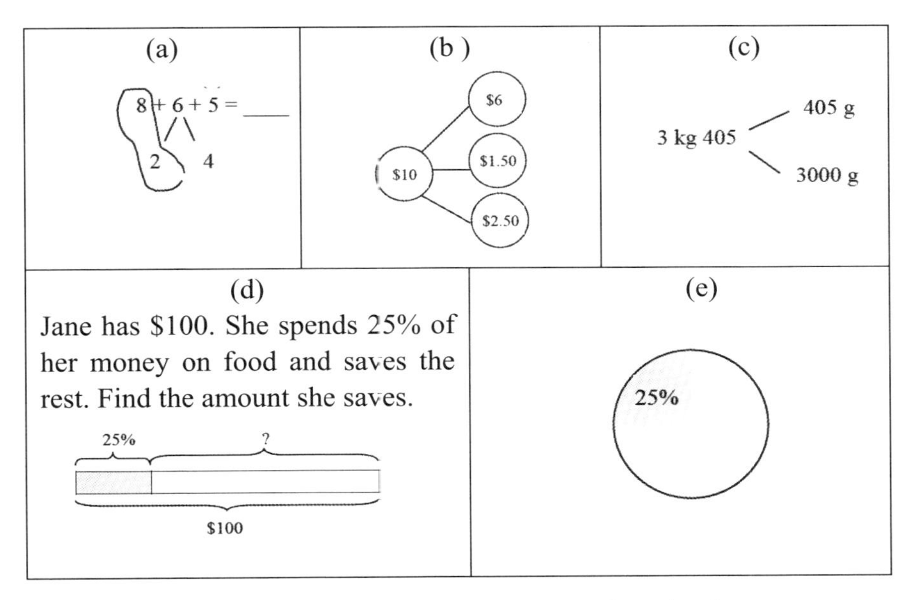

Figure 4. Part-whole relationship across grades and topics

Part-whole relationship helps students decide "when to use each operation – whatever method of computation may be appropriate" (Ashlock, 2006, p. 3) in Task 6 and 9. The value of total-part-part meanings extends to conversion of units (Figure 4c) and to upper primary topics, such as, percentages (Figure 4d), pie-chart, which shows a part of something relates to the whole (Figure 4e).

Idea of partitioning is also important in many mathematical topics such as fractions. Task 7 requires Primary 2 students to partition a given whole into several equal parts and used a part to name it a fractional part of the whole. Awareness of partitioning is one of the ideas necessary for the understanding of length measurement (Lehrer, 2003 as cited in Szilágyi, Clements & Sarama, 2013, p. 583). For example, partitioning is also required in Task 5 and 9 for students to read the scale in the ruler (Primary 2) and weighing scale (Primary 3) respectively.

3.1 *Multiplicative thinking*

Multiplicative thinking is the third Big idea in Siemon's et al. (2012) Big ideas in Numbers. Multiplicative thinking was also to some extent "considered the 'biggest' of the 'big number ideas'" in their list (Hurst and Hurrell, 2014, p. 13). "Multiplicative thinking involves recognising and working with relationships between quantities" (Siemon, 2013, p. 41). "The development of multiplicative thinking required that students construct and coordinate three aspects of multiplicative situations (groups of equal size, number of groups and the total amount" (Siemon and Breed, 2006, p. 4). Acquiring an understanding of an equal-grouping structure is at the core of multiplicative thinking (Killion & Steffe, 2002).

"Students who are not well established with multiplicative thinking do not have the foundational knowledge and skills needed to participate effectively in further school mathematics" (Siemon, Breed, Dole, Izard, & Virgona, 2006 as cited in Hurst & Hurrell, 2014, p. 9). That is because knowledge of division and of fractions relies very much on multiplicative thinking (Hurst & Hurrell, 2014). Multiplicative thinking is also fundamental to the development of many other important mathematical concepts and understandings such as algebraic reasoning, proportional reasoning, rates and ratios, measurement, and statistical sampling (Mulligan & Watson, 1998; Siemon, Izard, Breed, & Virgona, 2006). Multiplicative thinking, in particular, "supports efficient solutions to more difficult problems involving multiplication and division, fractions, decimal fractions, ratio, rates and percentage" (Siemon, 2013, p. 41). Although not discussed in this chapter, "multiplicative thinking can be addressed through the use of multiplicative arrays" (Hurst & Hurrell, 2014, p. 9) and multiplicative arrays are considered to be powerful ways to represent multiplication.

One of the ways to engage students in multiplicative thinking is when they "recognise a range of problems involving multiplication and/or division including direct and indirect proportion" (Siemon et al., 2012, p 28). That is, what is required is the development of the ability to apply these facts to a variety of situations which are founded on multiplication. Tasks 3 and 4 required the students to engage in multiplicative thinking - construct word problems for division and solve the word problems (Task 3 for Primary 1 students); communicate personal meanings of multiplication and division (Task 4 for Primary 2 students). The team recognized the multitude of connections between the notion of division and multiplication and other ideas such as fraction (Cheng & Zhao, 2011), ratio, proportion in the upper primary mathematics, hence, when designing tasks for division and multiplication for their Primary 1 and 2 students, the teachers wanted to (i) lay strong foundation for multiplicative thinking for their lower primary students (ii) make such connections explicit for themselves. This agrees with Big idea thinking that it has the capacity to develop teacher knowledge - Ma's (1999) knowledge packages and concept knots. Such deep and connected knowledge empowered the teachers to focus on relational understanding (Skemp, 1976) instead of instrumental understanding. By focusing on Big idea of multiplicative thinking in numbers led them to be more "effective concept-based teaching rather than a reliance on teaching procedures" (Hurst & Hurrell, 2014, p. 14).

The team believed that multiplicative thinking is more than just remembering and utilizing multiplication facts. The team also felt strongly that "multiplicative thinking is substantially more complex than additive thinking and may take many years to achieve (Vergnaud, 1983; Lamon, 2007)" (as cited by in Siemon, 2013, p. 41). Task 8 was designed to cultivate Primary 3 students' multiplicative thinking further by engaging in both additive and multiplicative thinking within a problem situation - Multiplicative thinking to find the number of wheels for the car and to find the number of wheels for the trucks; Additive thinking to add the total number of wheels for both types of vehicles. Task 8 used trucks and cars (*discrete countable objects*) that the children were familiar with to think multiplicatively.

Task 10 was designed for Primary 3 students to engage students in multiplicative thinking involving *discrete* countable objects as well as *measurement models*. Number line is an example of *measurement model*, that is, "lengths, rather than *discrete,* countable objects, are used to represent quantities in the problem" (Cathcart et al., 2006, p. 132). The *discrete countable objects*, the trees, and *measurement model*, the 'number line' where the trees are positioned.

Using the above two considerations, teachers will be able to continue to develop meaningful tasks to continue to develop students' multiplicative thinking (in this case, multiplication and division) for upper primary students. Two examples are shown below.

- A delivery company charges $5 for parcels delivered on time and $3.50 for parcels delivered late. In January the company collected $7298. For every 19 parcels delivered, 4 were delivered late and 15 were delivered on time.
  (a) What was the total number of parcels delivered late?
  (b) How much less money did the company collect in January because of parcels delivered late? (PSLE specimen paper from SEAB, 2009)
- A total of 18 lightbulbs are set up at an equal distance apart along three sides, AB, BC and CD of a rectangle platform. The figure shows part of the set-up. The breadth of the platform is 260 cm. What is the length of the platform? (PSLE question from SEAB, 2016)

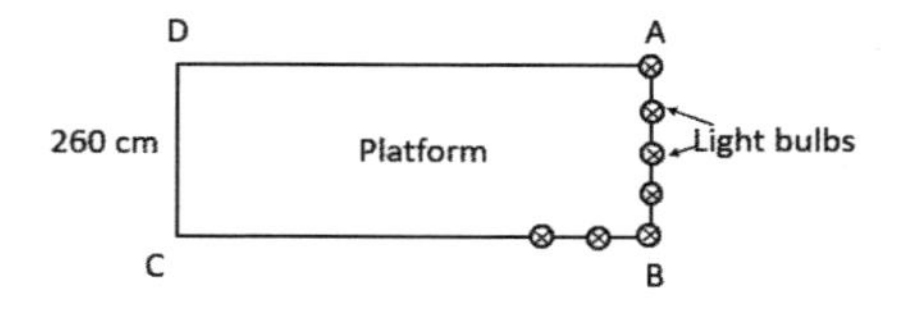

Figure 5. Diagram reproduced from SEAB (2016)

4 Implementation of Task to Engage Students

4.1 *Primary 1 students*

Our reflection with the teachers when they worked with Primary 1 students helped us realized that the design of good tasks may not always result in ways that they were engineered. The Primary 1 learners who attempted the journal writing prompts faced the challenge of expressing themselves in the prompts and they require much help from the teachers. The students needed first "to be able to communicate mathematics with the teacher freely and confidently" (Bin Amir & Fan, 2001, p. 4) for the teachers to develop their ability to express and represent their mathematical ideas and thinking. Hence, the team found themselves providing more opportunities during classroom instruction for their Primary 1 students to articulate and represent their mathematical ideas and reasoning through materials, drawings, words and mathematical symbols.

To provide greater connections among the various representations, the learning experiences for the teaching of specific concepts, for example, multiplicative thinking in building the 4 times table from Cheng (2014), was shared with the team. The representations and connections among the various representations needed to be emphasized to help make Big ideas of multiplicative thinking, part-whole relationships, partitioning etc. stick with the students. Figure 6a shows the worksheet designed by teachers for their Primary 3 students to think multiplicatively when building the 6 times table by making explicit connections between the pictorial and abstract representations. Students then reason out the solutions to the tasks in Figure 6b by building from the mathematical idea and reasoning developed in Figure 6a.

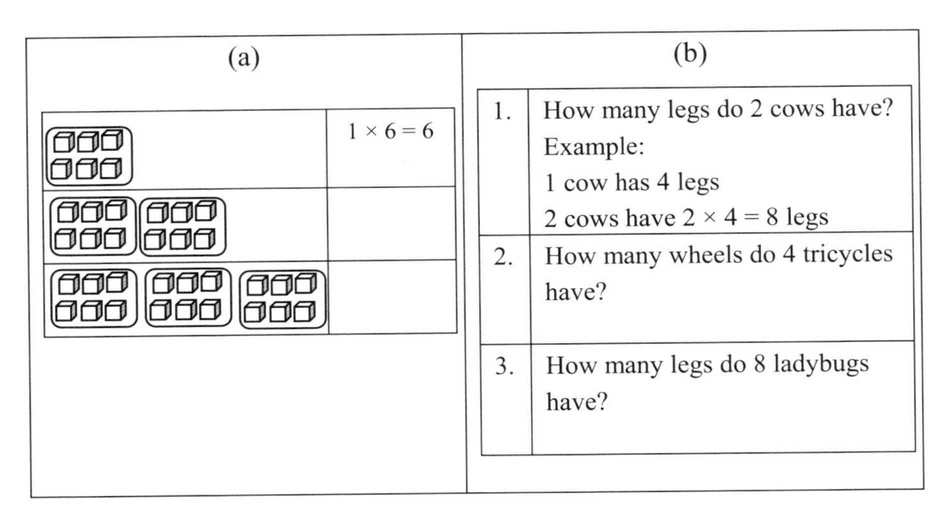

Figure 6. Explicit connections between representations

For the Primary 1 young learners, their motivation towards learning mathematics were the teachers' utmost consideration. Because the young learners are still developing their language and literacy skills, the teachers needed to provide adequate scaffold for the Primary 1 to express their mathematical ideas and reasoning. "Some of the students may have strong mathematical knowledge and ability but do not have sufficient facility to express their understanding" (Krpan, 2001, p.13). Hence, it is vital that scaffold ensures cognitive and emotional support and eventually transfer of responsibility (Hughes, 2015; Neitzel & Stright, 2003, 2004). The following three stages of writing were used by the teachers for the Primary 1 students:

- Stage 1: Wrote as a class using a journal writing prompt (e.g. Figure 7a) – provided the vicarious experience (Bandura, 1997) for the students. Students listened and observed how and why representations of the 2 division situations were constructed.
- Stage 2: Students wrote a similar prompt in pairs (e.g. Figure 7b) – provided cognitive and emotional support. Feedback was given to the pair-writing.
- Stage 3: Students wrote a similar prompt independently (e.g. Figure 7c) – Transfer of responsibility to individual learners to develop independence and control.

Further scaffolding were provided to primary 1 students who still struggled with expressing their mathematical ideas and reasoning at the end of the 3 stages of implementation of the journal writing prompts (see Figure 8).

(a) Stage 1 Write together

Mrs Tan has 20 sweets. She packs them equally into 5 bags. How many sweets are there in each bag? Draw how you figure it out.

There are _____ sweets in each bag.

Mrs Tan has 20 sweets. She packs 5 sweets in each bag. How many bags did she use? Draw how you figure it out.

Compare the 2 problems. How are they different?

(b) Stage 2 Write in pairs

Use the cubes below to write a division word problem. Solve the problem. Draw how you figure it out.

(c) Stage 3 Write independently

Write a division word problem based on the picture below. Solve the problem. Draw how you figure it out.

Figure 7. Journal writing prompts - implementation for Primary 1 students

Write or draw what multiplication means to you.
You may use these words to help you.

equal group total number

Figure 8. Helping words in journal writing prompt.

4.2 *Primary 2 and 3 students*

The Primary 2 teachers worked on some of the journal writing prompts as a class (Stage 1) before providing a similar but more challenging prompt for the students to work on individually (Stage 3). Figure 9a & 9b show examples of the journal writing prompts designed by the Primary 2 teachers to develop part-whole relationship. The students in Primary 3 were expected to work on the journal writing prompts independently (Stage 3).

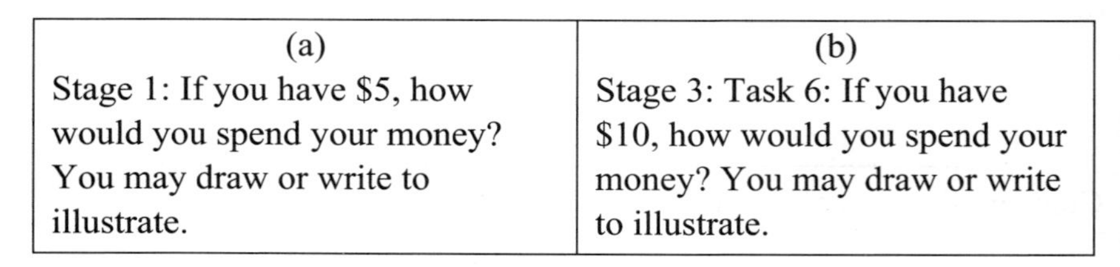

(a)	(b)
Stage 1: If you have $5, how would you spend your money? You may draw or write to illustrate.	Stage 3: Task 6: If you have $10, how would you spend your money? You may draw or write to illustrate.

Figure 9. Journal writing prompts implementation for Primary 2 students.

5 Implications for Designing Journal Writing Prompts

Similar to Leong, Cheng, Toh, Kaur and Toh (2019), the teachers referred to some base materials (instructional materials that teachers refer to when they prepare their instructional materials) when preparing the journal writing prompts. They studied the journal writing prompts in the students' workbook and came up with additional content and process prompts to fill up gaps in the students' learning. The prompts the teachers designed reflects the Big ideas they had in mind for their learners. That is, the teachers were very clear of the intended goals of the journal writing prompts in order to maximise the learning opportunities afforded by the prompts.

The design of the "tasks do not exist separately from the pedagogies associated with their use" (Sullivan, Knott & Yang, 2015, p. 84). Because the journal writing prompts for each topic focus on specific Big ideas, in the teaching of each topic, the teachers had to make explicit links between representations, and highlight critical ideas in their instruction to prevent the development of any misconceptions. Thus, by focusing on Big ideas, the journal writing prompts acted as a springboard for classroom instruction to be very clear on the overarching goal of a topic, across topics and grade levels. Figure 10 summarises the process the teachers engaged in when designing and enacting the journal writing prompts.

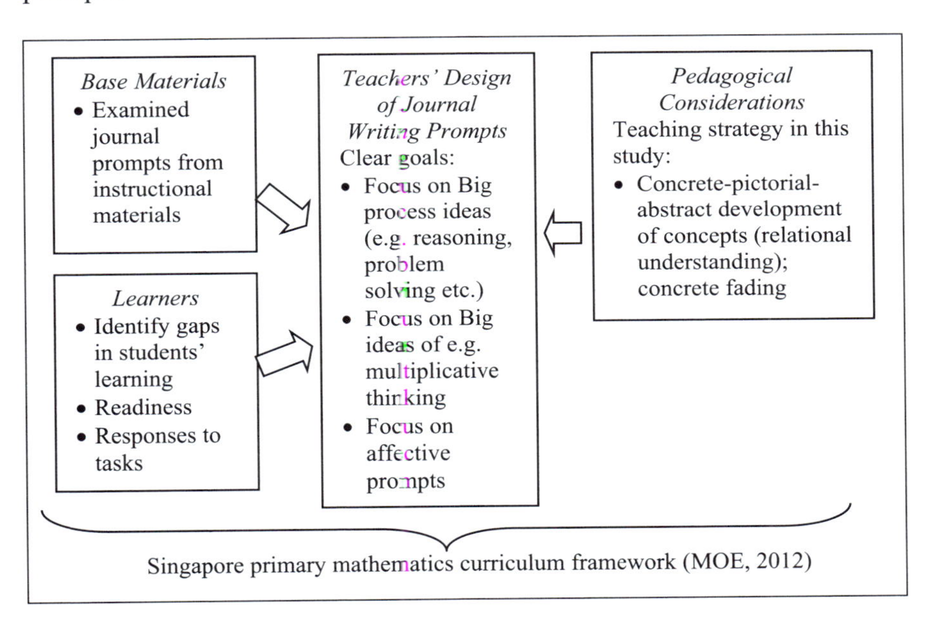

Figure 10. Process in designing journal writing prompts in this study adapted from Leong et al. (2019)

The teachers provided several reasons why their effort in this study supported students' development of mathematical processes (e.g. reasoning, communication and connections) (i) worksheet that helped them draw explicit connections between the pictorial to abstract representation (ii) worksheet that encouraged pattern recognition and generalisation e.g. identify the "groups" and "what is to be found" in the problem in Figure 6a. (iii) space and time for verbal 'mathematical' communication in the mathematics classroom e.g. have students to explain why the operation "+" or "×" is used in a given problem situation. As such, the teachers could see for themselves the struggles that students have as they progress from additive to multiplicative thinking and design appropriate learning experiences to help students transit to multiplicative thinking.

Acknowledgement

The authors wish to thank the teachers and the school who participated in the research to design journal writing prompts for the cultivation of mathematical thinking and reasoning in the primary mathematics classrooms.

References

Ashlock, R. B. (2006). *Error patterns in computation: Using error patterns to improve instruction.* Upper Saddle River, NJ: Pearson/Merrill Prentice Hall

Askew, M. (2013). Big ideas in primary mathematics: Issues and directions. *Perspectives in Education, 31*(3), 5-18.

Bandura, A. (1997). Self-efficacy: The exercise of control. New York: W. H. Freeman.

Bin Amir, Y., & Fan, L. (2002). *Exploring how to implement journal writing effectively in primary mathematics in Singapore.*

Borasi, R., & Rose, B. J. (1989). *Educational Studies in Mathematics, 20*(4), 347-365.

Cathcart, W., Pothier, Y., Vance, J., & Bezuk, N. (2006). *Learning mathematics in elementary mathematics and middle schools* (4th ed.). Columbus, OH: Merrill Prentice Hall.

Charles, R. I. (2005). Big ideas and understandings as the foundations for elementary and middle school mathematics. *Journal of Mathematics Education Leadership, 7*(3), 9-24.

Cheng, L. P. (2014). Teaching for mathematical abstraction. In P. C. Toh, T. L. Toh & B. Kaur (Eds.), *Learning experiences to promote mathematics learning* (pp. 131-152). Singapore: World Scientific.

Cheng, L. P. & Zhao, D. S. (2011). Making connections. *Mathematics Teaching, 223*, 23-25.

Collars, C., Koay, P. L., Lee, N. H., & Tan, C. S. (2014a). *Shaping maths activity book 2A*. Singapore: Marshall Cavendish Education.

Collars, C., Koay, P. L., Lee, N. H., & Tan, C. S. (2014b). *Shaping maths activity book 2B*. Singapore: Marshall Cavendish Education.

Collars, C., Koay, P. L., Lee, N. H., & Tan, C. S. (2015). *Shaping maths activity book 3B*. Singapore: Marshall Cavendish Education.

Dougherty, B. J. (1996). The write way: A look at journal writing in first-year algebra. *The Mathematics Teacher, 89*(7), 556-560.

Fong, H. K., Ramakrishnan, C., & Lau, B. P. W. (2013). *My pals are here! Maths workbook 1B*. Singapore: Marshall Cavendish Education.

Fong, H. K., Ramakrishnan, C., & Choo, M. (2014). *My pals are here! Maths workbook 2B*. Singapore: Marshall Cavendish Education.

Hughes, C. H. (2015). The transition to school. *The Psychologist, 28*, 714-717.

Hurst, C. (2015). New curricula and missed opportunities: Crowded curricula, connections, and big ideas. *International Journal for Mathematics Teaching and Learning*. Retrieved from https://www.cimt.org.uk/ijmtl/index.php/IJMTL

Hurst, C., & Hurrell, D. (2014). Developing the big ideas of number. *International Journal of Educational Studies in Mathematics, 1*(2), 1-18.

Killion, K., & Steffe. L. P. (2002). Children's multiplication. In D. L. Chambers, (Ed.), *Putting research into practice in the elementary grades: Readings from Journals of the NCTM* (pp. 90-92). Madison, Wisconsin: Wisconsin Centre for Education Research.

Krpan, C. M. (2001). *The write math: Writing about math in the classroom.* Parsippany, NJ: Dale Seymour Publications.

Lamon, S. J. (2007). Rational numbers and proportional reasoning: Towards a theoretical framework for research. In F. K. Lester (Ed.), *Second handbook of research on mathematics teaching and learning* (pp. 629-667). Charlotte, NC: Information Age Publishing.

Lehrer, R. (2003). Developing understanding of measurement. In J. Kilpatrick, W. G. Martin, & D. E. Schifter (Eds.), *A research companion to principles and standards for school mathematics* (pp. 179-192). Reston, VA: National Council of Teachers of Mathematics.

Leong, Y. H., Cheng, L. P., Toh, W. Y., Kaur, B., Toh, T. L. (2019). Making things explicit using instructional materials: A case study of a Singapore teacher's practice. *Mathematics Education Research Journal, 31(1),* 47-66.

Lord, E., & Kiddle, A. (2016). What do we really mean when we talk about big ideas in mathematics? *Mathematics Teaching, 252*, 30–31.

Ma, L. (1999). *Knowing and teaching elementary mathematics.* Mahwah, NJ: Erlbaum.

Ministry of Education (MOE). (2012). *Primary mathematics teaching and learning syllabus 2013*. Singapore: Author.

Mulligan, J. T., & Watson J. (1998). A developmental multi-modal model for multiplication and division. *Mathematics Education Research Journal, 10(2), 61-86.*

National Council of Teachers of Mathematics (NCTM). (2014). *Principles to action: Ensuring mathematical success for all*. Reston, VA: NCTM.

Neitzel, C., & Stright, A. D. (2003). Mothers' scaffolding of children's problem solving: Establishing a foundation of academic self-regulatory competence. *Journal of Family Psychology, 17*(1), 147-159.

Neitzel, C., & Stright, A. D. (2004). Parenting behaviours during child problem solving: The roles of child temperament, mother education and personality, and the problem-solving context. *International Journal of Behavioral Development, 28*(2), 166-179.

Schifter, D., & Fosnot, C. T. (1993). *Reconstructing mathematics education: Stories of teachers meeting the challenge of reform*: New York: Teachers College Press.

Schifter, D., Russell, S. J., & Bastable, V. (1999). Teaching to the big ideas. In M. Solomon (Ed.), *The diagnostic teacher: Revitalizing professional development* (pp. 22-47). New York: Teachers College Press.

Siemon, D. (2013). Launching mathematical futures: The role of multiplicative thinking. In S. Herbert, J. Tillyer, & T. Spencer (Eds.), *Proceedings of the 24th biennial conference of the Australian Association of Mathematics Teachers: Mathematics: Launching Futures* (pp. 36-52). Adelaide: AAMT.

Siemon, D., Bleckly, J., & Neal, D. (2012). Working with the Big Ideas in Number and the Australian Curriculum: Mathematics. In B. Atweh, M. Goos, R. Jorgensen, & D. Siemon, (Eds.), *Engaging the Australian National Curriculum: Mathematics: Perspectives from the field* (pp. 19-45). Online Publication: Mathematics Education Research Group of Australasia.

Siemon, D., & Breed, M. (2006, December). *Assessing multiplicative thinking using rich tasks.* Paper presented at the annual conference of the Australian Association for Research in Education, Australia. Retrieved from https://aare.edu.au/data/publications/2006/sie06375.pdf

Siemon, D., Izard, J., Breed, M., & Virgona, J. (2006). The derivation of a learning assessment framework for multiplicative thinking. In J. Novotna, H. Moraova, M. Kratka, & N. Stehlikova (Eds.), *30th Conference of the International Group for the Psychology of Mathematics Education* (Vol. 5, pp. 113-120). Prague: PME.

Singapore Examinations and Assessment Board (SEAB). (2016). *PSLE Mathematics Examination Questions 2015-2017*. Singapore: Educational Publishing House.

Singapore Examinations and Assessment Board (SEAB). (2009). *PSLE Mathematics Examination Questions 2008-2012*. Singapore: Educational Publishing House.

Skemp, R. (1976). Relational understanding and instrumental understanding. *Mathematics Teaching, 77*, 20-26.

Stein, M. K., Grover, B. W., & Henningsen, M. (1996). Building student capacity for mathematical thinking and reasoning: An analysis of mathematical tasks used in reform classrooms. *American Educational Research Journal, 33*(2), 455-488.

Sullivan, P., Knott, L., & Yang, Y. (2015). The relationships between task design, anticipated pedagogies, and student learning. In A. Watson & M. Ohtani (Eds.), *Task design in mathematics education: An ICMI study 22* (pp. 83-114). Cham: Springer International Publishing.

Szilágyi, J., Clements, D. H., & Sarama, J. (2013). Young children's understandings of length measurement: Evaluating a learning trajectory. *Journal for Research in Mathematics Education, 44*(3), 581-620.

Vergnaud, G. (1983). Multiplicative structures. In R. Lesh & M. Landau (Eds.), *Acquisition of mathematics concepts and processes* (pp. 127–173). New York: Academic Press.

Chapter 18

Enhancing Students' Reasoning in Mathematics: An Approach using Typical Problems

Jaguthsing DINDYAL

One way to enhance students' learning of mathematics is to develop their reasoning skills. While reasoning is one of the most important aspects of mathematics, it remains an elusive concept for many teachers and is quite often implicit rather than explicit in their day-to-day teaching. It is important for teachers to have an explicit focus on reasoning, for which, they need to identify key skills that enhance students' reasoning. In this chapter, I enumerate some of the most important skills that students need to develop to demonstrate their reasoning skills and I also focus on an approach that uses typical problems to generate tasks, either by modifying the typical problem or using the typical problem as a source for an entirely new problem that can help students develop their reasoning skills.

1 Introduction

In secondary school mathematics, students may be lured by the procedural aspects and focus on learning only the "how" at the expense of the "how and why". While this instrumental approach (see Skemp, 1987) may serve an immediate purpose, it would hold hidden traps for the future, as it denies the majority of students, including low performing students, an opportunity to see mathematics from a higher vantage point. The "why" is important in developing students' reasoning so that they can see the connections between the different concepts they learn in mathematics. "Promoting students' reasoning helps the teacher assess students' understanding and plan appropriate instruction" (Carroll, 1999, p. 248).

Also, Learning Experiences (LEs) are an integral part of the secondary mathematics syllabus. The Ministry of Education in Singapore (MOE) highlighted that one of the generic LEs that focuses on the development of good learning habits and skills should give students opportunities to discuss, articulate, and develop ideas to develop *reasoning skills* (MOE, 2012). For many teachers, reasoning is implicit in their teaching of mathematics rather than explicit. An explicit focus on reasoning by teachers is essential for helping students to develop their reasoning skills. However, such a focus on reasoning in the mathematics class requires teachers to identify what the essential reasoning skills are. In this chapter, first, I will focus on what reasoning is and what some reasoning skills that students need to demonstrate are. I will then illustrate how to use "typical problems" that are easily accessible to teachers, to create tasks that enhance students' reasoning in mathematics so as to meet the dual purpose of developing students' procedural skills as well as conceptual fluency that is suitable for students across all performance levels.

How can we develop students' reasoning skills? First, and foremost, what do we mean by reasoning in mathematics?

2 Reasoning

Without reasoning, there is no mathematics. The National Council of Teachers of Mathematics (NCTM, 2000) emphasized that, "Reasoning is an integral part of doing mathematics (p. 262). Hanna (2014) described reasoning as the common human ability to make inferences deductive or otherwise. On the other hand, Lithner (2008) described reasoning as, "the line of thought adopted to produce assertions and reach conclusions" (p. 257), whereas Hanna and Lithner make no statement about the domain in which the reasoning takes place or about the validity of the reasoning, On the other hand, O'Daffer and Thornquist (1993) described mathematical reasoning as a part of mathematical thinking that involves forming generalizations and drawing *valid* conclusions about ideas and how they are related. These descriptions of reasoning and/or mathematical reasoning bring to the fore certain key ideas that reasoning involves using some kind of *inferences* to form some kind of *generalizations*. O'Daffer and Thornquist have also highlighted the two key types of reasoning in mathematics: *inductive reasoning* and *deductive reasoning*. However, other types of reasoning have also been surfaced in the mathematics education literature, for example: proportional reasoning, analogical reasoning, spatial reasoning, quantitative reasoning and some domain specific types of reasoning such as algebraic reasoning, geometric reasoning or statistical reasoning. This still begs the question, "How do we develop students' reasoning skills?"

One important idea put forward by NCTM (2000) was that, "Reasoning mathematically is a habit of mind, and like all habits, it must be developed through consistent use in many contexts" (p. 56). One way to look at strategies to develop students' reasoning skills is to identify the skills that students need to demonstrate when they are reasoning. In an earlier document, NCTM (1989) highlighted some of the skills connected to reasoning in mathematics that the students in grades 5 to 12 were required to develop. Amongst others, the NCTM (1989) document mentioned that students should be able to recognize and apply deductive and inductive reasoning; make and test conjectures; formulate counterexamples; follow logical arguments; judge the validity of arguments; and construct simple valid arguments. If teachers have to develop these skills in students, then they should be able to create new tasks, modify or adapt existing tasks that can help students develop those reasoning skills. Typical problems are very flexible in meeting these requirements.

3 Typical Problems

Teachers use a lot of typical problems in their day-to-today teaching. Typical problems are described as:

> … standard examination-type questions or textbook-type questions which focus largely on developing procedural fluency and at times, conceptual understanding ... These questions can be solved by students in a much shorter time than challenging tasks, and are used frequently in mathematics lessons. Given the omnipresence of such questions in textbooks and other curriculum materials, we see typical problems as an untapped resource, which can be used to orchestrate learning experiences on a day-to-day basis. Using tasks developed from typical problems to orchestrate learning experiences would position mathematical learning experiences as an integral part of our mathematics lessons, and not just reserved for the occasional "enrichment" lessons. (Choy & Dindyal, 2017, p. 158)

A typical problem can be contextual as described below:

> Mary puts six identical red balls and four identical blue balls in a bag. She then takes out two balls at random from the bag, without replacement. Find as a fraction in its simplest form that she draws one red ball and one blue ball.

Typical problems can also be without any specific context as shown below:

> Factorize $x^2 - 3x - 4$.

Typical problems can be modified or reformulated to test a wider range of skills (see Dindyal, 2018) but can also become the source for generating new problems relevant for class discussions on a particular topic. Typical problems have been generally used to develop procedural skills but it is possible to use typical problems for developing conceptual fluency. If we apply Gibson's (1986) ideas to typical problems and consider the teacher to be the observer, then we can state that: (1) an affordance for using a typical problem exists relative to the action and capabilities of the teacher, (2) the existence of the affordance is independent of the teacher's ability to perceive it, and (3) the affordance does not change as the needs and goals of the teacher change. Affordances in relation to an observer could be positive or negative which in our context may lead to productive or less productive use of the problems in class by the teacher. Gibson also stated that "To realize the positive affordance of something, magnify its optical structure to that degree necessary for the behavioral encounter (p. 124)". As such, the affordances of typical problems can be perceived only when these problems are reformulated in different ways by the teacher. The typical problem then becomes the source of a new problem or new idea that the teacher can use in class to orchestrate the lesson and go beyond the routine use of developing procedural skills.

4 Use of Typical Problems to Develop Reasoning Skills

In what follows, I highlight a few ways in which a typical problem can be used to further students' reasoning skills. The typical problem can be modified or *can become the source of an entirely new problem* that can elicit the required reasoning skills from students. I will sequentially look how typical problems can be used to help students develop the following: forming conjectures; generating counterexamples; using inductive reasoning; using deductive reasoning; verifying conclusions; judging the validity of arguments; and constructing valid arguments.

4.1 *Conjectures*

Basically, a conjecture is a statement that is put forward for reasoning. It is not yet proved, for if it is proved, then it will be a theorem. The National Council of Teachers of Mathematics (NCTM, 2000) described a conjecture as an informed guessing, which is a major pathway to discovery. The NCTM document emphasized that teachers can help students make conjectures by asking questions such as: What do you think will happen next? What is the pattern? Is this true always or sometimes? Just making the conjecture is not enough. "Students at all grade levels should learn to investigate their conjectures using concrete materials, calculators and other tools, and increasingly through the

Table 1

Some examples of conjectures based on typical problems

Typical problem	**A conjecture based on the typical problem**	**Ask for an investigation of the conjecture**
Find the area of a triangle having base 10 cm and height 8 cm.	Each triangle has only one base and one height.	Is this statement true?
Name a quadrilateral with only 2 lines of symmetry.	Quadrilaterals can only have an even number of lines of symmetry.	Is this statement true or false? Give reasons.
How many faces has a cube?	A solid with six faces is a cube.	True or false? Give reasons for your answer.
The sum of three consecutive even numbers is 60. Find the numbers.	The sum of two even numbers is even.	Justify your answer.
Write down a set of five numbers for which the standard deviation is zero.	The standard deviation of a set of numbers can never be negative.	Why? Explain.
Draw any quadrilateral. Measure its exterior angles.	The sum of the exterior angles of a quadrilateral is 360°.	Can you explain why?
Find the probability of getting at least one head when two coins are tossed.	The probability of an event cannot be less than zero.	Explain why.
Find the equation of the perpendicular bisector of the line segment joining (1, 3) and (5, –1).	In the Cartesian plane, the product of the gradients of perpendicular lines is –1.	Explain why this conjecture is false.

grades, mathematical representations and symbols" (NCTM, 2000, p. 57). Given the importance of conjectures, teachers should provide students opportunities to make conjectures. Typical problems can be used to elicit numerous conjectures. Some examples are provided in Table 1.

As can be seen above, typical problems can become the source of several conjectures. The formation of conjectures is an important goal in the learning of mathematics. Subsequently, students can then be asked to verify whether their conjectures are true or false. At this stage some relevant explanation should suffice. It is important to include some conjectures which can be shown to be true and some which are false. A single typical problem can be the source of several conjectures. The teacher can formulate or help students formulate those that would be most relevant for the lesson that he or she is teaching.

4.2 *Counterexamples*

It is important for students to know that if statements in mathematics are true, then they are universally true and apply to all cases within the domain of discussion. A related idea which students should be aware of is that a single counterexample is enough to justify that a statement is false. Counterexamples are very useful in showing that conjectures could be false.

Typical problems can become the source of some conjectures and students can be asked to generate suitable counterexamples as illustrated in Table 2 below.

The previous section focused on the idea of conjecture as an important aspect which can help in developing the reasoning of students. One way to show that a conjecture is false is to find a suitable counterexample. For example, $\frac{a}{a} = 1$ seems to be true for $a \in R$ but in fact is a false statement as the statement does not hold for $a = 0$. In this case, the student who finds $a = 0$, has found a suitable counterexample. Thus encouraging students to find counterexamples can be a goal in itself to help students develop their reasoning skills. As shown below, typical problems have the potential for helping students to generate counterexamples.

Table 2

Some possible counterexamples generated from typical problems

Typical Problem	Some related conjectures	Possible Counterexample
Find the perimeter and area of a rectangle with sides of lengths 8 cm and 15 cm.	The perimeter of a figure cannot have the same numerical value as its area.	A square with side 4 cm.
Write $(1+\sqrt{2})^2+(1-2\sqrt{2})^2$ in the form $a+b\sqrt{2}$.	The sum of two irrational numbers is irrational.	$\sqrt{2}+(-\sqrt{2})=0$
The sum of three consecutive even numbers is 60. Find the numbers.	The sum of two odd numbers is odd.	$3+5=8$

4.3 *Inductive reasoning*

Polya (1957) defined inductive reasoning as “the process of discovering general laws by the observation and combination of particular instances” (p. 114). On the other hand, O’Daffer and Thornquist (1993) stated that, “inductive reasoning is a mathematical reasoning process in which information about some members of a set is used to form a generalization about other or all members of the set” (p. 43). Inductive reasoning is used a lot in mathematics teaching. “Students can use inductive reasoning to search for mathematical relationships through the study of patterns” (NCTM, 2000, p. 262). For example, if the student is asked to find the next two terms in the sequence: 3, 7, 11, 15, ___, ___, then the student will have to note a pattern from the first four elements of the sequence. The student can note that each element, after the first, is obtained by adding a constant 4 to the preceding term and so the two terms would have to be 19 and 23, respectively. This type of reasoning is inductive in nature as the student is reasoning based on some information from the previous members of the set.

The typical problems, shown in Table 3 below, can be used to reinforce students' ability to reason inductively. The main strategy can be to help students identify a pattern. Suppose a student is asked to find the number of squares in on a regular 8×8 chess board. The student may proceed by finding the number of squares on a 1×1, 2×2, 3×3, and 4×4 grid sequentially and note a pattern such as 1, $1 + 4$, $1 + 4 + 9$, $1 + 4 + 9 + 16$ and come to the conclusion that the total number of squares on the chess board is $1 + 4 + 9 + \ldots + 64 = 204$. Again, the type of reasoning used by the student in this example is inductive in nature.

Table 3
Some typical problems modified to elicit inductive reasoning

Typical Problem	**Modified problem to elicit inductive reasoning**
Ten persons meet at a party and shake hands with each other only once. How many handshakes are there?	A party is attended by some persons. Each one shakes hands with the others only once. Find the number of handshakes if there are (i) 2 persons, (ii) 3 persons, (iii) 4 persons, (iv) 5 persons, and (v) n persons.
Find the number of diagonals in a convex pentagon.	Find the number of diagonals in a convex polygon of (i) 4 sides, (ii) 5 sides, (iii) 6 sides, and (iv) n sides.
Write down the sum of the interior angles of a polygon of 10 sides.	Write down the sum of the interior angles of a polygon of (i) 4 sides, (ii) 5 sides, (iii) 6 sides, and (iv) n sides.
Write down the sum of the exterior angles of a quadrilateral.	Write down the sum of the exterior angles of a polygon of (i) 4 sides, (ii) 5 sides, (iii) 6 sides, and (iv) n sides.

While inductive reasoning is very important in the teaching and learning of mathematics, some caution is in order. Consider this situation in a class where a teacher leads students to measure the external angles of a triangle, after which the students conclude that the sum of the exterior angles of a triangle is 360°. While the generalization is correct for all triangles, this method does not constitute a proof. Also, a student may note that the diagonals in a square and a rhombus bisect the opposite angles and may conclude incorrectly that the diagonals of a rectangle bisect the opposite angles in the figure. In general, generalizations formed by inductive reasoning are not always true. Another point to note here is that although conclusions from inductive reasoning are not always true, proof using mathematical induction is a valid method of proof and is deductive in nature.

4.4 *Deductive reasoning*

Mathematics is essentially deductive in nature. The process of deduction involves reasoning logically from generalized statements or premises to conclusions about particular cases (Greenes & Findell, 1999). Unlike inductive reasoning which lead to conclusions that are not always valid, deductive reasoning guarantees that the conclusions are true if the starting premises are true and the logic used has no flaw in it. For example, a student who knows that a parallelogram is a quadrilateral in which the opposite sides are parallel can conclude that a square has two pairs of opposite sides parallel and therefore is a parallelogram. This type of reasoning is deductive in nature.

Students use deductive reasoning in mathematics often without being aware that they are doing so. Nickerson (2004) distinguished between two types of reasoning in mathematics: (1) reasoning to arrive at a conclusion, and (2) reasoning to justify a conclusion. Students are mostly involved in reasoning to arrive at a conclusion when deductively solving problems to get a viable solution and in reasoning to justify a conclusion when proving or showing that a statement is true using deductive reasoning. Typical problems can be used as a source of new tasks for students to develop their deductive reasoning skills as illustrated in Table 4 below.

Table 4
New problems from typical problems to elicit deductive reasoning

Typical Problem	**New problem asking for a justification or proof**
Mary says that the triangle with angles $2x°$, $3x°$ and $4x°$ is acute-angled. Is she correct?	Show that the angles in a triangle add up to 180°.
Find the sum of the interior angles in a polygon of 10 sides.	Show that the sum of the interior angles in a polygon of n sides is $(n - 2) \times 180°$.
Use the formula to find the solution of the equation $2x^2 - 5x - 12 = 0$ to 2 d.p.	Derive the formula for finding the roots the quadratic equation $ax^2 + bx + c = 0, a \neq 0$.
Tom says that the area of a square is equal to half the product of the lengths of its diagonals. If a square has diagonals of lengths 10 cm, what would be its area according to Tom?	Tom says that the area of a square is equal to half the product of the lengths of its diagonals. Is Tom correct? Justify your answer.
Find the curved surface area of a cone with base radius 8 cm and height 10 cm.	Derive a formula for finding the curved surface area of cone of base radius r and height h.
A rectangle has sides of lengths 6 cm and 8 cm. Find the length of one of its diagonal.	Show that the diagonals of a rectangle are equal in length.

It is to be noted that all theorems are proved using deductive reasoning. Given the importance of deductive reasoning in mathematics, students should be encouraged to engage in tasks that elicit the students' deductive reasoning. NCTM (2000) has elaborated that increasingly over the grades students should learn to make deductive arguments based on the mathematical truths they are establishing in class.

Three other activities that follow from deductive reasoning are as important. NCTM (1989) stated that, amongst others, the assessment of students' ability to reason mathematically should provide evidence that they can use deductive reasoning to verify conclusions, judge the validity of arguments, and construct valid arguments.

4.4.1 Verify conclusions

To help students verify conclusions, we can create a scenario using a typical problem as a starting point. Suppose the typical problem is "Find the sum of the interior angles in a polygon of 10 sides." The teacher can say, "Tom has deduced that the sum of the interior angles of a polygon of 10 sides is $8 \times 180° = 1440°$". Explain how Tom arrived at this conclusion or the teacher can ask, "Can you verify if Tom's conclusion is correct or not?" Another example could be "Solve the quadratic equation $x^2 - 4x + 3 = 0$, using factorization." A scenario the teacher can create could be, "Mary says that the same values are obtained if the quadratic formula is used. Can you verify if Mary is correct?" Students should be given both conclusions which are valid and some which are invalid to verify.

4.4.2 Judge the validity of arguments

Students should be able to judge the validity of arguments in mathematics leading to a conclusion at the level they are studying. Caroll (1999, p. 253) has stated that, "One way of engaging students in the reasoning process is to have them examine and explain an error". Again, the starting point could be a typical problem. For example, "Solve the quadratic equation $x^2 - 4x + 3 = 0$." The teacher can then provide a solution and ask specific questions regarding the validity of the conclusion. For example, the teacher can say that Kevin gave the following solution:

$$x^2 - 4x + 3 = 0$$
$$(x - 1)(x - 3) = 0$$
$$x - 1 = 0 \text{ and } x - 3 = 0$$

$$x = 1 \text{ and } x = 3$$

The teacher can then ask students to comment on whether this solution is correct or not. Clearly, in this case, the solution to the problem is incorrect as x cannot be simultaneously 1 and 3. "In order to evaluate the validity of proposed explanations, students must develop enough confidence in their reasoning abilities to question others' mathematical arguments as well as their own" (NCTM, 2000, pp. 345-346)

4.4.3 Construct valid arguments

Mathematicians spend most of their time trying to construct valid arguments to prove new theorems. Typical problems can be again used as starting points. For example, the typical problem could be "Find the area of an isosceles triangle that has sides of lengths 6 cm, 8 cm and 8 cm." This problem can become the source of some related problems whereby students will have to give valid arguments. For example, the teacher can ask the students (1) to show that the angles opposite to the congruent sides in an isosceles triangle are equal, or (2) to show that the bisector of the angle between the congruent sides passes through the midpoint of the third side. The students in such cases will have to construct a sequence of valid arguments to justify whether these statements are true or not.

5 Conclusion

The ability to form conjectures, the ability to find counterexamples, the ability to use inductive and deductive reasoning, as well as the ability to verify conclusions, the ability to judge the validity of arguments and the ability to construct valid arguments are essential skills that students need to master to enhance their overall reasoning skills. These aforementioned skills are not specifically stated in common textbooks and are often implicit in the teaching of mathematics. Teachers need to be familiar with these skills so that they can develop the skills in their students. Tate and Johnson (1999) stated that, "one indicator of teacher quality is how well the teacher understands students' thinking and reasoning about mathematics and how best to extend their thinking and reasoning" (p. 225). The examples above amply demonstrate that typical problems can become a primary source of tasks that can help students develop their students' reasoning skills. The advantage of using typical problems is that they are readily available in textbooks, ten-year series of past examination papers, and common assessment books. Teachers do not have to create new tasks from scratch and as Carroll (1999) stated, "relatively short questions can assess and promote reasoning without placing a huge burden on the teacher" (p. 255). Also, it is to be noted that the tasks have to be implemented appropriately in class. Teachers must always ask students to explain their reasoning for any conclusions they make when learning mathematics.

References

Carroll, W. M. (1999). Using short questions to develop and assess reasoning. In L. V. Stiff & F. R. Curcio (Eds.), *Developing mathematical reasoning in grades K-12, 1999 Yearbook* (pp. 247-255). Reston, VA: NCTM.

Choy, B. H., & Dindyal, J. (2017). Snapshots of productive noticing: Orchestrating learning experiences using typical problems. In A. Downton, S. Livy, & J. Hall (Eds.), *Proceedings of the 40th annual conference of the Mathematics Education Research Group of Australasia: 40 years on: We are still learning!* (pp. 157-164). Melbourne: MERGA.

Dindyal, J. (2018). Affordances of typical problems. In P. C. Toh & B. L. Chua (Eds.), *Mathematics instruction: Goals, tasks and activities* (pp. 33-48). Singapore: World Scientific.

Gibson, J. J. (1986). *The theory of affordances: The ecological approach to visual perception* (pp. 127 - 143). Hillsdale, New Jersey: Lawrence Erlbaum Associates, Publisher.

Greenes, C., & Findell, C. (1999). Developing students' algebraic reasoning abilities. In L. V. Stiff & F. R. Curcio (Eds.), *Developing mathematical reasoning in grades K-12, 1999 Yearbook* (pp. 127-137). Reston, VA: NCTM.

Hanna, G. (2014). Mathematical proof, argumentation, and reasoning. *Encyclopedia of mathematics education*, 404-408.

Lithner, J. (2008). A research framework for creative and imitative reasoning. *Educational Studies in Mathematics*, 67(3), 255-276.

Ministry of Education (2012). *Ordinary-level & normal (academic)-level mathematics teaching and learning syllabus*. Singapore: Author.

National Council of Teachers of Mathematics (1989). *Curriculum and evaluation standards for school mathematics*. Reston, VA: NCTM

National Council of Teachers of Mathematics (2000). *Principles and standards for school mathematics*. Reston, VA: NCTM

Nickerson, R. S. (2004). Teaching reasoning. In J. P. Leighton & R. J. Steinberg (Eds.), *The nature of reasoning* (pp. 410-442). Cambridge, UK: Cambridge University Press.

O'Daffer, P. G., & Thornquist, B. A. (1993). Critical thinking, mathematical reasoning, and proof. In P. S. Wilson (Ed.), *Research ideas for the classroom: High school mathematics* (pp. 39-56). Reston, VA: NCTM.

Polya, G. (1957). *How to solve it: A new aspect of mathematical method* (2nd ed.). Princeton, NJ: Princeton University Press.

Skemp, R. R. (1987). *The psychology of learning mathematic*s. Hillsdale, NJ: Lawrence Erlbaum Associates.

Tate, W. F, & Johnson, H. C. (1999). Mathematical reasoning and educational policy: Moving beyond the politics of dead language. In L. V. Stiff & F. R. Curcio (Eds.), *Developing mathematical reasoning in grades K-12* (1999 Yearbook, pp. 221-233). Reston, VA: NCTM.

Chapter 19

Big Ideas from Small Ideas

LEONG Yew Hoong

In this chapter, I illustrate – using one case – how we can explore big ideas by starting with 'small ideas', as small as ideas derived from a few exercise items taken from one page of a textbook. But this is only the beginning of a virtually unending journey towards increasingly bigger ideas. The methodology of inquiry illustrated is recommended as a course of analysis for mathematics teachers who is attracted to the same pursuit.

1 What are "Big Ideas"?

This is usually the first big question we ask. We can take reference from curriculum panels of other jurisdictions that have gone ahead of the Singapore experience; in particular, we can start with this definition taken from a series of books commissioned by the National Council of Teachers of Mathematics (NCTM, 2015):

> Big ideas are mathematical statements of overarching concepts that are central to a mathematical topic and link numerous smaller mathematical ideas into coherent wholes. (p. viii)

This is helpful, but once we begin, as teachers, to start thinking of how we can plan to teach for big ideas, we run into roadblocks that can be presented in the form of these questions:

- What count as big ideas – how big is big? What does a "topic" mean? In concrete terms, is "Prime Factorisation" a topic, or must it be couched in something bigger, like "Highest Common Factor and Lowest Common Multiple", or even bigger, like "Integers"?
- How do I know that I got it 'right'? I may have identified a big idea for a topic and others identify others – is there a big idea for a topic that is more "central" than others?

These are important questions. And, there is no agreement to the exact answer to these questions (e.g., Charles, 2005; Siemon, Bleckly, & Neal, 2012). For this chapter, I do not intend to start by addressing these questions. But I do hope that the readers will find that they are at least partially addressed by the time we reach the end of the chapter. At this juncture, I like to insert this observation that we do not need to be totally clear about what big ideas are before we commit to the big idea of teaching mathematics for big ideas. There are two terms in the quotation that render teaching for big ideas particularly attractive – "link" and "coherent". These emphasise the importance that when we teach mathematics, we are not merely covering a sequence of disconnected contents or techniques; rather, we are also interested in presenting ideas that are beyond the particularities of specific cases or instances and that tie them together under an overarching umbrella. Teaching for big ideas is thus a pedagogical vision of teachers and students jointly engaging in regularly asking the question, "So what are the mathematical big ideas that tie these stuff that we have been doing?"

And this question actually motivates the approach I take for the remaining of this chapter: We can explore big ideas by starting (not with asking what these big ideas are but) with smaller more familiar objects.

2 What is the Big Idea behind these Questions?

We can start with objects as small as exercise items commonly found in textbooks. To illustrate, I extract these items from Discovering Mathematics (Chow, 2014) – a secondary mathematics textbook used in

Singapore (and it is assumed the sequence of items is similar for other Singapore textbooks):

Simplify these expressions.

(i) $\dfrac{6a^2b^3}{15ab^4}$

(ii) $\dfrac{c^2}{c^2 - cd}$

(iii) $\dfrac{(m-2)^2}{m^2 - m - 2}$ (p. 81)

First, we discuss the common ways in which the solutions of these items are presented to students. For (i), the expression can be simplified – usually through separate consideration of the numbers and the algebraic portions – using 'cancellations' of the original symbols, replacing them where appropriate with the simplified symbols, resulting in $\dfrac{2a}{5b}$; for (ii), there is an intermediate step of factorising the denominator before doing the cancellation of "c" to obtain the answer of $\dfrac{c}{c-d}$; similarly, the factorisation of the denominator for (iii) is done before the cancellation of $(m-2)$ to result in $\dfrac{m-2}{m+1}$.

If one adopts the instructional approach as described above, it would seem that the overarching commonality across the items is "cancellation". So, is "cancellation" the big idea in this 'topic'? But even a cursory reflection would cause one to doubt this conclusion – because, as is stated in the earlier NCTM quote, "big ideas are *mathematical* statements ..." Surely, there is nothing approximately mathematical about the act of cancellation in itself. [Of course, the weakness of fronting cancellation as the main focus goes beyond the fact that it does not reveal anything mathematical underlying the simplifying processes. See, for example, Lee and Wheeler (1989). Practically, it can breed errors of arbitrary cancellation as any experienced teachers would testify

– such as, in the case of (ii), it is possible for students to arbitrarily cancel c^2 in both the numerator and the denominator, leaving $\frac{1}{-cd}$.]

If cancellation should not be the big idea here, what is? Perhaps it is "factorisation". After all, we did factorisation for (ii) and (iii). This is certainly a better big idea than cancellation because factorisation is at least a *mathematical* concept. But upon closer examination, there is more to merely factorisation – recall that in this context, the factorisation is for the purpose of simplifying the fractions. So we continue to drill deeper in search of greater clarity of the big idea in this context. The following line of presentation of (ii) may show the lead to a sharper big idea:

$$\frac{c^2}{c^2 - cd} = \frac{c \times c}{c \times (c - d)} = \frac{c}{c} \times \frac{c}{c - d} = 1 \times \frac{c}{c - d} = \frac{c}{c - d}$$

From the above, we can see that factorisation forms the first step – and a crucial one – that enables us to 'see' factors both in the numerator and the denominator that are common, and because they are equal to one (in this case, $\frac{c}{c} = 1$), and $1 \times A = A$, we can 'simplify' by not including the "1 ×" in the final expression. Thus, to refine, the big idea may be phrased as: "To simplify fractions, we factorise to look for common factors in both the numerator and the denominator". To show that this is a feasible big idea across all the three items (and for this matter, all other items under the same topic on "simplifying algebraic fractions"), I illustrate similar lines of presentation for the other two items:

$$\frac{6a^2b^3}{15ab^4} = \frac{2 \times 3 \times a \times a \times b \times b \times b}{3 \times 5 \times a \times b \times b \times b \times b} = \frac{3}{3} \times \frac{a}{a} \times \frac{b^3}{b^3} \times \frac{2 \times a}{5 \times b} = 1 \times \frac{2a}{5b} = \frac{2a}{5b}$$

$$\frac{(m-2)^2}{m^2 - m - 2} = \frac{(m-2) \times (m-2)}{(m-2) \times (m+1)} = \frac{m-2}{m-2} \times \frac{m-2}{m+1} = 1 \times \frac{m-2}{m+1} = \frac{m-2}{m+1}$$

[A case can be made that in all the three items, since the result $\frac{ac}{bd}=\frac{a}{b}\times\frac{c}{d}$ is repeatedly used, that this too is a big idea in this situation. This would certainly be the case if we undertake the analysis without taking into consideration its local instructional context. Here, the context is teaching algebraic fractions at the secondary level. $\frac{ac}{bd}=\frac{a}{b}\times\frac{c}{d}$ is a result that is taught at the Primary level in Singapore schools. Thus, within this current context of analysis, it is a key resource that is applied in the working process, not a big idea in the sense of one that a secondary teacher would highlight as a main mathematical point to be learnt by the students at this level.]

When we look back at the process of searching for the big idea undergirding these items, we realise that these practices are helpful: (i) go beyond the conventional working steps (such as cancellation) to look for the actual *mathematical* basis for these steps (in this case, $1 \times A = A$); (ii) go beyond the working steps of one item to look for the common overarching mathematical structure (in this case, the role factorisation plays) from which each item is merely an example of.

This thought experiment also illustrates the practical advantages of teaching for big ideas (instead of presenting techniques as item type-specific) even when these big ideas are not so 'big' – merely across a few exercise items: (i) Students need only remember one (or two) big ideas – especially when it is reinforced through numerous examples – better than a bag of arbitrary unrelated rules that are item type-specific. (ii) Students are given the opportunity to make mathematical sense of the process; over time, students can indeed choose to shorten the process – that is, skip some steps – but when needed especially in cases when they are stuck, the sense-making can allow them to recover the skipped steps to overcome the obstacles. (iii) When they can fall back to the mathematical sense of the procedure, they are less likely to make mistakes that are more commonly associated with the quick cancellation method.

But this exploration of big ideas need not stop at this point. Using the metaphor of camera zoom lens, we can zoom out further to consider if there are related bigger idea(s) beyond the boundaries of this topic of "simplifying algebraic fractions".

3 What is the Big Idea across Neighbouring Topics?

As guided by the Singapore Secondary Mathematics syllabus, and followed closely by most textbooks, the neighbouring topics that are usually covered chronologically after "simplifying algebraic fractions" is the "multiplication and division of algebraic fractions", and in some cases, "addition and subtraction of algebraic fractions".

Is the big idea of "To simplify fractions, we factorise to look for common factors in both the numerator and the denominator" that we identified earlier of any relevance to these neighbouring topics, or do we consider them altogether as entirely separate once we cross the boundary of topics?

We can continue the method of analysis illustrated in the previous section: by putting exemplary items under these topics alongside each other to consider their underlying commonalties. We do so first for the topic on multiplication and division of algebraic fractions:

(iv) $\dfrac{6x^2y}{16y-8x}\times\dfrac{24y-12x}{4xy^2}$

(v) $\dfrac{x^2-3x-10}{x^2+3x+2}\div\dfrac{4x^2-20x}{2x^2+5x+3}$

It becomes clear, after the initial working steps of (iv) and (v), that the subsequent work is essentially identical to the simplification process elaborated for (i) – (iii) and is thus ellipted (as represented by "…"):

$$\frac{6x^2y}{16y-8x}\times\frac{24y-12x}{4xy^2}=\frac{6x^2y(24y-12x)}{4xy^2(16y-8x)}=\frac{12\times 6x^2y(2y-x)}{8\times 4xy^2(2y-x)}=\cdots$$

$$\frac{x^2-3x-10}{x^2+3x+2}\div\frac{4x^2-20x}{2x^2+5x+3}=\frac{x^2-3x-10}{x^2+3x+2}\times\frac{2x^2+5x+3}{4x^2-20x}=\cdots$$

Although the topic of "multiplication and division of algebraic fractions" may at first give the impression that the focus in the topic is on the operations of multiplication and division, the above shows that the weight of the 'algebraic work' still lies in the subsequent simplification. [Again, here the results of $\frac{a}{b}\times\frac{c}{d}=\frac{ac}{bd}$ for (iv) and for (v) are resources we draw upon from previously-taught topics in earlier years]. We may now place the big ideas of these adjacent topics together for easier comparison:

- For "simplification ..." topic: we factorise to look for common factors in both the numerator and the denominator.
- For "multiplication and division ..." topic: form products of fractions, then factorise to look for common factors in both the numerator and the denominator.

Clearly, the common big idea across these topics is "factorise to look for common factors in numerator and denominator". Pushing on, we consider further another neigbouring topic on "addition and subtraction of algebraic fractions". Using the same method, we highlight this prototypical example in this topic:

(vi) $\frac{x+y}{x-y}+\frac{x^2-4y^2}{x^2-y^2}-\frac{x-3y}{x+y}$

Again, we see here that factorisation is a key step in the process of combining the given fractions. Here, although the goal of factorisation differs from the earlier topics – while the previous topics look for common factors in the numerator and the denominator, for addition and subtraction, the focus is on looking for common factors in the numerator – there is nevertheless a common overarching idea behind the techniques employed in all these topics: "We factorise to enable us to see better ...", as illustrated in a different representation of (vi) above:

$$\frac{x+y}{x-y}+\frac{x^2-4y^2}{x^2-y^2}-\frac{x-3y}{x+y}=\frac{x+y}{x-y}+\frac{(x-2y)(x+2y)}{(x-y)(x+y)}-\frac{x-3y}{x+y}=\cdots$$

The above step presents more clearly the factors needed in the denominators to achieve "same denominator" in order to form a single fraction. [Again, it is acknowledged that there are other skills involved to complete the solution. The focus of our inquiry is on ideas that are central and are in common across neighbouring topics, not on the detailed analyses of each item].

Taking stock of our inquiry so far, we see that when we start with exercise items within a topic, upon careful study of their common mathematical basis, we can extract some 'small ideas'. In particular, from the topic on "simplifying algebraic fractions', we note that "we factorise to help us see the common factors in the numerator and the denominator". We then zoom-out from the topic to its neighbouring topics to consider if this big idea within a smaller domain can be somewhat generalized to a wider region of topics. As it turns out, while the setting in each topic is different – no more restricted to the purpose of looking for common factors in the numerator and denominator – the 'core' and more general idea remains: "We factorise to enable us to see better …". This more generalised statement that is adjusted for relevance for a wider range of neighboring topics can then be taken – for our purpose – as one big idea for these topics.

This realisation is potentially powerful: We now not only have a big idea for a topic; we also have a 'bigger' idea for a whole cluster of neighbouring topics which include this earlier topic, with the former big idea nested within the latter big idea. Returning to our camera zoom metaphor, we have a big idea for a zoomed-out picture of a bigger topic; but as we zoom in further to small topics within the bigger topic, we adjust the big idea by adding in more fine-grained details that take into considerations the specifics of the narrower topic. This way of thinking has pedagogical implications: it allows teachers to focus on one (or two) main ideas for a whole class of related topics, and present smaller topics within this set as specific applications of these big ideas. Doing so enables students to see these topics as tightly connected fundamentally – and hence heighten sense-making, instead of viewing the learning of mathematics as isolationist learning of a bag of unrelated rules for altogether different settings.

4 What is the Big Idea beyond Neighbouring Topics?

Pressing on, we can further ask if the big idea "we factorise to enable us to see better ..." is merely restricted to the big topic of "algebraic fractions"? Not really. I present three other places in the Singapore Syllabus where this big idea is again conspicuous.

First, within the 'near vicinity' (usually taught in the same year, if not the same semester) of algebraic fractions is the topic of "solving quadratic equations". It is clear that one key technique of solving quadratic equations is "we factorise to enable us to see [in this case, the zeroes of the quadratic equation, and hence the solution]".

Second, in a usually later treatment under the topic of "Sketching of quadratic graphs" in Secondary Three – while "Algebraic fractions" is usually done is Secondary Two – students will be required to sketch quadratic graphs, for example, of $y = x^2 + 3x - 4$. One of the ways is to re-present the quadratic function in a factorised form: $y = (x + 4)(x - 1)$. Why? This way, "we factorise to enable us to see better [in this case, the x-intercepts of the graph, which in turn, leads us to calculate the line of symmetry as x = mean of these values. These are substantial steps towards the sketching task]".

Third, in a usually earlier treatment under "prime factorisation" done in Secondary One, a textbook item such as this presents an interesting discussion:

> Find the smallest value of n such that the LCM of n and 15 is 45. (Yeap, 2013, p. 22)

In this case, since the numbers are small, the solution can be obtained by guess and check. But the re-presentation of the numbers in (prime) factorised form presents new ways of looking at the numbers and hence new insights:

$$45 = 3 \times 3 \times 5$$
$$15 = \quad 3 \times 5$$
$$n = 3 \times 3 \ldots$$

Presented this way, for the LCM (n, 15) to be 45, n must at least consist of the product of primes 3×3 in its prime factorisation. Again, this illustrates the recurrence of the big idea of "we (prime) factorise to enable us to see better [in this case, the primes needed to satisfy the LCM condition]". It is also interesting to connect prime factorisation as yet another form of factorisation, apart from the algebraic factorisation that we have focused mostly on in our analysis up to this point. It reminds us of yet another way of linking up and generalising big ideas or concepts in mathematics – "factorisation" means breaking up into factors, and it can take different special forms across the syllabus, such as prime factorisation, quadratic factorisation, etc.

The three other instances where the big idea of "we factorise to enable us to see better ..." show us that a big idea can be further zoomed-out beyond the confines of a big topic within a particular year level: we can look out for other similar occurrences of this big idea across the whole spectrum of the mathematics syllabus. Again, doing so has pedagogical implications: when students are presented with a big idea repeatedly across the year levels of study, not only are these repeated reinforcements to consolidate learning, it presents to them mathematics as it ought to be – mathematical ideas and methods are closely connected.

5 What is an even Bigger Big Idea?

Continuing with our method of inquiry, which is basically one of starting with a small area of mathematical inquiry – down to some usual textbooks questions from one exercise section of a topic – and then repeatedly asking "what is the overarching mathematical idea?" as we zoom out more and more to bigger regions. But it seems we have reached almost the limits of the secondary mathematics syllabus, can one zoom out further? Yes, in a different direction. Notice that "factorise" in this case can be seen as a special case of a more general mathematical procedure, which is "use another representation". In each of the situations that I described above, we can see (prime) factorisation as another way of representing the objects – be it algebraic expressions or numbers – so that we can better see different things that the original representation does not afford so easily. Doing the word replacement gives us this more general statement which is an even bigger idea, "We *use another representation* to enable us to see better …"

Illustrating this bigger idea, we can zoom back in to the quadratic expressions, for example, $x^2 + 3x - 4$. Apart from the factorised form, which affords us better access to seeing the roots of the equation $x^2 + 3x - 4 = 0$, is there another way to represent the expression that will help us see other essentials better? Indeed, the form, after the process of completing the square, yields $\left(x+\frac{3}{2}\right)^2 - \frac{25}{4}$. This form helps us see better the minimum value of the function and the x-value that corresponds to that minimum. This example shows us that "factorisation" – the main consideration of the earlier sections of this chapter – is but one of the many alternative representations that help us see better; thus while the big idea that "factorisation helps us to see better" is concretely relevant across a big region of secondary mathematics, the bigger idea that "using alternative suitable representations helps us see better" cuts at the core across all of mathematics. In fact, if one considers the various techniques of symbol manipulations within secondary algebra and trigonometry (e.g., expressing a single fraction into partial fractions, using trigonometric identities to convert one trigonometric function to another), they belong within the ambit of this same big idea.

This big idea is so 'big' that even whole chapters in the secondary mathematics textbooks are devoted to it: "Graphs" is an example. Graphs can be viewed as "another representation to enable us to see better [in this case, functions – whether linear, quadratic, or others]". I would imagine questions such as this is very hard for students to comprehend without resorting to using graphs as an alternative representation:

Find the range of values of k for which $kx^2 + 1 > 2kx - k$ for all real values of x. (Ho, Khor, & Yan, 2013, p. 22)

The usual stumbling block is the meaning of "$kx^2 + 1 > 2kx - k$ (or $kx^2 - 2kx + 1 + k > 0$), *for all real values* of x". But we can 'see' it more clearly when this expression is translated into the equivalent form graphically – that of the quadratic curve being above the x-axis.

Thus, there are numerous instantiations of this big idea of "using another representation …" throughout the mathematics syllabus. If we examine closely, we will chance upon them at every turn. This can make a difference pedagogically: when students are regularly reminded that, especially when we are stuck at a problem, we can employ the mathematical tools we have learnt to represent the situation differently – it can give us fresh insights into a way out of being stuck.

Summarising the journey that we have taken in Sections 2 – 5 of this chapter, I have attempted to illustrate just one case of how we can explore "big ideas" by starting with 'small ideas' derived from a page of exercise questions in a familiar textbook. Depending on how "big" we want to go, we can keep zooming-out from this small region to increasingly bigger topical landscape all the way to the whole of the secondary mathematics syllabus. This ability to see the big idea with different levels of specificity across different grain sizes does two things for us as teachers: (1) at the broadest level, it helps us see the connections across the whole syllabus and thus provides us many opportunities to make these links for our students as we teach them in a year level (or better, across year levels); but at the same time (2) at each appropriate level of zoom, we can specify concretely how the big idea can be adjusted particularly to the demands of the more 'local' situation. This flexibility of adjustment in seeing the general-and-the-particular is also a necessary disposition for successful learning of mathematics. Also, alongside these considerations, I have argued that these realisations are not merely theoretical good-to-know things; rather, there are real implications in the teaching of mathematics: as teachers, when we are fascinated by the overarching concepts tying many parts of the syllabus together, we will emphasise them in class; students will then experience mathematics less as learning isolated bits of rules-to-follow but more authentically as a coherent system based on sense-making.

I should also at this point clarify that the way I develop from small ideas to increasingly bigger ideas is but *one* line of inquiry. The same starting point of "algebraic fractions" can lead to other lines of inquiry and hence other big ideas. For example, we can look at multiplication, division, addition and subtraction of fractions from the big idea of mathematical binary operations – and again, we can find many points in the syllabus where operations are repeatedly emphasised not just in real number systems, but even in other mathematical systems (e.g., vectors, matrices).

6 Conclusion: Back to "What are big ideas?"

In Section 1 of this chapter, I began with these questions:

- What count as big ideas – how big is big? What does a "topic" mean? In concrete terms, is "Prime Factorisation" a topic, or must it be couched in something bigger, like "Highest Common Factor and Lowest Common Multiple", or even bigger, like "Integers"?

If we follow the methodology of zooming-out at various levels of grain sizes that I have illustrated, the exact definition of "big idea" and "topic" becomes irrelevant. The focus shifts rather to the match between the 'bigness' of the idea to the level of zoom appropriate for the scale of content coverage. To reiterate with the contents of our earlier discussion, "we *factorise* ..." is appropriate within the local region of "algebraic fractions", but when considering wider regions cutting across traditional 'topics', this 'big idea' is evidently over-specific, and a more general bigger idea of "we *use another representation* ..." becomes a more appropriate overarching big idea.

I conclude with an attempt to answer this question that I wish more teachers would ask, "How can I develop the expertise to explore increasingly big ideas from small ideas, as illustrated in this chapter?" To many readers, the methodology that I delineate here may not come natural to us. For example, we may have been teaching simplification of fractions by direct cancellation without exploring the deeper mathematical basis for a long time now – to the extent that, unless pointed out by others, we would have been 'blind' to its limitations from the big idea perspective. In other words, developing big ideas from small ideas is more difficult than it looks – at least, it may not be achieved merely through self-reflection.

I think, practically, we can start this enterprise as a joint effort among mathematics teachers from the same school. In a 'safe environment', we can discuss big ideas following the line of inquiry that I have proposed in this chapter. In a group, we can point out one another's blind spots as we work towards increasingly greater clarity as to the "big ideas" of a particular region of content coverage. Since it is *mathematical* big ideas that we are exploring, it is also very helpful to have a 'content strong' member in your group to provide the disciplinary expertise. For most schools in Singapore, there is already a ready structure to support this enterprise – the weekly Professional Learning Community (PLC) groups. It is perhaps worthwhile to re-style PLCs of mathematics departments into places where teachers jointly study the big ideas of the Singapore mathematics syllabus, one at a time.

Acknowledgement

I thank the mathematics teachers from Orchid Park Secondary School and Ahmed Ibrahim Secondary School whom I worked closely in developing replacement units for algebraic fractions. Much of what is written here are taken from our discussions.

References

Charles, R. I. (2005). Big ideas and understandings as the foundations for elementary and middle school mathematics. *Journal of Mathematics Education Leadership, 7*(3), 9-24.

Chow, W. K. (2014). *Discovering Mathematics 2A* (2nd ed.). Singapore: Star Publishing.

Ho, S. T., Khor, N. H., & Yan, K. C. (2013). *Additional Mathematics 360.* Singapore: Marshall Cavendish.

Lee, L., & Wheeler, D. (1989). The arithmetic connection. *Educational Studies in Mathematics, 20*(1), 41-54.

National Council of Teachers of Mathematics. (2015). *Developing Essential Understanding: An NCTM series*. VA: Author.

Siemon, D., Bleckly, J., & Neal, D. (2012). Working with the Big Ideas in number and the Australian curriculum: Mathematics. In B. Atweh, M. Goos, R. Jorgenson, & D. Siemon (Eds.), *Engaging the Australian national curriculum: Mathematics – Perspectives from the field* (pp. 19-45). Online Publication: Mathematics Education Research Group of Australasia.

Yeap, B. H. (2013). *New Syllabus Mathematics 1.* Singapore: Shinglee.

Chapter 20

Enhancing Mathematics Teaching with Big Ideas

Cynthia SETO CHOON Ming Kwang PANG Yen Ping

Teachers need to organize the teaching of a topic around big ideas to help students develop a deeper and more robust understanding of mathematics. This enables students to see mathematics as a coherent and integrated whole, not merely isolated pieces of knowledge. Big ideas express ideas which are central to mathematics and connect ideas from different strands and levels. Engaging teachers to construct a web of interconnected ideas (WICI) helps them to have a greater awareness of how concepts are built in preceding and succeeding grade levels and content strands. We propose the use of Content Representation in Mathematics (M-CoRe) as a way for teachers to think about the big ideas associated with teaching a given topic. M-CoRe sets out two critical aspects of teaching: big idea(s) in the content and big idea(s) in teaching to develop students' understanding of that content. Enhancing mathematics teaching with big ideas is discussed through WICI and M-CoRe for two topics, namely Mixed Numbers at Primary 4 and Simple Probability at Secondary 2.

1 Introduction

The Singapore Mathematics Curriculum underscores the importance of Mathematics in providing the "foundation for many of today's innovations and tomorrow's solutions" (Ministry of Education, 2018, p. S1-2). In particular, it recognizes that a strong foundation in mathematics is fundamental in many Smart Nation initiatives which depend heavily on computational power and mathematical insights. The learning experiences that students should have in order to develop critical mathematical processes, such as reasoning and communication, to support the development of 21st century competencies, have been implemented in the 2013 syllabuses (Ministry of Education, 2012). Building on this implementation, one of the key shifts in the revised 2020 Singapore Mathematics syllabuses is to teach towards big ideas and make visible the central ideas, coherence and connections across topics, and continuity across levels.

This key shift requires teachers to facilitate students to see and make connections among mathematical ideas within a topic or between topics across levels or strands. Kaur, Wong and Chew (2018) highlighted the importance of the role of teachers to support meaningful connections between procedures, concepts and contexts so as to achieve the goals stipulated by the Singapore mathematics curriculum. Gleaning insights from this finding, we consider that an organized knowledge base is paramount for teachers to be able to explain the connections or guide students to uncover these connections for themselves by asking questions about related small ideas in mathematics. This is also supported by Silverman and Thompson (2008) who asserted that "an expert teacher, then, is one with organized knowledge bases that can be quickly and easily drawn upon while being engaged in the act of teaching" (p. 501).

2 Teaching Mathematics though the Lens of Pedagogical Content Knowledge (PCK)

Mathematics teachers need a broad and deep knowledge base for teaching mathematics, which includes knowledge of the subject matter, pedagogy, students and student learning (NCTM 1991; Shulman, 1986). The nature, depth and organization of teacher knowledge influences teachers' presentation of ideas, capacity to help students connect mathematical ideas and flexibility in responding to students' questions (Ball, 1988).

A study by Ma (1999) shows that Chinese teachers often view a piece of knowledge as a part of a larger context of knowledge as they have the belief that mathematics is a set of relationships between concepts, facts, and procedures. Calling this kind of knowledge as 'knowledge package', Ma elaborated that organizing mathematics knowledge in a coherent way reflects the Chinese teachers' understanding of mathematics topics. This mathematics knowledge of teachers is pedagogical content knowledge (PCK), which is defined by Shulman (1986) as "the ways of representing and formulating the subject that makes it comprehensible to others" (p. 9). As an accepted academic construct, PCK has been closely linked to views associated with the professional knowledge bases of teachers. PCK is a unique knowledge base for teaching that distinguishes the expert teacher in the subject area (for example, Mathematics Master Teacher) from the subject expert (Mathematician).

PCK encapsulates the notion of big ideas. Content shapes pedagogy (Loughran, Milroy, Berry, Gunstone, & Mulhall, 2001). The amalgam of a teacher's pedagogy and understanding of content is the foundation of one's PCK such that PCK influences a teacher's teaching in ways that will best engender students' learning of mathematics for understanding. As such, building teachers' capacity to view content from the lens of big ideas of mathematics will empower them to make use of connections and links within and between such big ideas to make them explicit to students.

The need for teachers to have a sense of how mathematics is a coherent and connected whole has been consistently emphasized by scholars (Boaler & Humphreys, 2005; Sullivan, 2011). Studies by scholars (Askew, Brown, Rhodes, Wiliam, & Johnson, 1997; Charles, 2005; Hattie, 2003; Ma, 1999) recognized that the most effective mathematics teachers are the ones who make connections and support their students in making connections in mathematics. Recent research has shown that teachers experience difficulties in articulating the big ideas that inform their teaching and teachers' understanding of key mathematical ideas will impact their selection and use of appropriate tasks in their lessons (Clarke, Clarke, & Sullivan, 2012; Roche, Clarke, Clarke, & Sullivan, 2014).

A teacher's development of PCK is a complex process which is determined by the content to be taught, the context in which the content is taught and the way the teacher reflects his/her teaching experiences. To capture and portray experienced science teachers' PCK, Loughran, Mulhall, and Berry (2004) and Loughran, Berry, and Mulhall (2006) developed Content Representations (CoRe) for specific science topics. A CoRe represents the particular content/topic of science teaching. It is developed by asking teachers to think about what they consider to be the 'big ideas' associated with teaching a given topic for a particular grade level(s) based on their experience of teaching that topic. These 'big ideas' refer to the science ideas that the teacher(s) see as crucial for students to develop their understanding of the topic. In some cases, a big science teaching idea may be the same as a big science idea but the two are not necessarily synonymous as the interaction between content and teaching impacts how teachers conceptualize these big science teaching ideas.

In addition, eight pedagogical prompts in CoRe probe deeper understanding of the content; not only how it might be taught, but also, how it might (or might not) be learned. The eight pedagogical prompts are:

1. What do you intend the students to learn about this idea?
2. Why is it important for students to know this?
3. What else do you know about this idea (that you do not intend students to know yet)?
4. Difficulties/limitations connected with teaching this idea
5. Knowledge about students' thinking which influences your teaching of this idea
6. Other factors that influence your teaching of this idea
7. Teaching procedures (and particular reasons for using these to engage with this idea)
8. Specific ways of ascertaining students' understanding or confusion around this idea (include likely range of responses)

Although CoRe has been used as a way of empowering science teachers to actively develop their PCK in a specific content area, we posit that the pedagogical prompts in CoRe are relevant for our mathematics teachers. For example, we find that Prompt 4 on 'Difficulties/limitations connected with teaching this idea' and Prompt 5 on 'Knowledge about students' thinking which influences your teaching of this idea' resonate with the 'Knowledge of Content and Student (KCS)' in the domain map for mathematical knowledge for teaching (MKT) by Hill, Ball and Schilling (2008) which outlines the mathematical knowledge that teachers use in classrooms to produce instruction and student learning. They defined KCS as "content knowledge intertwined with knowledge of how students think about, know or learn this particular content" (p. 375). These two prompts in CoRe draw teachers' attention to both the specific content and something particular about students as highlighted in MKT.

We also find parallels between the horizon knowledge in MKT and Prompt 3 on 'What else do you know about this idea (that you do not intend students to know yet)?' Ball, Thames and Phelps (2008) highlighted the importance of horizon knowledge, awareness of how mathematics topics are related over the span of mathematics included in the curriculum, for teachers to make instructional decisions to set the mathematical foundations for what their students will learn at later years. We find great relevance in Prompt 3 of CoRe to facilitate our mathematics teachers to see connections between topics across levels or strands which is a key focus in the revised 2020 syllabuses.

Based on studies by scholars discussed above, we advocate the importance of equipping teachers with the knowledge base to support their students in connecting mathematics ideas for deep conceptual understanding. In Section 3, we propose engaging teachers to construct a web of interconnected ideas (WICI) to empower themselves to see the links and make connections among the various mathematical ideas. This is discussed in the context of a Primary 4 topic on Mixed Numbers and a Secondary 2 topic on Simple Probability to help teachers teach towards the big ideas of equivalence and measures respectively. For this chapter, our definition of big ideas is based on the 2020 Singapore Mathematics Syllabuses (Ministry of Education, 2018) which states that "big ideas express ideas that are central to mathematics. They bring coherence and connect ideas from different strands and levels." (p. S2-3) To facilitate teachers to think about a mathematics topic through big ideas, Section 4 will detail the use of CoRe in Mathematics (m-CoRe) to develop teachers' conceptual understanding of the content for teaching, which will lead to deepening teachers' PCK for quality student learning.

3 Web of Interconnected Ideas (WICI)

Concept maps are commonly used to help teachers see links and make connections among the various mathematical ideas. For example, Novak and Gowin (1984) engaged teachers in constructing 'concept maps' of understanding numbers in their study of effective teachers of numeracy in primary schools. Hough, O'Rode, Terman, and Weissglass (2007) studied how concept maps can also be used with prospective teachers to help them think about connections in mathematics. According to Davis and Simmt (2006), access to a web of interconnected ideas that constitute a concept is essential for teaching.

We believe that teachers with a coherent and connected knowledge base will be more effective in facilitating their students to see the connections among mathematical ideas so as to organize their knowledge better. When students connect mathematical concepts, they reduce the amount of mathematics to be learnt. This deepens their understanding of basic concepts which in turn increases their ability to apply mathematics within any context. Constructing the WICI makes the knowledge base visible to the teacher as well as communities of teachers. WICI serves as a basis for teachers to refine and strengthen their PCK. In Section 3.1 and 3.2, we present two WICIs (Mixed Numbers and Simple Probability) and discuss the connections between related concepts and big ideas in mathematics based on the 2020 Singapore Mathematics Syllabuses.

3.1 *WICI for Mixed Numbers*

From Primary 1 to 3, students learn that adding whole numbers will give another whole number. Through topics such as number bonds and adding whole numbers, students appreciate the part-whole relationships in these mathematics concepts which are taught at various levels. At Primary 4, students are introduced to the idea that a mixed number has the same value as the sum of a whole number and a proper fraction.

To provide students with the lens to look at the big idea of equivalence, teachers need to highlight the relationship that a mixed number expresses the 'equality' of the sum of a whole number and a proper fraction. Figure 1 presents a WICI showing the various topics connected to a mixed number. Although a mixed number looks different from the sum of a whole number and a proper fraction (mathematical objects), they are numerically the same. It expresses the same part-whole relationship (equivalence criterion). This equivalence criterion can be expressed as an addition relationship as well as a division relationship. At Primary 5, in the topic 'Concept of Fraction as Division', students will learn that a division algorithm can be used to obtain a mixed number when a greater whole number is divided by a smaller whole number. In highlighting the equivalence criterion, students will appreciate the connections between the quotient, divisor, remainder and the mixed number.

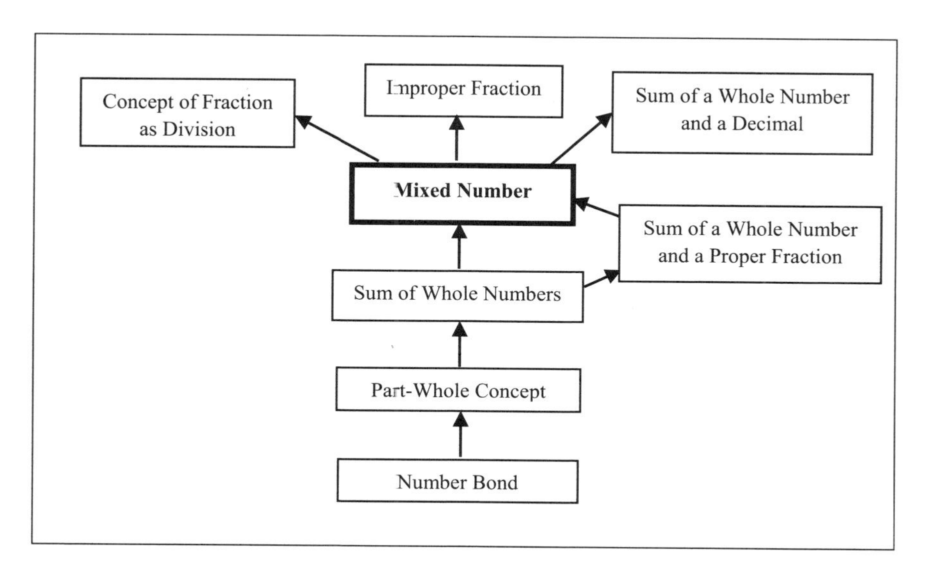

Figure 1. WICI for Mixed Numbers

Encouraging teachers to construct a WICI, such as Figure 1, will empower them to be more prepared to facilitate students to see the central idea of equivalence that is coherent among the mathematics concepts / topics learnt at various levels and content strands. With such awareness, teachers are more likely to find opportune points to highlight the idea of equivalence that connect these concepts of mixed numbers and addition of whole numbers. This understanding is also fundamental for students to learn decimals as the sum of a whole number and a decimal. Knowing that mixed numbers are linked to the concept of fractions and also division is important as it will influence teachers' instructional decisions to lay the mathematical foundations for what students will learn at later years (Ball et al., 2008). In articulating what they know about the connections across topics through constructing a WICI, we believe teachers will be more prepared to guide their students to uncover these connections and continuity across topics and levels.

3.2 *WICI for Simple Probability*

The WICI in Figure 2 shows the various pieces of knowledge that are connected to Simple Probability. It illustrates the concepts in preceding and succeeding grade levels and content strands when Secondary 2 students are introduced to the concept of probability as a measure of chance of events that are random or uncertain. Awareness of this connected knowledge of mathematical concepts empowers teachers to be more intentional in exploiting students' prior knowledge of fractions, decimals, percentages and ratios (learnt at primary levels) when they learn to compute probabilities by counting and comparing the ratios of equally likely outcomes. This also helps students to appreciate the curriculum continuity. Teachers are encouraged to draw on students' intuitive sense of chance to help them to appreciate that numbers are used to quantify the likelihood of various real-world phenomenon so that they can be analysed, compared or ordered. Explicit links to the topic of Proportion are essential for students to understand that Probability uses proportion to measure chance and has a finite range from 0 to 1. Making such connections enables students to see mathematics as a coherent whole. Providing just-in-time knowledge from the topic of Set Theory will help students make sense of the term 'sample space' and associated set notations (such as { }, n(S), n(E) and E′). Since Set Language and Notation is formally taught in Secondary 4, it is important for teachers to realize that these notations and academic vocabulary, used to define and quantify probability, will make little sense to Secondary 2 students if they are not explicitly explained.

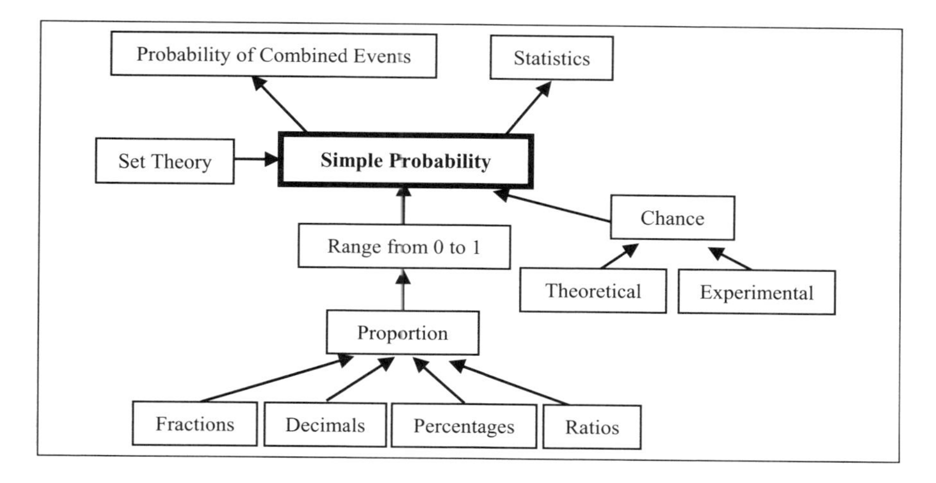

Figure 2. WICI for Simple Probability

To promote deeper understanding of the big ideas about measures, teachers can highlight that there are other measures which students have learnt in other levels or topics. For example, they have learnt about grams, kilograms (units) and how one unit of these measures look like. Just like having a sense of one unit of measure allows one to make reasonable estimates of the measure, the study of probability provides us with the means to understand likelihood, randomness and uncertainty that pervade our everyday life.

4 Content Representation in Mathematics (M-CoRe)

With the call for teaching towards big ideas in the 2020 Singapore Mathematics Syllabuses, we recognize the need to build teachers' capacity to link the 'what', 'why' and 'how' of the content to be taught with what teachers consider to be important in shaping students' learning. Shulman (1987) stipulates that "to teach is first to understand" and that "teaching begins with a teacher's understanding of what is to be learned and how it is to be taught" (p. 14). Premised on this conviction and the relevance of the eight pedagogical prompts in CoRe (discussed in Section 2) to frame teachers' thinking about big ideas and their teaching, we formulate a template for Content Representation in Mathematics (M-CoRe) as shown in Tables 1 and 2.

Based on the work of Loughran et al. (2001), the eight pedagogical prompts are presented in the first column in M-CoRe to probe deeper understanding of the content and conceptualizing of teaching in ways to enhance student learning. The top row of M-CoRe are the big ideas to be interrogated against the list of prompts. This is to make explicit the concepts and issues related to the content to be taught and to provide direction(s) and route(s) of engagement to develop conceptual understanding of mathematics.

For teachers to teach towards big ideas, we reiterate the importance of knowing the big idea(s) in the content and the big idea(s) in teaching so that they will be able to foreground big ideas in ways that will enhance student learning. To support this ambitious goal of teaching towards big ideas, the authors developed the M-CoRe for Mixed Numbers and Simple Probability as presented in Table 1 and Table 2 respectively.

4.1 *M-CoRe for Mixed Numbers (Primary 4)*

Mixed Numbers is often regarded as an easy topic to teach and very often it is taught procedurally. As we work on the M-CoRe, we have a deeper appreciation of the importance of making the big idea of equivalence in Mixed Numbers explicit. From the WICI on Mixed Numbers (Figure 1), teachers will be more intentional to connect the concept of mixed numbers to students' understanding of adding whole numbers based on part-whole relationships. They will also be more prepared to help students see the distinction in these two instances.

Besides using fraction discs to show the whole and fractional parts, it is important to get students to make connections between the manipulatives, symbols of mixed numbers and also the representations on the number line. In abstracting the symbol of mixed numbers, students also need to know that there are multiple modes of representing mixed numbers, for example, $1\frac{3}{10}$ *l* of water can be shown as in Diagram 1 or Diagram 2 in Table 1.

Table 1
M-CoRe for Mixed Numbers (Primary 4)

	Big Idea A	**Big Idea B**
Important math ideas/ concepts	A mixed number has the same value as the sum of a whole number and a proper fraction. It has a value greater than 1.	Mixed numbers can be represented with concrete/real-life objects, in pictorial (e.g. number line) or abstract forms.
What do you intend the students to learn about this idea?	• Express the sum of a whole number and a proper fraction as a mixed number.	• Use fraction discs/number line to represent and interpret fractions greater than one whole as mixed numbers.
Why is it important for students to know this?	• To write fractional parts that is greater than 1 whole • Mixed numbers are used in real-life situations to express a quantity e.g. $1\frac{1}{2}$ pizza, $3\frac{1}{4}$ *kg* of rice.	• To interpret the fraction notation of mixed numbers. $1\frac{3}{4}$ is made up of 1 and $\frac{3}{4}$. (1 whole and 3 quarters)

	Big Idea A	Big Idea B
What else do you know about this idea (that you do not intend students to know yet)?	• Calculations involving division do not always have whole number solutions. For example, while $8 \div 4$ will have a whole number solution, $7 \div 4$ will have a remainder 3. The remaining 3 is not able to be shared out or is not enough to form another group of 4. The remainder can be expressed as $\frac{3}{4}$. So $7 \div 4 = 1\frac{3}{4}$.	
Difficulties/ limitations connected with teaching this idea	Students need to discern that • A whole number can be made up of 2 parts: a whole number and a whole number to give another whole number e.g. $2 + 3 = 5$. • Fractions can be made up of 2 parts: a whole number and a fraction to give a mixed number e.g. $2 + \frac{1}{3} = 2\frac{1}{3}$.	• Students often do not see a fraction as a number. It is challenging for students to plot a fraction, such as $\frac{3}{5}$ or $1\frac{3}{5}$, on a number line.
Knowledge about students' thinking which influences your teaching of this idea	• Students have worked with fractions less than or equal to 1 in Primary 2 and 3. • Students may consider fractions as two whole numbers separated by a line and use the rules of adding and subtracting whole numbers to fractions. e.g. $1 + \frac{1}{3} = \frac{2}{3}$ or $\frac{2}{4}$.	
Other factors that influence your teaching of this idea	• Students are generally fearful of fractions. One of the reasons is the whole number bias, which causes students to have difficulties conceptualising whole numbers as decomposable units.	

	Big Idea A	**Big Idea B**
Teaching procedures (and particular reasons for using these to engage with this idea)	• Show examples of mixed numbers using fraction strips or discs or paper cut-outs of objects such as slices of cheese, pizza etc. (make connections with real-life application of mixed numbers). Ask students to write down the mixed numbers represented by these area models. • Move students to the pictorial stage of their learning through the use of questioning and prompts such as - How many wholes are there? - How many equal parts of the whole are shaded? - When you add the wholes and the shaded parts, what do you get? 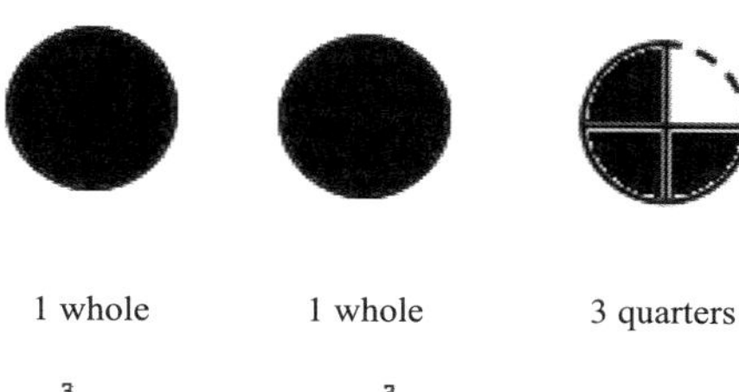 ($2\frac{3}{4}$ is made up of 2 and $\frac{3}{4}$) • Use different shapes as wholes to reinforce the learning of mixed numbers (Dienes' Variability Principles) and to prevent students from having the misconception of relating mixed numbers to only one particular shape. • Guide students to read fraction scales beyond 1. Use questions such as ✓ Look at the number 1 on the number line below. If we divide the space between a whole number and the next number on a number line into 5 equal parts, what fraction does each of the parts represent? ✓ If we count on from the next marking, it will be 1 and $\frac{1}{5}$. How do we write it as a mixed number? ✓ How do you write the mixed number to represent P? 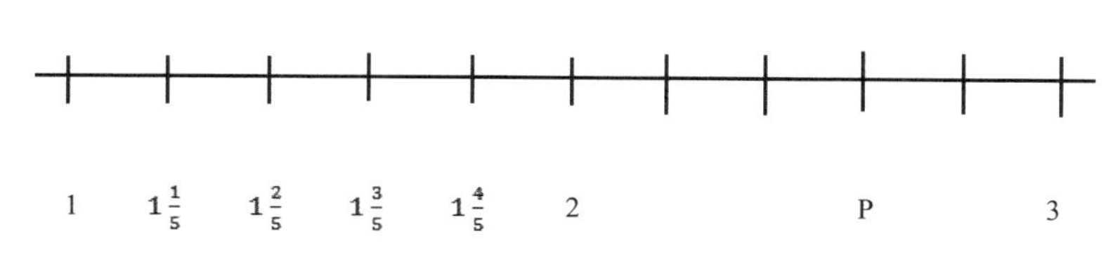	

	Big Idea A	Big Idea B
Specific ways of ascertaining students' understanding or confusion around this idea (include likely range of responses)	• Ask students to give an example of a mixed number used in everyday situations. • Students are able to explain to their partner what a mixed number means and to illustrate it with a diagram or a number line. • Students are able to show different ways of pictorial representations as below to indicate $1\frac{3}{10}$ *l* of water and they could also relate it to a number line.	

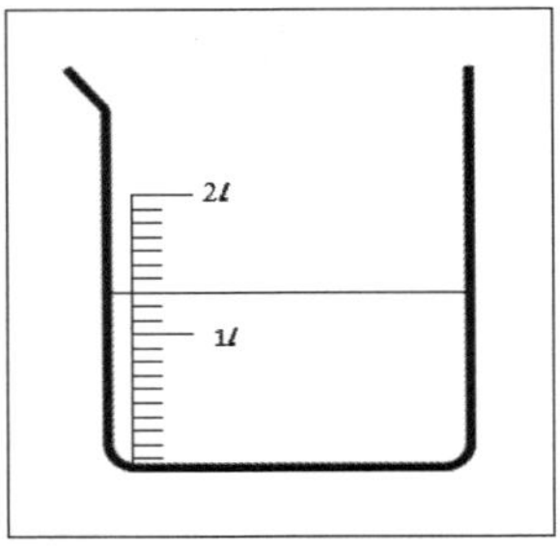

Diagram 1

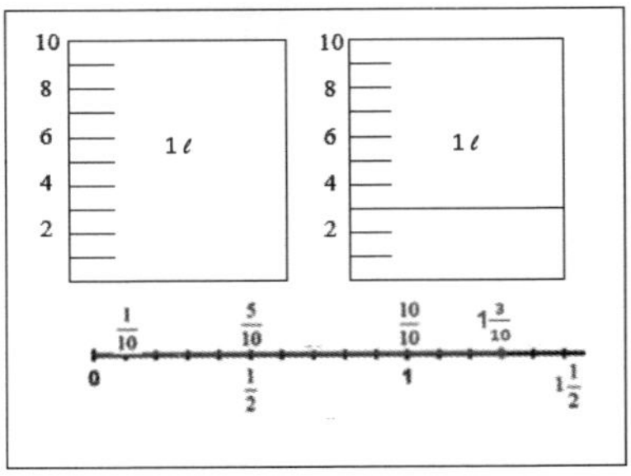

Diagram 2

4.2 *M-CoRe for Simple Probability (Secondary 2)*

Responding to the eight prompts in M-CoRe as shown in Table 2 serves as a guide for teachers to think through the lesson and unit planning for the topic on Simple Probability. When drawing up the M-CoRe in relation to the WICI presented in Figure 2, teachers will be more aware of the opportunity to tap on students' schema of the concept of chance, derived from their everyday experiences, and are sensitized to the need to provide just-in-time knowledge from the topic of Set Theory. Portraying experienced teachers' PCK also illuminates opportunities to highlight the big idea of measures during lesson enactment. Teachers should engage students in discussions to compare experimental and theoretical values of probability using probability experiments and/or computer simulations.

Table 2

M-CoRe for Simple Probability (Secondary 2)

	Big Idea A	Big Idea B
Important math ideas/ concepts	• Probability is a measure of chance (of events that are random or uncertain). • Numbers are used to measure or quantify an attribute of various real-world or mathematical objects, so that they can be analysed, compared and ordered. Examples of measures include length, area, volume, money, mass, time, temperature, speed, angle, probability, mean, etc.	• Probability is the study of likelihood or chance. • A random experiment is a process in which the result of the process cannot be predicted with certainty. • From conducting a random experiment, one can determine the experimental probability.
What do you intend the students to learn about this idea?	• Probability uses proportion to measure chance. • Probability of an event E, P(E), in a random experiment with equally likely outcomes is defined as the ratio (proportion) of $\frac{\text{Number of outcomes favorable to the event E}}{\text{Total number of possible outcomes}}$ • This ratio satisfies the following basic properties: 1. For any event E, $0 \leq P(E) \leq 1$. 2. For an event E that will definitely happen, $P(E) = 1$. 3. For an event E that is impossible to happen, $P(E) = 0$. 4. E' is called the complement event of E. $P(E') = 1 - P(E)$. • The closer this measure is to 1, the higher the chance of a phenomenon happening. This measure has a finite range and doesn't involve units.	• The experimental probability from a random experiment is not necessarily equal to the theoretical probability. • As more trials are conducted, the experimental probability generally gets closer to the theoretical probability.
Why is it important for students to know this?	• There are numerous applications of probability in real life such as in weather, insurance, sports and card games. Students' ability to understand probability as a measure of chance will enable them to make sense of information involving probability. • Students can make sense of probability through the lens of measures – What property is it measuring? What method is used to obtain the measure? What are the units? Does the measure have a finite range? What are the special values of the measure?	• The probability of any outcome of a random phenomenon is the proportion of times the outcome would occur in a very long series of repetitions. • The study of probability is thus based on the ideas of randomness and large numbers.

	Big Idea A	Big Idea B
What else do you know about this idea (that you do not intend students to know yet)?	• For combined events, diagrams (e.g. possibility diagrams, tree diagrams) can be used to visualise the possible combinations of outcomes. • Computation of the probabilities of combined events such as mutually exclusive events and independent events involve the addition and multiplication of probabilities.	• In addition to theoretical probability, we can also assign probabilities empirically, that is, from our knowledge of numerous similar past events.
Difficulties/ limitations connected with teaching this idea	• Students may be familiar with different forms of measurements but not with the big idea of measures.	• Secondary 2 is the first time students formally learn Probability so they will not know the academic vocabulary and notations related to notations from Set Theory commonly used in this topic.
Knowledge about students' thinking which influences your teaching of this idea	• Students usually have an intuitive sense of chance and should be familiar with common sayings such as '100% certain' and '0% chance' in real-life. • Students are already familiar with different forms of measurements (e.g. length, area, volume, money, mass, time, temperature, speed, angle and mean), so probability as a measure of chance builds on these prior conceptions of measures.	• Students may think that 20 or 50 are already considered to be large numbers in an experiment, so learning experiences involving large numbers in an experiment should be introduced as part of the teaching of this topic.
Other factors that influence your teaching of this idea	• Students are able to convert between percentages, fractions and decimals but may not be aware that these are all ratios under the broader concept of proportion.	• Students may sometimes state the number of favourable outcomes as the probability of an event. For example, students may mistakenly say that the probability of obtaining an even number when a die is rolled is 3 instead of $\frac{3}{6}$.

	Big Idea A	Big Idea B
Teaching procedures (and particular reasons for using these to engage with this idea)	• Facilitate class discourse to bring out how quantifying probability as a measure helps us compare, analyse and order the chance of a phenomenon happening, as opposed to using descriptive means (likely, unlikely, etc.) • Discuss the concept of chance using everyday events/occurrences.	• Enact learning experiences such as conducting simple experiments of single events to help students learn to compute probabilities by counting and comparing ratios of equally likely outcomes, e.g. conduct a coin-tossing experiment using both physical coins and an online coin toss simulator to generate an increasingly larger number of coin tosses and get students to observe how the value of the experimental probability of obtaining a 'Head' or 'Tail' changes. • Compare and discuss experimental and theoretical probabilities based on random experiments, so students can discern their differences and come to appreciate the importance of large numbers in the study of probability.
Specific ways of ascertaining students' understanding or confusion around this idea (include likely range of responses)	• Facilitate classroom discourse for students to explain and clarify their thinking.	• Questions on simple probability for students to work on. • Get students to design and conduct a probability experiment to verify a probability statement.

5 Implications for Developing Mathematics Teachers' PCK

Teachers' knowledge of practice is largely tacit (Polanyi, 1962). Although Cooper, Loughran, and Berry (2015) viewed PCK as a tacit form of teachers' professional knowledge that is topic, person, and context specific, their extensive research also indicated that teachers hold a shared knowledge (or collective PCK) around key notions of teaching and learning particular science topics. When teachers use the pedagogical prompts in M-CoRe as specific cues to discuss teaching, they access each other's content-specific pedagogical reasoning which provides insights into their tacit knowledge of practice. For example, developing the M-CoRe for Mixed Number and Simple Probability have enabled us to 'concretize' teaching and learning issues and make visible the 'what', 'why' and 'how' in teaching a specific mathematics topic. Through this shared knowledge, we deepen our understanding of the content which is the foundation to enhance our mathematics teaching.

The process of reflecting and framing these ideas may be challenging and this is consistent with studies by Cooper et al. (2015) who found that individually, teachers often struggle to construct big ideas and to answer the prompts in CoRe. Working in small groups helps teachers to better articulate their ideas as a result of discussing and testing each other's thinking. They also found that it was helpful to have a facilitator to stimulate and focus the discussion. As such, we recommend teachers to work collaboratively to construct WICI and M-CoRe either in their professional learning community (within school) or networked learning community (across schools).

6 Conclusion

A greater awareness of how concepts are built in preceding and succeeding grade levels and content strands as depicted in Figure 1 for Mixed Numbers and Figure 2 for Simple Probability will support teachers to be more adept in finding opportune points to bring out the big ideas that connect the various topics. With their big ideas in content interrogated against a list of pedagogical prompts as presented in Tables 1 and 2, teachers will have a deeper understanding that activities or tasks should be linked to the relevant big ideas and key skills. Together, the big ideas in the content and the big ideas in teaching empower teachers to exude the richness of connections, coherence, continuity and centrality of mathematics through emphases placed on their teaching actions and explanations.

In conclusion, we reiterate that enhancing mathematics teaching with big ideas starts with teachers making sense of the interconnected knowledge base of mathematics, such as by constructing a WICI and finding opportunities to articulate their understanding and teaching of specific mathematics through the use of prompts in CoRe. In doing so, we believe teachers will deepen their PCK and enhance their professional knowledge of practice to help students see mathematics as a connected and coherent whole, and not isolated pieces of knowledge.

References

Askew, M., Brown, M., Rhodes, V., Wiliam, D., & Johnson, D. (1997). *Effective teachers of numeracy: Report of a study carried out for the teacher training agency*. London: King's College, University of London.

Ball, D. L. (1988). Unlearning to teach mathematics. *For the learning of mathematics, 8*(1), 40-81.

Ball, D. L., Thames, M. H., & Phelps, G., (2008). Content knowledge for teaching: What makes it special? *Journal of Teacher Education, 59,* 389-407.

Boaler, J., & Humphreys, C. (2005). *Connecting mathematical ideas: Middle school video cases to support teaching and learning*. Portsmouth, NH: Heinemann.

Charles, R. I. (2005). Big ideas and understandings as the foundations for elementary and middle school mathematics. *Journal of Mathematics Education Leadership, 7*(3), 9-24.

Clarke, D. M., Clarke, D. J., & Sullivan, P. (2012). Important ideas in mathematics: What are they and where do you get them? *Australian Primary Mathematics Classroom, 17*(3), 13-18.

Cooper, R., Loughran, J. J., & Berry, A. (2015). Science teachers' PCK: Understanding sophisticated practice. In A. Berry, P. Friedrichsen, & J. Loughran (Eds.), *Re-examining pedagogical content knowledge in science education* (pp. 60-74). New York: Routledge.

Davis, B., & Simmt, E. (2006). Mathematics-for-teaching: An ongoing investigation of the mathematics that teachers (need to) know. *Educational Studies in Mathematics, 61*, 293-319.

Hattie, J. A. C. (2003, October). *Teachers make a difference: What is the research evidence*? Paper presented at the ACER Research Conference: Building Teacher Quality: What does the research tell us? Melbourne, Australia. Retrieved from http://research.acer.edu.au/research_conference_2003/4/

Hill, H. C., Ball, D. L., & Schilling, S. G. (2008). Unpacking pedagogical content knowledge: Conceptualizing and measuring teachers' topic-specific knowledge of students. *Journal for Research in Mathematics Education, 39*, 372-400.

Hough, S., O'Rode, N., Terman, N., & Weissglass, J. (2007). Using concept maps to assess change in teachers' understanding of algebra: A respectful approach. *Journal of Mathematics Teacher Education, 10*, 23-41.

Kaur, B., Wong, L. F., & Chew, C. K. (2018). Mathematical tasks enacted by two competent teachers to facilitate the learning of vectors by grade ten students. In P. C. Toh & B. L. Chua (Eds.), *Mathematics instruction: Goals, tasks and activities* (pp. 49-66). Singapore: World Scientific.

Loughran, J. J., Milroy, P., Berry, A., Gunstone, R., & Mulhall, P. (2001). Documenting science teachers' pedagogical content knowledge through PaP-eRs. *Research in Science Education, 31*, 289-307.

Loughran, J. J., Mulhall, P., & Berry, A. (2004). In search of pedagogical content knowledge in science: Developing ways of articulating and documenting professional practice. *Journal of Research in Science Teaching, 41*(4), 370-391.

Loughran, J. J., Berry, A., & Mulhall, P. (2006). *Understanding and developing science teachers' pedagogical content knowledge.* Dordrecht: Sense Publishers.

Ma, L. (1999). *Knowing and teaching elementary mathematics*. Mahwah, NJ: Erlbaum.

Ministry of Education. (2012). *Primary mathematics teaching and learning syllabus*. Singapore: Curriculum Planning and Development Division.

Ministry of Education. (2018). *2020 secondary mathematics syllabuses (draft)*. Singapore: Curriculum Planning and Development Division.

National Council of Teachers of Mathematics. (1991). *Professional standards for teaching mathematics*. Reston, VA: NCTM.

Novak, J. D., & Gowin, D. B. (1984). *Learning how to learn.* New York, NY: Cambridge University Press.

Polanyi, M. (1962). *Personal knowledge: Towards a post-critical philosophy*. London: Routledge & Kegan Paul.

Roche, A., Clarke, D. M., Clarke, D. J., & Sullivan, P. (2014). Primary teachers' written unit plans in mathematics and their perceptions of essential elements of these. *Mathematics Education Research Journal, 26*(4), 853-870.

Shulman, L. S. (1986). Those who understand: Knowledge growth in teaching. *Educational Researcher, 15*(2), 4-14.

Shulman, L. S. (1987). Knowledge and teaching: Foundations of the new reform. *Harvard Educational Review, 57*(1), 1-22.

Silverman, J., & Thompson, P. W. (2008). Towards a framework for the development of mathematical knowledge for teachers. *Journal of Mathematics Teacher Education, 11*, 499-511. http://dx.doi.org/10.1007/s10857-008-9089-5

Sullivan, P. (2011). Teaching mathematics: Using research-informed strategies. *Australian Education Review, 59*, 57-62.

Chapter 21

Making Vertical Connections when Teaching Towards Big Ideas

LOW Leng   WONG Lai Fong

In the 2020 Secondary Mathematics Syllabus (Draft), the big ideas "are central to the discipline and bring coherence and connections between different topics so as to develop in students a deeper and more robust understanding of mathematics and better appreciation of the discipline". This chapter outlines teaching towards the big ideas by making connections across different topics using mathematical tasks. These tasks relate to topics taught at different levels and are suitable for application within the transition phase in the 4 or 5 years in the secondary mathematics curriculum, serving as a bridging function for the purpose of vertical connectivity. An example of how such a task looks like will be discussed in this chapter.

1 Introduction

1.1 *Big ideas in mathematics*

A Big Idea is a statement of an idea that is central to the learning of mathematics, one that links numerous mathematical understandings into a coherent whole (Charles, 2005). In the field of cognitive science, the understanding of big ideas leads to knowledge use more flexibly, improves problem solving, helps in sense-making and facilitates knowledge transfer (Niemi, Vallone, & Vendlinski, 2006). Fosnot (2007) opined that big ideas are not only deeply connected to the structures of mathematics but also feature in shifts on learners' reasoning, such as shifts in perspective, in logic and in the mathematical relationships they set up.

Singapore mathematics curriculum adopts a spiral approach with an implicit link within topics and across levels. The emphasis of the mathematical processes – Reasoning, Communication & Connections, Applications & Modelling, Thinking Skills & Heuristics – in the Singapore school mathematics curriculum (Ministry of Education, 2013), is crucial for developing and making explicit links between and within the big ideas. However, the enacted curriculum is often presented in a linear and silo manner and this does not explicitly show the links and connections advocated in the big ideas of mathematics. Thus, in the 2020 revised secondary mathematics syllabus, there is an explicit emphasis on teaching towards big ideas, making links and connections. Eight big ideas are identified in the syllabus documents, relating to the four themes – Properties & Relationships, Representations & Communications, Operations & Algorithms, Abstractions & Applications. Understanding the big ideas allows for making meaningful connections to facilitate deeper understanding and appreciation of the nature of mathematics.

One of the aims of the mathematics syllabus in the 2020 revised documents is to enable students to "connect ideas within mathematics and between mathematics and other subjects through applications of mathematics" (Ministry of Education, 2018). Teachers must make mathematics meaningful and relevant to the learners, connecting learning of mathematics to the everyday experiences rather than merely emphasizing procedural demonstration by working through selected examples and expecting learners to merely practise by applying the procedures.

1.2 *Integration within and across topics*

In what ways can teachers make such connections explicit? The basic premise is that teachers think of mathematics as a connected and linked subject and in ways other than is presented in the current curriculum structure. This may include making the links across the various topics, between the three content strands and real world problem solving where the day to day experiences of the learners is also connected to the learning of mathematics (Adler, Pournara, & Graven, 2000). As recommended in the syllabus documents, teachers can teach towards big ideas by finding an opportune point to bring out each big idea and by integrating into the current sequence of instructions or lessons and within this sequence. While making the connections within or across the topics, the 'vertical' connections can also be added. If connections across different topics and year levels can be used as the basis for teaching, the curriculum will become more connected and this will encourage more sense making in our learners. A focus on big ideas can assist teachers to understand how mathematical topics are connected across the levels rather than be considered in isolation.

Siemon, Bleckly, and Neal (2012) pointed out the need for big ideas to be thought of in ways that are different from the present way of organizing curricula. According to Hurst (2015), the present way can be represented in Figure 1, where content is "pigeonholed" according to the three areas: content strands, topics and year levels. It seems that these three areas often do not overlap and are isolated. If 'big idea' thinking is used to organize content, Figure 2 represents a curriculum design with connections across content strands, topics and year levels. The curriculum will be more connected and a connected curriculum will more likely help students make sense of what they are learning. While the horizontal arrow shows the connections between mathematical ideas across the content strands, for example, fractions, decimals, percentages and ratio and proportion, the vertical arrow shows the development of key mathematical ideas across the year levels. An example of such a vertical connection is the big idea of proportionality that threads through the topics "Ratios" in secondary 1, "Similarity" in secondary 2 and "Trigonometric ratios" in secondary 3.

The mathematical processes that develop competencies in abstracting and reasoning, representing and communicating, applying and modelling have to be emphasized by teachers when facilitating mathematical tasks and classroom discourse in the classroom for a deeper understanding of the mathematics taught and to make the connections explicit.

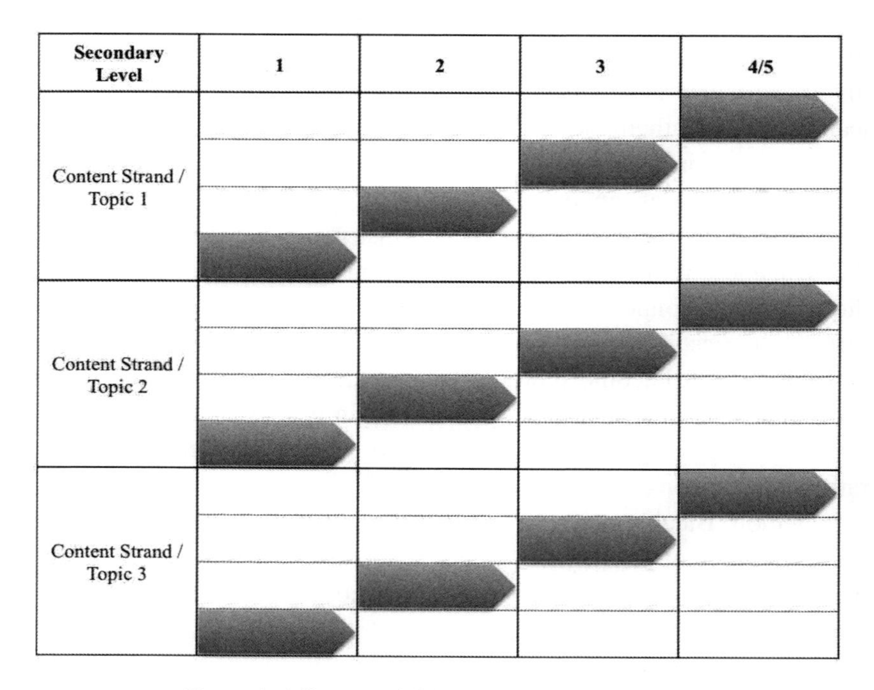

Figure 1. A linear model (adapted from Hurst, 2015)

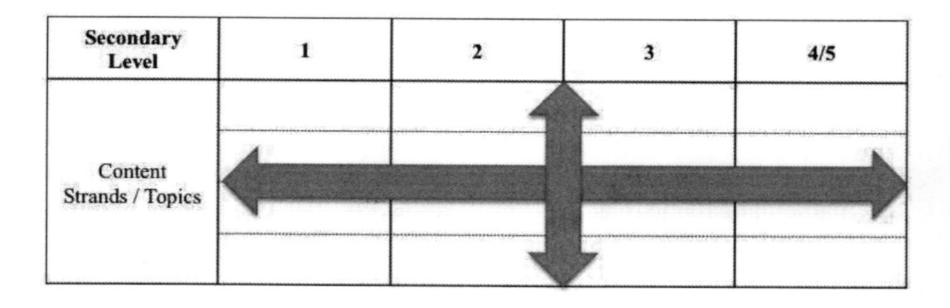

Figure 2. A more connected model based on big ideas and key concepts (adapted from Hurst, 2015)

1.3 *Use of mathematical tasks*

Mathematical tasks are central to students' learning because "tasks convey messages about what mathematics is and what doing mathematics entails" (National Council of Teachers of Mathematics (NCTM), 1991, p. 24). Students are often engaged in activities using tasks to learn some aspect of mathematics, be it a concept, skill or mathematical process. By working through well-designed tasks, students can expand their thinking about mathematical ideas and their approaches to solving mathematical problems (Garofalo & Trinter, 2012). They can also appreciate the value of approaching the tasks from different perspectives and of using different representations, concepts and skills. In particular, NCTM (2000) advocates that "much of the power of mathematics comes from being able to view and operate on objects from different perspectives" (p. 361). When working on their assigned tasks, students should be encouraged to respond to them in as many ways as they can. Appropriate use of these tasks challenges learners' intuitions while broadening and deepening their knowledge of mathematical concepts through the linking of multiple content areas across levels. Designing mathematical tasks that relate to topics across the different levels may serve as a bridging function for the purpose of vertical connectivity. An example of such a task will be discussed in the next section.

2 Using a Common Mathematical Task to Make Connections

2.1 *An example of a common task posed at different levels*

Consider the following mathematical task.

Find the height of the flag pole.

This is a mathematical task that teachers typically engage students after they have learnt the trignometric ratios or the angle of elevation. In this section, the authors are suggesting how this one task can be implemented across three levels, namely secondary 1 to 3 (or grades 7 to 9), to teach towards the big idea of proportionality.

Proportionality is a central theme that develops through many topics in the secondary mathematics curriculum, including ratio, rates, percentage, proportions, similarity, scaling, linear functions, trigonometry and probability. Although the mechanics of setting up an equation of two equivalent ratios with one missing term is generally straightforward, the notion of proportionality is subtle and students usually solve problems using algorithmic approaches that have little or no meaning to them instead of applying strategies or reasoning on proportional relationships (Ben-Chaim et al., 1998; Cramer & Post, 1993). Hence, the mathematical task shared in this section can be used to help students to reason about and make sense of proportional relationships, and, more importantly, to make connections the mathematical ideas or concepts they learn over the levels or years.

Figure 3 shows how the task to find the height of the flag pole is posed to the students at secondary 1 and the topic it relates to at that level. The task requires students to set up an equation of two equivalent ratios with one missing term, namely the height of the flag pole. Students' proportional thinking is required to make comparisons that are not dependent on specific numerical values, and such qualitative thinking encourages students to establish appropriate parameters for their models.

Topic: Ratios
Instructions:
1. Write down your height.
2. Now, take a picture of you standing next to the flag pole.
3. Describe and explain how you can use this picture to get an estimate of the height of the flag pole. You may want to draw a suitable diagram to make your explanation clearer. State any assumption(s) made.

Figure 3. Task posed at secondary 1

Figure 4 shows how the same task is posed to the students at secondary 2. The task now requires students to identify the pair of similar triangles. Students learn that similar triangles have sides that are proportional and use this knowledge to determine the height of a flagpole by setting up an equation using corresponding sides of the two triangles. The task allows students to focus on underlying proportional relationships rather than solely applying quantitative procedures.

Topic: Similarity
Instructions:
1. Write down your eye's height.
2. Now, place a mirror on the ground at __ metres from the foot of the flag pole. Looking at the mirror, walk backwards, in line with the flag pole and the mirror, till you see the tip of the flag pole in the mirror. Measure the horizontal distance between you and the mirror.
3. Describe how you can use this distance to get an estimate of the height of the flag pole. You may want to draw a suitable diagram to make your explanation clearer. State any assumption(s) made.

Figure 4. Task posed at secondary 2

Figure 5 shows how the task to find the height of the flag pole is posed to the students at Secondary 3. Using the given distance between an observer and the flag pole, and the angle of elevation, that is, the measure of the angle between the horizontal and the observer's line of sight obtained from their clinometer, students are required to set up an equation using the tangent ratio to solve for the height of the flag pole.

Topic: Trigonometry
Instructions:
1. Write down your eye's height.
2. Now, download a clinometer app on your smartphone.
3. Describe how you can use the clinometer to get an estimate of the height of the flag pole. You may want to draw a suitable diagram to make your explanation clearer. State any assumption(s) made.

Figure 5. Task posed at secondary 3

2.2 Making connections across levels

Students' understanding and progression in mathematics are not measured in terms of completed tasks – a datum that teachers must be mindful of, especially if the students merely respond to tasks without contacting the mathematical ideas and thinking as intended. While tasks influence learners by directing their attention to particular aspects of content and by specifying ways of processing information (Doyle, 1983, p. 161), the sequence of tasks may not develop the students' understanding of the big idea if the teacher does not make meaningful connections between the topics across the levels and the big idea overt.

When working on the task at secondary 1, students are expected to use ratios to set up an equation. Students' proportional thinking is required to make comparisons that are not dependent on specific numerical values, and the idea of proportionality (and also the idea of equivalence) can be weaved in when the teacher orchestrates a discussion on the two possible equations set up using different pairs of ratios shown in Figure 6. The teacher can also motivate students to look forward to the learning of the concept of scales in the following year.

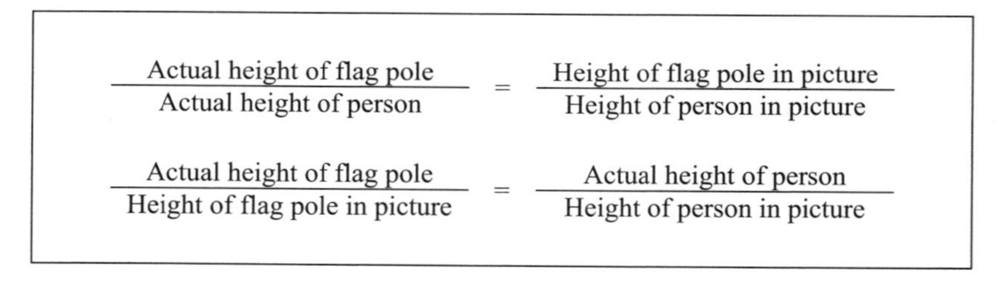
$$\frac{\text{Actual height of flag pole}}{\text{Actual height of person}} = \frac{\text{Height of flag pole in picture}}{\text{Height of person in picture}}$$

$$\frac{\text{Actual height of flag pole}}{\text{Height of flag pole in picture}} = \frac{\text{Actual height of person}}{\text{Height of person in picture}}$$

Figure 6. Two possible equations set up using ratios

When students work on the same task at secondary 2, using the concept of similar triangles this time, the fundamental idea of proportionality involved in similarity is again emphasized. Students can also use a different approach by identifying the similar triangles created by the person and his/her shadow, and the flag pole and its shadow, when the measurements are taken at the same time of day. The teacher can also motivate students to look forward to the learning of similarity tests in the following year when their intuitive notions about similarity can be proven. In addition, the teacher should orchestrate a discussion on the similarities and differences between the approaches used in the current and previous years, weaving in the idea of diagrams that serve to communicate properties of the objects and facilitate problem solving.

At secondary 3, students should understand how the trigonometric ratios are derived from the properties of similar triangles, and their understanding is reinforced by them working on the same task. Another discussion on the similarities and differences between the approaches used in the current and previous years, leading to the idea of proportionality, must be orchestrated, and if possible, creating an opportunity to weave in another idea of models as abstractions of real-world situations or phenomena using mathematical objects and representations.

The above three tasks involve mathematical concepts and skills that are typically included in secondary mathematics curriculum. They provide students opportunities to work on the same task and reason from different mathematical ideas in unique ways. At each level, what was previously the focus of attention becomes knowledge that is thought about and used at the next level, through informal and formal reasoning, and leading eventually to the big idea. Learning can be further enhanced if students are provided with guiding questions to aid them in reflecting upon the completion of each task, and hopefully these questions can generate more ideas for contemplation. Students who solve these tasks and others like them are subsequently motivated to think more flexibly when solving other problems, to try alternative methods, for example, to consider algebraic and geometric representations and solutions, and to evaluate their solution methods and intuitions.

Learning is neither uniform nor uni-directional. According to Mason and Johnston-Wilder (2006), "learners realize that they did not understand what they thought they understood... suddenly realize that there are gaps in their knowledge... and find there are wrinkles in their understanding that they have been overlooking" (p. 121). Working the same task but using different approaches each year enables students to reinforce, construct and consolidate their learning by looking back at what they have learnt or understood, looking at what they are learning, and looking forward to what they will be learning. Learning advances when students revert to familiar objects and situations as new experience emerges, use and link ideas, strategies and techniques from earlier topics, and eventually generalize and reconstruct ideas for themselves.

3 Issues and Challenges

We must be mindful of the fact that students respond to instruction very differently, depending on the structure and demands shaped by tasks enacted in the classroom (Shimizu, Kaur, Huang, & Clarke, 2010), and the task with which "students actually engage may or may not be the same task that the teacher announced at the outset" (Stein, Grover & Henningsen, 1996, p. 462). Furthermore, as Christiansen and Walther (1986) pointed out, "even when students work on assigned tasks supported by carefully established educational contexts and by corresponding teacher-actions, learning as intended does not follow automatically from their activity on the tasks" (p. 262). The nature of tasks often changes as they pass from one phase to another, and in this case, from one topic and level to another. Thus, there are some issues and challenges that we need to address.

3.1 Teachers' content knowledge

Teaching towards big ideas has the potential to help the students organise their knowledge better and give a sense of coherence to their learning. This means that teachers need to have a strong understanding of fundamental mathematics, as described by Ma and Kessel (2003), in terms of four key features: connectedness, multiple perspectives, recurring basic ideas, and longitudinal coherence; instead of merely in terms of facts, skills and procedures.

Our key message is for teachers to think about the way they view content knowledge and the pedagogies accompanying it. The way we view our curriculum and teach topics in silos may not be adequate for purposeful learning and teachers need to actively seek links and connections within and between concepts and explicitly point out how the links are connected.

This means teachers may need to apply their mathematical knowledge in different ways. Teachers with good 'horizon content knowledge' (Ball, Thames, & Phelps, 2008) will be able to go back and forth in a mathematical idea and think of ways to help the learners relate to topics they would encounter in the future. Over a period of time, the way teachers think about their subject and the way students relate to it will change. This is shared by Siemon et al. (2012):

> Effective teachers recognise the connections between different aspects and representations of mathematics. They ask timely and appropriate questions, facilitate and maintain high-level conversations about important mathematics, evaluate and respond to student thinking during instruction, promote understanding, help students make connections, and target teaching to ensure key ideas and strategies are understood. (p. 21)

3.2 Teachers' facilitation of classroom discourse

Just as it is crucial for the teacher to select appropriate mathematical tasks to help students make connections in the learning of mathematical ideas, the teacher also plays an important role as a facilitator in the classroom discourse so as to further student learning. When the teacher orchestrates classroom discourse, opportunities are provided for students to communicate using the language of mathematics. When the teacher asks questions, the big ideas and the connections have to be clearly articulated and made visible to the students. The more the students hear an idea, the more likely they will be able to internalize it so as to advance their learning.

Teachers should attend to students' thinking and sense-making by asking questions to elicit their thinking. For instance, in using the above task of finding the height of the flag pole to teach towards the big idea of proportionality, teachers may use the following questioning prompts to facilitate classroom discourse to make connections to mathematical ideas:

1. Read the problem. What is the problem about?
2. What information is given/do you need/is important?
3. What assumptions are made and why are they made?
4. What mathematics topics/concepts can you use to solve this problem?
5. Are there alternative methods to solve this problem?
6. Compare your current method with that/those used in your previous level(s). Which method do you prefer? Why?
7. How do you know that your solution is correct and you have arrived at the correct answer?
8. Compare your current solution with that/those obtained in your previous level(s). Which solution is more accurate? Why?
9. How is this problem similar or different from what you have already solved?

The big idea of proportionality that threads through the three topics of ratio, similarity and trigonometric ratios has to be explicitly articulated by the teacher (or the students) and not to be assumed that students will pick up the connection along the way. Teachers have to understand the importance of each topic and how that topic can build mathematical understanding and leverage on mathematical tasks to make connections and support transfer.

The notion of the teacher factor in respect to the teacher's content knowledge and facilitation of classroom discourse entails the essential of a strong content knowledge and a pedagogical shift for the teachers so as to present mathematics as a coherent and connected body of knowledge and a way of thinking for the students to develop a deep understanding of the subject.

4 Conclusion

Learning is meaningful when the learners can make sense of and use the knowledge and skills intended in the curriculum. Well-designed tasks can be used to encourage thinking about the big ideas that weave across mathematics and that are pivotal in students' mathematics learning.

This chapter also highlights the importance of the role of teachers as they enact the mathematical tasks in order to support meaningful connections between procedures, concepts and contexts, besides providing opportunities for students' engagement in key practices such as reasoning and problem solving. Teachers play an important role in the implementation of the intended curriculum, incorporating integrated mathematical tasks to make the connections that can turn a curriculum of skills, processes and knowledge that seems isolated into a coherent and purposeful understanding of mathematics. If students develop a more connected view of mathematics as modelled by their teachers, they will deepen their mathematical knowledge and will more likely be able to apply it in a range of contexts.

We hope that our idea on the use of task to make vertical connections across levels will initiate rich conversation among teachers. When teachers are thinking about big ideas, they become more conscious of the tasks to choose for students and the connections to highlight in and through their teaching. The big ideas need to be explicit, not implicit. Looking at the big ideas that thread through the various topics in different levels enable both teachers and students see what have been learnt, being learnt and will be learning. In fact, the big idea of teaching towards big ideas is beautifully highlighted by Boaler, Munson, and Williams (2018):

> We are now in the 21st century, student do not need to be trained to be calculators – we have technology for this – but they do need to experience mathematics as a beautiful and connected subject of enduring big ideas. Students who learn through big ideas and connections enjoy mathematics more, understand mathematics more deeply and are better prepared to tackle the big complex problems and discoveries that they will meet in their lives. (p. 100)

References

Adler, J., Pournara, C., & Graven, M. (2000). Integration within and across mathematics. *Pythagoras, 53*(12), 2-13.

Ball, D. L., Thames, M. H., & Phelps, G. (2008). Content knowledge for teaching: What makes it special? *Journal of Teacher Education*, *59*(5), 359-407.

Ben-Chaim, D., Fey, J. T., Fitzgerald, W. M., Benedetto, C., & Miller, J. (1998). Proportional reasoning among 7th grade students with different curricular experiences. *Educational Studies in Mathematics*, *36*(3), 247-273.

Boaler, J., Munson, J., & Williams, C. (2018). *What is mathematical beauty? Teaching through big ideas and connections.* California: Youcubed. Retrieved from https://bhi61nm2cr3mkdgk1dtaov18-wpengine.netdna-ssl.com/wp-content/uploads/2017/08/What-Is-Mathematical-Beauty-1.pdf

Charles, R. I. (2005). Big ideas and understandings as the foundation for elementary and middle school mathematics. *Journal of Mathematics Education Leadership*, *7*(3), 9-24.

Christiansen, B., & Walther, G. (1986). Task and activity. In B. Christiansen, G. Howson, & M. Otte (Eds.), *Perspectives on mathematics education* (pp. 243-307). Dordrecht, Reidel.

Cramer, K., & Post, T. (1993). Proportional reasoning. *The Mathematics Teacher*, *86*(5), 404-407.

Doyle, W. (1983). Academic work. *Review of Educational Research*, *53*(2), 159-199.

Fosnot, C. T. (2007). Our teaching and learning philosophy. In C. T. Fosnot (Ed.), *Investigating multiplication and division: Grade 3–5* (pp. 13-15). Portsmouth, NH: Heinemann.

Garofalo, J., & Trinter, C. P. (2012). Tasks that make connections through representations. *Mathematics Teacher*, *106*(4), 302-307.

Hurst, C. (2015). New curricula and missed opportunities: Crowded curricula, connections, and 'big ideas'. *International Journal for Mathematics Teaching and Learning*. Retrieved from https://www.cimt.org.uk/ijmtl/index.php/IJMTL

Ma, L., & Kessel, C. (2003). *Knowing mathematics*. Boston: Houghton Mifflin.

Mason, J., & Johnston-Wilder, S. (2006). *Designing and using mathematical tasks*. UK: Tarquin Publications.

Ministry of Education. (2013). *O-Level mathematics teaching and learning syllabus*. Singapore: Curriculum Planning and Development Division.

Ministry of Education. (2018). *2020 secondary mathematics syllabuses (draft)*. Singapore: Curriculum Planning and Development Division.

National Council of Teachers of Mathematics (NCTM). (1991). *Professional standards for teaching mathematics*. Reston, VA: NCTM.

National Council of Teachers of Mathematics (NCTM). (2000). *Principles and standards for school mathematics*. Reston, VA: NCTM.

Niemi, D., Vallone, J., & Vendlinski, T. (2006). The power of big ideas in mathematics education: Development and pilot testing of POWERSOURCE assessments (CSE Report 697). *National Center for Research on Evaluation, Standards, and Student Testing (CRESST)*.

Shimizu, Y., Kaur, B., Huang, R., & Clarke, D. (2010). The role of mathematical tasks in different cultures. In Y. Shimizu, B. Kaur, R. Huang, & D. Clarke (Eds.), *Mathematical tasks in classrooms around the world* (pp. 1-14). Sense Publishers.

Siemon, D., Bleckly, J., & Neal, D. (2012). Working with the big ideas in number and the Australian Curriculum: Mathematics. In B. Atweh, M. Goos, R. Jorgensen, & D. Siemon (Eds.), *Engaging the Australian National Curriculum: Mathematics – Perspectives from the field* (pp. 19-45). Online Publication: Mathematics Education Research Group of Australasia.

Stein, M. K., Grover, B. W., & Henningsen, M. (1996). Building student capacity for mathematical thinking and reasoning: An analysis of mathematical tasks used in reform classrooms. *American Educational Research Journal*, *33*(2), 455-488.

Contributing Authors

CHENG Lu Pien is a Lecturer in the Mathematics and Mathematics Education Academic Group at the National Institute of Education, Nanyang Technological University, Singapore. She received her PhD in Mathematics Education from the University of Georgia (U.S.) in 2006. She specializes in mathematics education courses for primary school teachers. Her research interests include the professional development of primary school mathematics teachers, task design in mathematics education and cultivating children's mathematical thinking in the mathematics classrooms.

CHOON Ming Kwang is a Master Teacher (Primary Mathematics) at the Academy of Singapore Teachers (AST). Prior to joining AST, Ming Kwang was a Lead Teacher in North Spring Primary School. He works closely with Teacher Leaders to champion quality learning in primary mathematics. He builds teacher capacity and encourages pedagogical innovations through mentoring Teacher Leaders. His interest in mathematical communication has led him to conduct several workshops on the use of 'Talk Moves' and 'Pentagon Questions' to engage students in clarifying their thinking and justifying their mathematical reasoning. Through workshops and networked learning communities, he seeks to nurture teachers to be reflective professionals who look into enhancing students' learning experiences in mathematics lessons.

CHOY Ban Heng, a recipient of the National Institute of Education (NIE) Overseas Graduate Scholarship in 2011, is currently an Assistant Professor in the Mathematics and Mathematics Education Department at

NIE, Nanyang Technological University, Singapore. Prior to joining NIE, he taught secondary school students mathematics for more than 10 years, and was the Head of Department (Special Projects) before he joined the Curriculum Planning and Development Division as a Curriculum Policy Officer. He holds a Doctor of Philosophy (Mathematics Education) from the University of Auckland, New Zealand, specialising in mathematics teacher noticing. Ban Heng was awarded the Early Career Award at the 2013 Mathematics Education Research Group of Australasia Conference in Melbourne for his excellence in writing and for presenting a piece of mathematics education research. He has a keen interest in several other related areas of research, including mathematics task design, mathematics classroom practices, teacher professional development and design-based research. Currently, Ban Heng is the Principal Investigator of a research which looks into the professional development of primary mathematics teachers; the research is funded by the MOE Academy Funding Grant.

CHUA Boon Liang is an Assistant Professor in mathematics education at the National Institute of Education, Nanyang Technological University, Singapore. He received his PhD in Mathematics Education from the Institute of Education, University College London, UK. His research interests include students' and teachers' mathematical reasoning and justification, the design of mathematical tasks to probe and promote students' and teachers' understanding of mathematical concepts, and the teaching and learning of algebra, particularly in the area of pattern generalisation.

DINDYAL Jaguthsing completed his PhD at Illinois State University in USA. He is currently an Associate Professor in the Mathematics and Mathematics Education Academic Group at the National Institute of Education (NIE), Nanyang Technological University, Singapore. He has prior experience in teaching mathematics at the secondary level and has taught mathematics education courses to both primary and secondary preservice and inservice teachers at NIE. He has worked on several research projects, the most recent one, being on teacher noticing. His interest areas include teacher education, the teaching and learning of algebra and geometry, mathematical tasks, problem solving, and teacher noticing. He has published several papers, book chapters and conference papers. Daya is a member of the International Group for the Psychologlogy of Mathematics Education (PME), the National Council of Teachers of Mathematics (NCTM), and the Mathematics Education Research Group of Australasia (MERGA). He is on the Editorial Board of the *International Journal for Mathematics Teaching and Learning*, a member of the International Advisory Board of the *Journal for Research in Mathematics Education* (JRME), and an Associate Editor of the *Mathematics Education Research Journal* (MERJ).

Chris HURST is a lecturer in primary mathematics education at Curtin University in Perth and he was awarded his PhD in 2007. He is an active member of the Mathematics Education Research Group of Australasia (MERGA) and regularly presents at national and international conferences. His current research interests are children's multiplicative thinking, teacher content knowledge, and the big ideas of mathematics. Dr Hurst was recently involved with a national research project titled Building an Evidence Base for Best Practice in Mathematics Education and has published his research work from various projects in national and international journals. He also regularly presents research-based professional learning for primary teachers.

Berinderjeet KAUR is a Professor of Mathematics Education at the National Institute of Education, Nanyang Technological University, Singapore. She holds a PhD in Mathematics Education from Monash University in Australia. She has been with the Institute for the last 30 years and best described as the doyenne of Mathematics Education in Singapore. In 2010, she became the first full professor of Mathematics Education in Singapore. She has been involved in numerous international studies of Mathematics Education and was the Mathematics Consultant to TIMSS 2011. She was also a core member of the MEG (Mathematics Expert Group) for PISA 2015. She is passionate about the development of mathematics teachers and in turn the learning of mathematics by children in schools. Her accolades at the national level include the public administration medal in 2006 by the President of Singapore, the long public service with distinction medal in 2016 by the President of Singapore and in 2015, in celebration of 50 years of Singapore's nation building, recognition as an outstanding educator by the Sikh Community in Singapore for contributions towards nation building.

Vincent KOH Hoon Hwee is the Head of Department for Mathematics at a local Primary School and has been teaching Mathematics at the primary school level for more than 15 years. The school had collaborated with NIE in a research project to engage students in mathematical processes through journal writing. He believes in inculcating the joy of learning in the teaching of mathematics. He is currently exploring ways to deepen and engage students in mathematical activities to help them in their conceptual understanding. One of the approaches he has implemented in the school is in the using of storytelling to teach mathematics.

KOH Khee Meng is a former Professor (1972–2014) in the Department of Mathematics at the National University of Singapore (NUS) and an Adjunct Professor (2015-2018) at the Singapore University of Technology and Design. He is currently teaching at the National Institute of Education, Nanyang Technological University, Singapore. He graduated from Nanyang University in 1968 and obtained his M.Sc. and Ph.D. from the University of Manitoba in Canada in 1969 and 1971 respectively. Among several other significant appointments, Professor Koh was the chairman of the Singapore International Mathematical Olympiad Committee (1991-1993), a council member of the Institute of Combinatorics and Its Applications (International) (1995-1997) and the president of the Singapore Mathematical Society (1996-1998). For his fine teaching, he has won numerous Teaching Awards, culminating in the NUS Outstanding Educator Award in 2011. Professor Koh specializes in Graph Theory, and he is widely regarded as one of the few foremost Graph Theorists in Singapore and the ASEAN region. He, together with his research students, has published more than 160 research articles in international scientific journals. Professor Koh is also the co-author of more than five books. In recognition of his accomplishments and contributions, he was honored with the NUS Outstanding Science Alumnus Award in 2014.

LEONG Yew Hoong is an Associate Professor in the Mathematics and Mathematics Education Academic Group of the National Institute of Education, Nanyang Technological University, Singapore. He is committed to research that influences classroom practice. Around this focus, his research interests and publications in mathematics education include: realistic ambitious pedagogy, low-progress learners, and teacher professional development.

Pearlyn LIM Li Gek is currently a PhD student with the National Institute of Education (NIE), Nanyang Technological University, Singapore. Her research interests include fostering critical thinking in mathematics lessons and the use of interesting approaches like drama and magic tricks in the teaching of mathematics. She was formerly the head of the mathematics department at a primary school. She was also a teaching fellow at NIE for six years and co-authored the New Syllabus Primary Mathematics textbooks.

LOW Leng has taught mathematics in secondary schools for 27 years and is currently a Master Teacher at the Academy of Singapore Teachers. Her interest in catering to the different needs of the students saw her partnering teachers in networked learning communities and in professional development courses. As a Master Teacher, she strives to engage in continual professional growth through meaningful collaboration with teachers so as to deepen students' conceptual understanding of mathematics and to nurture in them the joy of learning. She has presented in both local and international conferences in the learning of mathematics and mentoring.

PANG Yen Ping is a Master Teacher (Secondary Mathematics) at the Academy of Singapore Teachers. Prior to her appointment as a Mathematics Master Teacher, she was a Lead Teacher at Gan Eng Seng School for seven years. She holds a Master of Teaching and is passionate about mentoring teachers to enhance their pedagogical content knowledge. With her conviction to make teaching and learning of mathematics fun and engaging for both students and teachers, Yen Ping seeks to promote quality teaching in the mathematics fraternity. Her research interests include constructivist pedagogy, contrasting examples instructional approach, mathematical reasoning and motivation.

Cynthia SETO is a Principal Master Teacher (Mathematics) at the Academy of Singapore Teachers (AST), Ministry of Education. As the Principal Master Teacher and Cluster Head for Mathematics at AST, she leads a team of Maths Master Teachers to spearhead curricular and pedagogical initiatives. She holds a PhD in mathematics education and her dissertation is on classroom learning environment and networked learning communities for teachers. With about forty years of service in education, she has taught all levels of mathematics from Secondary Two to Primary One. Besides authoring several book chapters, she has presented and published papers at international conferences. She was also invited to host workshops at several international conferences. As the appointed Singapore Governing Board member of SEAMEO Regional Centre for QITEP in Mathematics, Cynthia also contributes to the advancement of Mathematics Education in South East Asia.

Padmanabhan SESHAIYER, a professor of mathematical sciences at George Mason University in Fairfax, Virginia, works in the broad areas of computational mathematics, mathematical modeling, teacher professional development and STEM education. As the director of the Center for Outreach in Mathematics Professional Learning and Educational Technology (COMPLETE), he has initiated a variety of educational programs in K-16 to foster the interest of students and teachers in STEM at all levels.

Jennifer SUH is a professor of mathematics education at George Mason University in Fairfax, Virginia. She is interested in deepening teachers' understanding of the learning trajectories in mathematics through lesson study. Currently, she serves as the Co-Director of the COMPLETE center.

TAY Eng Guan is an Associate Professor and Head of the Mathematics and Mathematics Education Academic Group of the National Institute of Education (NIE), Nanyang Technological University, Singapore. Dr Tay specializes in Graph Theory and Mathematics Education and has had papers published in international scientific journals in both areas. He is co-author of the books *Counting*, *Graph Theory: Undergraduate Mathematics* and *Making Mathematics Practical*. Dr Tay has taught in Singapore junior colleges and served a stint in the Ministry of Education. He has also worked as sub-dean of the degree programmes at NIE.

TOH Pee Choon received his PhD from the National University of Singapore in 2007. He is currently an Assistant Professor at the National Institute of Education, Nanyang Technological University, Singapore. A number theorist by training, he continues to research in both Mathematics and Mathematics Education. His research interests in Mathematics Education include problem solving, proof and reasoning, and the teaching of mathematics at the undergraduate level.

TOH Tin Lam is an Associate Professor and currently the Deputy Head of the Mathematics and Mathematics Education Academic Group in the National Institute of Education, Nanyang Technological University, Singapore. He obtained his PhD from the National University of Singapore in 2001. He publishes papers in international scientific journals in both Mathematics and Mathematics Education, and is doing research in both areas. He has been involved in several large scale mathematics education research projects on mathematical problem solving and comics for mathematics instruction.

TOH Wei Yeng Karen is a research assistant at the National Institute of Education, Nanyang Technological University, Singapore. She has a Bachelor of Science in Applied Mathematics from the National University of Singapore. Prior to joining the Institute, she was a secondary school mathematics teacher for seven years. She is interested in researching classroom practices and instructional materials of mathematics teachers.

TONG Cherng Luen is a research assistant at the National Institute of Education, Nanyang Technological University, Singapore. He has a Bachelor of Engineering from the Nanyang Technological University of Singapore. Prior to joining the Institute, he was a secondary school mathematics teacher for more than ten years. He is interested in the use of technology in capturing classroom activities and also quantitative analyses of data.

David WAGNER is a professor in the Faculty of Education at the University of New Brunswick. He is most interested in human interaction in mathematics and mathematics learning and the relationship between such interaction and social justice. This inspires his research which has focused on identifying positioning structures in mathematics classrooms by analyzing language practice, on ethnomathematical conversations in Aboriginal communities, and on working with teachers to interrogate authority structures in their classrooms. He has widely published and serves as Associate Editor of *Educational Studies in Mathematics*, on the editorial board of *Mathematics Education Research Journal*, on the board of the journal *For the Learning of Mathematics*, and on the Nonkilling Science and Technology Research Committee. He has taught grades 7-12 mathematics in Canada and Swaziland.

WONG Lai Fong has been a mathematics teacher for over 20 years. For her exemplary teaching and conduct, she was given the President's Award for Teachers in 2009. As a Head of Department (Mathematics) from 2001 to 2009, a Senior Teacher and then a Lead Teacher for Mathematics, she set the tone for teaching the subject in her school. Recipient of a Post-graduate Scholarship from the Singapore Ministry of Education, she pursued a Master of Education in Mathematics at the National Institute of Education. Presently, she is involved in several Networked Learning Communities looking at ways to infuse mathematical reasoning, metacognitive strategies, and real-life context in the teaching of mathematics. Lai Fong is active in the professional development of mathematics teachers and in recognition of her significant contribution toward the professional development of Singapore teachers, she was conferred the Associate of Academy of

Singapore Teachers in 2015 and 2016. She is currently seconded as a Teaching Fellow in the National Institute of Education, Nanyang Technological University, Singapore, and is also an executive committee member of the Association of Mathematics Educators.

YAP Von Bing obtained a B.Sc. (Hons) in Mathematics and a M.Sc. in Applied Mathematics, both from the National University of Singapore (NUS), and then a Ph.D. in Statistics from the University of California. Since 2004, he has been teaching at the Department of Statistics and Applied Probability in NUS, where he is an Associate Professor. His main interest is applied statistics, the interface between mathematical theory and science. Other fields that fascinate him include probability, molecular evolution, comparative genomics, ecology, and the philosophy and practice of teaching abstract concepts.

Joseph B. W. YEO is a lecturer in the Mathematics and Mathematics Education Academic Group at the National Institute of Education, Nanyang Technological University, Singapore. He is the first author of the *New Syllabus Mathematics* textbooks used in many secondary schools in Singapore. His research interests are on innovative pedagogies that engage the minds and hearts of mathematics learners. These include the use of an inquiry approach to learning mathematics (e.g. guided-discovery learning and investigation), ICT, and motivation strategies to arouse students' interest in mathematics (e.g. catchy maths songs, amusing maths videos, witty comics, intriguing puzzles and games, and real-life examples and applications). He is also the Chairman of Singapore and Asian Schools Math Olympiad (SASMO) Advisory Council, and the creator of Cheryl's birthday puzzle that went viral in 2015.

YEO Kai Kow Joseph is a Senior Lecturer in the Mathematics and Mathematics Education Academic Group at the National Institute of Education, Nanyang Technological University, Singapore. Before joining the National Institute of Education in 2000, he held the post of Vice Principal and Head of Mathematics Department in secondary schools. As a mathematics educator, he teaches pre- and in-service as well as postgraduate courses in mathematics education and supervises postgraduate students pursuing Masters degrees. His publication and research interests include mathematical problem solving at the primary and secondary levels, mathematics pedagogical content knowledge of teachers, mathematics teaching in primary schools and mathematics anxiety.

ZHU Ying is an Assistant Professor in the Mathematics and Mathematics Education Academic Group at the National Institute of Education, Nanyang Technological University, Singapore. She received an M.Sc. in Applied Statistics from University of Oxford and a Ph.D. in Statistics from University College London, UK. She also has industry experience at ACNielsen and PricewaterhouseCoopers as a statistician and a senior analyst. At NIE, she teaches statistics courses at both undergraduate and graduate levels. Her research interests include multivariate classification of high-dimensional data and image classification. She has been working on developing these methodologies with applications to biomedical sciences, as well as in statistics education.